Information Systems Engineering and Management 45

The book series "Information Systems Engineering and Management" (ISEM) publishes innovative and original works in the various areas of planning, development, implementation, and management of information systems and technologies by enterprises, citizens, and society for the improvement of the socio-economic environment.

The series is multidisciplinary, focusing on technological, organizational, and social domains of information systems engineering and management. Manuscripts published in this book series focus on relevant problems and research in the planning, analysis, design, implementation, exploration, and management of all types of information systems and technologies. The series contains monographs, lecture notes, edited volumes, pedagogical and technical books as well as proceedings volumes.

Some topics/keywords to be considered in the ISEM book series are, but not limited to: Information Systems Planning; Information Systems Development; Exploration of Information Systems; Management of Information Systems; Blockchain Technology; Cloud Computing; Artificial Intelligence (AI) and Machine Learning; Big Data Analytics; Multimedia Systems; Computer Networks, Mobility and Pervasive Systems; IT Security, Ethics and Privacy; Cybersecurity; Digital Platforms and Services; Requirements Engineering; Software Engineering; Process and Knowledge Engineering; Security and Privacy Engineering, Autonomous Robotics; Human-Computer Interaction; Marketing and Information; Tourism and Information; Finance and Value; Decisions and Risk; Innovation and Projects; Strategy and People.

Indexed by Google Scholar. All books published in the series are submitted for consideration in the Web of Science.

For book or proceedings proposals please contact Alvaro Rocha (amrrocha@gmail.com).

Toufik Marir · Maroua Bouzid ·
Rohallah Benaboud · Abdelhabib Bourouis
Editors

Proceedings of the 3rd International Conference on Computer Science's Complex Systems and Their Applications

ICCSA'2024

Springer

Editors
Toufik Marir
University of Oum El Bouaghi
Oum El Bouaghi, Algeria

Rohallah Benaboud
Research Laboratory on Computer Science
University of Oum El Bouaghi
Oum El Bouaghi, Algeria

Maroua Bouzid
Normandie Université de Caen Campus II
University of Caen
Caen Cedex, France

Abdelhabib Bourouis
Research Laboratory on Computer Science
University of Oum El Bouaghi
Oum El Bouaghi, Algeria

ISSN 3004-958X ISSN 3004-9598 (electronic)
Information Systems Engineering and Management
ISBN 978-3-031-90757-9 ISBN 978-3-031-90758-6 (eBook)
https://doi.org/10.1007/978-3-031-90758-6

This Springer imprint is published by the registered company Springer Nature Switzerland AG
The registered company address is: Gewerbestrasse 11, 6330 Cham, Switzerland

Foreword

We are delighted to present this book, which showcases the proceedings of the 3rd International Conference on Computer Science's Complex Systems and Their Applications (ICCSA'2024), held on November 13–14, 2024, at the University of Oum El Bouaghi in Algeria. The conference was organized by the Research Laboratory on Computer Science's Complex Systems (ReLa$(CS)^2$).

In this edition of ICSSA, the conference features a range of research topics, including applications of artificial intelligence in healthcare and engineering, advances in collective artificial intelligence, software engineering, and the integration of artificial intelligence within emergent technologies in networked systems. These topics reflect the diverse and evolving landscape of complex systems research, offering novel insights and applications that address real-world challenges. Indeed, applications of designing, understanding, or monitoring complex systems are met in particle physics to understand the forces that govern their behavior ; in healthcare to predict disease and assist medical diagnoses ; in environmental science to predict and monitor natural disasters, analyze climate data, and optimize energy; in biology to analyze large genomic data sets and understand the protein folding, in social science to understand human behavior, predict social trends and analyze the impact of government policies and in art and design to generate new forms of expression. Studying complex systems is the new trends of the AI revolution. Indicators of this new trend are the Nobel prize of 2024 in Biology and Physics concerning the understanding of natural complex systems (biology) and physical complex system (physics and chemistry) using AI technology.

In this context, the primary objective of this book is to highlight the latest advancements and discoveries in the development of complex systems within the field of computer science, particularly artificial intelligence. By providing a platform for researchers to share their work, we aim to foster dialogue and collaboration, facilitating new directions and innovations in this rapidly advancing field and shed the light on the new international research directions to motivate and initiate the development of new national and international projects.

Building on the success of previous editions held in 2016 and 2021, this year's conference saw a highly selective process, with only 25 papers accepted for presentation over two days. The high quality of these contributions was ensured by a rigorous double-blind review process, with each paper evaluated by at least three expert reviewers.

We are confident that this book will serve as a valuable resource and reference for researchers, students, and professionals engaged in complex systems research in computer science. We hope that you find the insights and research presented within both informative and inspiring.

Abdel-Illah Mouaddib
Farid Mokhati

Preface

We are honored to present the proceedings of the International Conference on Computer Science's Complex Systems and Their Applications (ICCSA'2024), held on November 13–14, 2024, at the University of Oum El Bouaghi, Algeria. This conference was organized by the Research Laboratory on Computer Science's Complex Systems (ReLa$(CS)^2$) in collaboration with the Faculty of Exact Sciences and Natural and Life Sciences, both at the University of Oum El Bouaghi.

ICCSA'2024 provided a unique platform for researchers, academics, and practitioners to explore, discuss, and exchange ideas on the latest advancements and challenges in the field of complex systems and their diverse applications in computer science. Topics included artificial intelligence and its applications, cybersecurity, software engineering, and more. Particularly, several papers focused on artificial intelligence techniques and their application in engineering and healthcare domains, reflecting the growing significance of these topics in today's rapidly evolving technological landscape.

The high-quality contributions to these proceedings reflect the rigorous peer-review process employed by the conference. Each paper was evaluated by at least three reviewers following a double-blind review policy. Only 33% of the submitted papers were accepted for presentation at the conference, ensuring a strong focus on relevance, originality, and scientific merit. The conference also featured keynote addresses by distinguished speakers from Canada, Portugal, and France, whose expertise and insights enriched the discussions and inspired participants.

The success of ICCSA'2024 would not have been possible without the efforts of many dedicated individuals. We extend our deepest gratitude to the program committee members and reviewers for their commitment to maintaining the highest standards of quality. We also thank the organizing committee, volunteers, and institutional partners for their invaluable support in bringing this conference to life. Additionally, ICCSA'2024 was supported by *the General Directorate of Scientific Research and Technological Development (DGRSDT) and the Thematic Agency for Research in Science and Technology (ATRST), both under the auspices of the Ministry of Higher Education and Scientific Research (Algeria).*

Finally, we express our heartfelt thanks to the authors and attendees for their active participation and valuable contributions. We hope that these proceedings will serve as a lasting reference for researchers and practitioners in the field of complex systems and beyond.

Thank you for being part of ICCSA'2024. We look forward to welcoming you to future editions of the conference!

November 2024

Toufik Marir
Chair, ICCSA'2024

Acknowledgments

We would like to express our sincere gratitude to all those who have contributed to the success of ICCSA'2024. Special thanks goes to the **Faculty of Exact Sciences and Natural and Life Sciences** and the **Thematic Agency for Research in Science and Technology (ATRST)**, whose financial support has been instrumental in making this conference possible. Their continued commitment to fostering scientific research and collaboration has greatly enriched this event.

We also extend our heartfelt appreciation to our esteemed keynote speakers, program committee members, reviewers, and all the authors for their valuable contributions. Without their dedication and expertise, this conference would not have been such a great success.

Organization

Program Committee Members

Abbas Cheddad	Blekinge Institute of Technology, Sweden
Abdeldjalil Ouahabi	University of Tours, France
Abdelhabib Bourouis (Chairman)	University of Oum El Bouaghi, Algeria
Abdelhamid Mammeri	National Research Council, Canada
Abdeljalil Gattal	University of Tebessa, Algeria
Abdel-Illah Mouaddib	University of Caen, France
Abdelouahab Moussaoui	University of Setif 1, Algeria
Abdelouahed Gherbi	ÉTS, University of Québec, Canada
Abderrahim Siam	University of Khenchela, Algeria
Boubaker Nour	Ericsson Research, Canada
Bouziane Brik	University of Burgundy, France
Brahim Nini	University of Oum El Bouaghi, Algeria
Chamseddine Talhi	ÉTS, University of Québec, Canada
Chawki Djeddi	University of Tebessa, Algeria
Cherif Taouche	University of Oum El Bouaghi, Algeria
Djamel Benmerzoug	University of Constantine 2, Algeria
Djamel Nessah	University of Khenchela, Algeria
Djamila Mechta	University of Setif 1, Algeria
Farid Mokhati	University of Oum El Bouaghi, Algeria
Farida Dahmani-Bouarab	University of Tizi Ouzou, Algeria
Fateh Boutekkouk	University of Oum El Bouaghi, Algeria
Flavien Balbo	Henri Fayol Institute, France
Hadj Ahmed Bouarara	University of Saida, Algeria
Hichem Haouassi	University of Khenchela, Algeria
Kamel Karoui	University of Carthage, Tunisia
Kenza Belhouchette	University of Oum El Bouaghi, Algeria
Kerkouche Elhillali	University of Jijel, Algeria
Khadidja Henni	Université TELUQ, Canada
Khaled Khalfaoui	University of Jijel, Algeria
Laid Kahloul	University of Biskra, Algeria
Mahieddine Djoudi	University of Poitiers, France
Mahmoud Emam	Faculty of Artificial Intelligence, Menoufia University, Egypt
Maroua Bouzid	University of Caen, France
Meriem Riahi	Higher Engineering School of El Manar, Tunisie

Mohamed Chaouki Babahenini	University of Biskra, Algeria
Mohamed Sedik Chebout	University of Oum El Bouaghi, Algeria
Mohamed Tayeb Laskri	University of Kofi Annan, Guinea
Mohammed el Habib Souidi	University of Khenchela, Algeria
Moufida Maimour	University of Lorraine, Nancy, France
Mounir Hemam	University of Khenchela, Algeria
Muhammad Shafiq	Cyberspace Institute of Advanced Technology, Guangzhou University, China
Nassim Seghir	University of Edinburgh, UK
Okba Kazar	United Arab Emirates University, UAE
Ouassila Hioual	University of Khenchela, Algeria
Rafik Menassel	University of Tebessa, Algeria
Reda Yaich	SystemX Institute, France
Rohallah Benaboud	University of Oum El Bouaghi, Algeria
Salah Merniz	University of Constantine 2, Algeria
Salim Zerrougui	University of Oum El Bouaghi, Algeria
Salima Ouadfel	University of Constantine 2, Algeria
Sofiane Zaidi	University of Oum El Bouaghi, Algeria
Sofiane Mounine Hemam	University of Khenchela, Algeria
Soufiane Boulehouache	University of Skikda, Algeria
Tahar Guerram	University of Oum El Bouaghi, Algeria
Toufik Maarouk	University of Khenchela, Algeria
Toufik Marir	University of Oum El Bouaghi, Algeria
Toufik Sari	University of Annaba, Algeria
Zianou Ahmed Seghir	University of Khenchela, Algeria

External Reviewers

Abd-ed-Daïm Tenachi	University of Oum El Bouaghi, Algeria
Abdelaali Bekhouche	University of Khenchela, Algeria
Abdelhamid Tioura	University of Khenchela, Algeria
Abdelheq Silem	University of Oum El Bouaghi, Algeria
Ahlem Meziani	University of Tebessa, Algeria
Ahlem Nasri	University of Oum El Bouaghi, Algeria
Asma Maziz	University of Oum El Bouaghi, Algeria
Atef Chorfi	University of Oum El Bouaghi, Algeria
Brahim Belgroun	University of Khenchela, Algeria
Brahim Djoudi	University of Constantine 1, Algeria
Hadia Cherrid	University of Constantine 1, Algeria
Hichem Rahab	University of Khenchela, Algeria
Kamel Tebessi	University of Oum El Bouaghi, Algeria

Khalid Ghoul	University of Oum El Bouaghi, Algeria
Lazhar Khamer	University of Souk Ahras, Algeria
Mihoub Mazouz	University of Ouargla, Algeria
Mohamed Amin Laouadi	University of Setif 1, Algeria
Mohamed Ben Othmane	University of Khenchela, Algeria
Ridha Mehalaine	University of Khenchela, Algeria
Yahia Menassel	University of Tebessa, Algeria
Youcef Himri	University of Bechar, Algeria
Zouheyr Tamrabet	National School of AI, Algeria

Contents

Machine Learning-Based Intrusion Detection Systems for the Internet of Drones: A Systematic Mapping Study

Mostafa Ogab[1(✉)], Sofiane Zaidi[1], Abdelhabib Bourouis[1], Carlos T. Calafate[2], and Ramzi Bouzoubia[1]

[1] Department of Mathematics and Computer Science, Research Laboratory on Computer Science's Complex Systems (ReLa(CS)2), University of Oum El Bouaghi, Oum El Bouaghi, Algeria
{mostafa.ogab,zaidi.sofiane,bourouis.abdelhabib, ramzi.bouzoubia}@univ-oeb.dz

[2] Department of Computer Engineering (DISCA), Universitat Politècnica de València, Valencia, Spain
calafate@disca.upv.es

Abstract. The increased deployment of drone-based solutions, and their integration into various sectors, necessitate robust security measures. Machine learning-based intrusion detection systems (IDS) have emerged as a promising solution to safeguard the Internet of Drones (IoD). In this paper, our objective is to systematically map the research landscape of machine learning-based IDS for the IoD, and to categorize studies based on machine learning models, IDS types, datasets, software environments, and attack classifications. This study aims to identify trends, gaps, and opportunities in this emerging field. To this end, we conducted a systematic mapping study, reviewing 12 studies published between 2014 and 2024. We categorized and analyzed the selected studies to present a comprehensive overview of the current state of research. Our approach incorporated elements from systematic review guidelines, particularly defining the search strategy and conducting a quality assessment. The analysis revealed diverse machine learning models used in the IoD context. The studies predominantly explore network-based IDS as the primary type of intrusion detection system. The software environments were primarily based on Python and its libraries, while the datasets employed varied widely. In terms of attack classification, multi-class classification was the main focus. This diversity in methodologies and focus areas indicates a lack of standardization, highlighting the need for more consistent approaches in future research. The findings highlight the significant potential of machine learning-based IDS to improve IoD security by effectively identifying and mitigating cyber threats.

Keywords: Internet of Drones · IoD · Intrusion Detection System · IDS · Machine Learning · Anomaly · Security · UAV

T. Marir et al. (Eds.): ICCSA 2024, ISEM 45, pp. 1–15, 2025.
https://doi.org/10.1007/978-3-031-90758-6_1

1 Introduction

The broad adoption and use of drones in many sectors have given rise to the Internet of Drones (IoD) [7], a network that enables seamless communication and coordination among drones. Drones, also known as Unmanned Aerial Vehicles (UAVs), are being used ever more for applications such as agriculture, military, search and rescue, delivery services, entertainment, healthcare and medical services, and traffic monitoring [18]. While the interconnectivity of drones within the Internet of Things (IoT) offers many benefits, such as improved operational effectiveness and enhanced capabilities, it also introduces significant security and privacy challenges. While drones communicate via the IoD, they also become vulnerable to various kinds of cyber attacks, such as jamming, spoofing, hijacking, and data breaches. These security weaknesses can result in significant ramifications, including compromised missions, infringements on privacy, and potential harm to individuals and property. Thus, it is crucial to emphasize the protection of IoD networks [17].

Intrusion Detection Systems (IDS) are crucial in identifying and mitigating security threats within the IoD. Specifically designed to monitor and analyze network traffic and system activity, they identify any warning signs of malicious behavior. IDS immediately alerts administrators when such activity is detected, enabling them to take the necessary actions against potential intruders. However, traditional IDS may not adequately handle the ever-changing and intricate nature of threats in IoD environments [5,9].

Researchers are now focusing on developing sophisticated IDS that leverage machine learning (ML) techniques to protect the Internet of Drones from cyber threats. ML algorithms can analyze large datasets, detect patterns, and quickly adapt to emerging risks, making them highly suitable for the dynamic security of the IoD. Hence, IDS may significantly improve their accuracy, minimize false positives, and efficiently deal with novel forms of attack [9].

Although previous studies have explored machine learning applications for IDS in conventional networks and IoT, several studies specifically address the unique challenges of the Internet of Drones. The dynamic and distributed nature of drone networks introduces specific vulnerabilities that traditional IDS models may not adequately address. Furthermore, additional research is needed on the comparative effectiveness of different machine learning algorithms in the context of IDS for the IoD. This gap in the literature necessitates a comprehensive analysis to identify the most suitable ML techniques for enhancing security of drone networks.

This study seeks to fill this gap by systematically mapping the current research landscape for ML-based IDS in the IoD. It aims to offer a thorough overview of the methodologies, algorithms, datasets, IDS types, attack classification, and software environments used in this domain by analyzing and categorizing existing related research from various databases. In particular, this study's main objective is to identify trends and gaps in the existing body of literature, providing insightful information for future research efforts and practical implementation. A comprehensive understanding of the latest advancements in

ML-based IDS for the IoD is required to design robust security measures to protect against the increasing range of cyberattacks aiming to compromise drone networks.

This paper is structured as follows: This section presents an introductory overview and examines relevant research. Section 2 thoroughly describes the review procedure adopted to conduct the systematic mapping study. Section 3 covers the selected research studies, answers the research questions, and discusses the investigation outcomes. Finally, Sect. 4 presents conclusions and highlights future research directions in the domain.

2 Methodology

The objective of this systematic mapping study is to provide an overview of the research on Machine Learning-Based Intrusion Detection Systems for the Internet of Drones (see Fig. 1). The study is based on the approach outlined by Petersen et al. [13], which provides a structured framework for conducting systematic mapping studies. The following steps were undertaken to ensure a rigorous and comprehensive review:

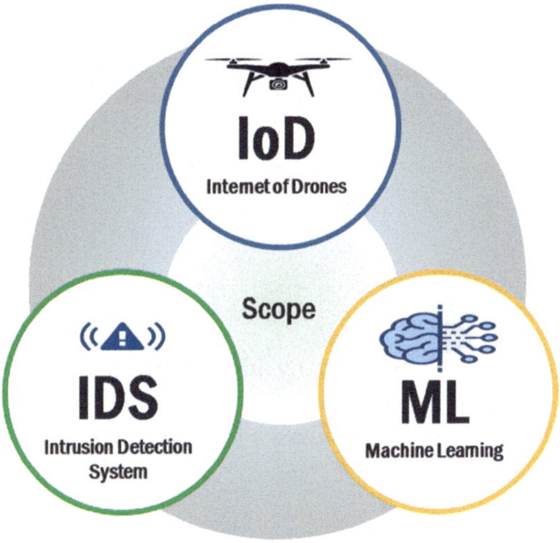

Fig. 1. Study Scope.

2.1 Research Questions

The first step involved formulating clear and specific research questions to guide the scope and focus of the study. The Research Questions (RQ) were designed to address key aspects of the research area, including machine learning models applied in the Internet of drones, intrusion detection systems types, datasets, attack classification, and software environment used. Table 1 presents RQs and their objectives. Based on this, the proposed RQs will be answered in Sect. 3.

Table 1. Research Questions.

RQ	Research Question	The aim of the question
RQ1	Which machine learning algorithms are applied in this IDS?	This question aims to identify the specific ML algorithms that have been utilized in developing IDS for the IoD. Understanding the algorithms applied will help in assessing their effectiveness and suitability for different types of intrusion detection.
RQ2	What types of IDS are implemented in this research?	This question seeks to categorize the types of IDS that have been developed and studied for the IoD. This includes Network-based, Host-based, and hybrid IDS, among others.
RQ3	Which datasets does this research use?	This question focuses on the datasets that researchers have used to train and evaluate ML-based IDS for the IoD. Knowing the datasets is crucial for understanding the context and limitations of the research findings.
RQ4	Which software environments do these experiments use?	This question aims to identify the software tools and platforms employed in the experimental setup of ML-based IDS research for the IoD. This includes programming languages, libraries, and frameworks.
RQ5	What attack classifications (binary or multi-class) are considered in this research?	This question examines whether the studies focus on binary classification (i.e., distinguishing between normal and malicious activities) or multi-class classification (i.e., identifying specific types of attacks) within their IDS implementations.

2.2 Search Strategy:

The literature search was conducted by identifying keywords and formulating search strings from research questions [13]. The search strings named Protocol were applied across multiple databases to ensure comprehensive coverage of relevant studies.

Databases: Scopus, IEEE Xplore, ScienceDirect, SpringerLink, ACM Digital Library, MDPI.

Keywords: "Machine Learning", "Deep Learning", "Anomaly", "Intrusion Detection System", "IDS", "Internet of Drones", "IoD", "Drone".

Protocol: ("Internet of Drone" OR Drone Or IoD) AND ("Intrusion Detection" OR IDS OR Anomaly) AND ("Deep Learning" OR "Machine Learning").

Keywords and search strings were formulated using Boolean operators to combine terms related to the scope. After applying the search protocol to several databases, we identified the various results outlined in Table 2. Conducted in 2024, this study utilized the reference management tool Mendeley to handle the large number of references and eliminate duplicates.

Table 2. Number of studies per database.

Database	Search results
Scopus	212
IEEE Xplore	3
ScienceDirect	86
SpringerLink	20
ACM Digital Library	6
MDPI	5

2.3 Study Selection

The study selection process was conducted in three phases: initial search, primary screening, and full-text review. The initial search yielded 332 papers, and after removing duplicates, 325 unique articles remained. During the primary screening phase, the titles and abstracts of these 325 articles were filtered considering the inclusion and exclusion criteria (see Table 3), resulting in 47 papers deemed potentially relevant. In the full-text review phase, the full texts of these 47 papers were thoroughly reviewed to assess their relevance and quality. This led to the exclusion of 35 papers due to a lack of specific focus on the research area or insufficient empirical evidence.

After this comprehensive review process, a total of 12 papers were selected as the final set of studies for inclusion in our systematic mapping. These papers represent the most relevant and high-quality research available on machine learning-based intrusion detection systems for the Internet of Drones, providing a solid foundation for our analysis.

2.4 Data Extraction

The data extraction process involved capturing relevant information from each included publication to provide a detailed overview of the research area suggested by this systematic mapping study. The following data points were extracted

Table 3. Inclusion and Exclusion Criteria.

	Criteria
Inclusion	Peer-reviewed journal articles, conference papers, and review articles. Studies specifically focused on ML-based IDS for the IoD. Articles published in English. Articles published between 2014 and 2024
Exclusion	Duplicated articles. Articles without available full texts. Non-peer-reviewed articles (e.g., editorials, opinion pieces). Articles that are not related to computer science.

from the research questions (RQs) mentioned at the beginning of this section: reference, year of publication, machine learning model applied, dataset utilized, software environment, IDS type, attack classification, and evaluation metrics used. Table 4 describes the outcomes of this process.

3 Results and Discussion

Statistics show that the distribution of publications across various databases indicates significant contributions from Scopus with twelve publications, followed by MDPI with five, ScienceDirect with three, SpringerLink with two, and IEEE Xplore with one publication (see Fig. 2). The temporal analysis reveals an increasing interest in recent years, starting in 2021 (see Fig. 3). The study includes a mix of venue types, predominantly eleven journal articles and a single conference paper (see Fig. 4), highlighting the preference for disseminating research findings through journals on the Internet of Drones (see Fig. 5).

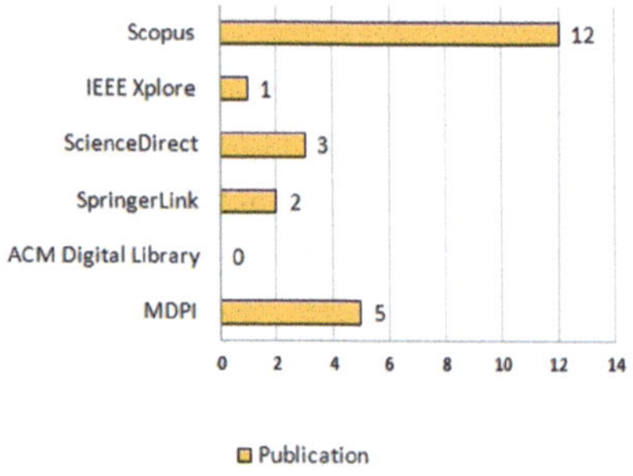

Fig. 2. Publications by Database.

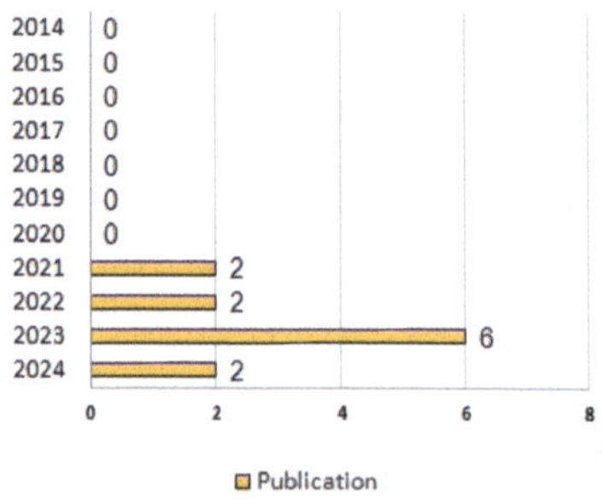

Fig. 3. Publications by Year.

Table 4. Analysis of the chosen studies.

Ref	Year	Model	Datasets	Software	IDS Type	Classification	Evaluation Metrics
Ramadan et al.[14]	2021	LSTM-RNN	KDDCup 99, NSL-KDD, UNSW-NB15, CICIDS2017, TON-IoT, WSN-DS	Python, TensorFlow, Spark	NIDS	Multi-class	• Accuracy • Precision • Recall • F1-Score • Critical Distance Diagrams (CDDs)
Perumalla et al.[11]	2021	Deep Neuro-Fuzzy Network	KDDCup 99	NS2	NIDS	Multi-class	• End-to-End Delay: Neptune attack - 1.253 s, Smurf attack - 5.117s • Precision: Neptune attack - 88.375%, Smurf attack - 87.631% • Recall: Neptune attack - 89.263%, Smurf attack - 89.893%
Al-Wesabi et al.[1]	2022	OPRFS-ODFNN (Optimal Deep Feed-forward Neural Network)	NSL-KDD	-	NIDS	Multi-class	• Accuracy: 99.54% • Precision: 91.33% • Recall: 84.52% • F1-Score: 86.20%
Alissa et al.[2]	2022	CSODAE-ID (Deep AutoEncoder)	NSL-KDD	Python	NIDS	Multi-class	• Accuracy: 99.12% • Precision: 98.25% • Recall: 98.24% • F1-Score: 98.24% • Selectivity: 99.40% • ROC Curve
Escorcia et al.[6]	2023	STFA-HDLID (Deep Belief Network)	TON-IoT, UNSW-NB15	Python	NIDS	Multi-class	• Accuracy: 99.51% and 98.85% • F1-Score • Sensitivity: 99.73% and 99.36% • Specificity • MCC
Ntizikira et al.[10]	2023	SP-IoUAV (CNN-LSTM)	CICIDS2017	Python, VSCode, OpenCV, PyTorch, NS3, OMNeT++	NIDS	Multi-class	• Accuracy: 99.98% • Precision: 99.93% • Recall: 99.92% • F1-Score: 99.92%
Heidari et al.[8]	2023	RBFNNs (Radial Basis Function Neural Networks)	UNSW-NB15, NSL-KDD, CICDDOS2019, CICIDS2017, AWID	Python, Keras, TensorFlow, Scikit-Learn	NIDS	Multi-class	• Specificity • F1-Score • Recall • Precision • Accuracy
Ashraf et al.[4]	2023	LSTM-RNN	CSE-CIC-IDS2018	-	NIDS	Multi-class	• Accuracy: 91.00% and 91.36% • F1-Score: 88.11% and 90.19% • Recall: 81.13% and 90.11% • Precision: 89.10% and 90.16%
Wu et al.[16]	2023	Q-TCID (Q-learning-based Two-layer Cooperative Intrusion Detection)	-	MATLAB	CIDS	Binary	• Attack Probability • Accuracy • Energy Consumption Rate • Per-Node Capture Rate
Perumalla et al.[12]	2023	OAOFS-MLIDS (XGBoost)	NSL-KDD, CICIDS2017	-	NIDS	Multi-class	• Accuracy: 99.53% and 99.59% • Precision: 85.73% and 95.44% • Recall: 92.91% and 96.79% • F1-Score: 88.57% and 96.09% • Geometric Mean • COH (Communication OverHead)
Tandra et al.[15]	2024	RegressionNet (LR and MLP)	NSL-KDD, KDDCup 99, STIN	Python, Keras, Scikit-Learn, Tensorflow	NIDS	Multi-class	• Accuracy: 99.89% • Average Accuracy: 97.90% • F1-Score: 99.80% • Recall: 99.80% • Precision: 99.80%
Alotaibi et al.[3]	2024	AMOA-DLID (Sparse AutoEncoder)	NSL-KDD	Python	NIDS	Multi-class	• Accuracy: 99.79% • F1-Score: 95.90% • Recall: 97.78% • Precision: 96.82% • AUC Score • ROC Curve • Precision-Recall Curve • Computational Time

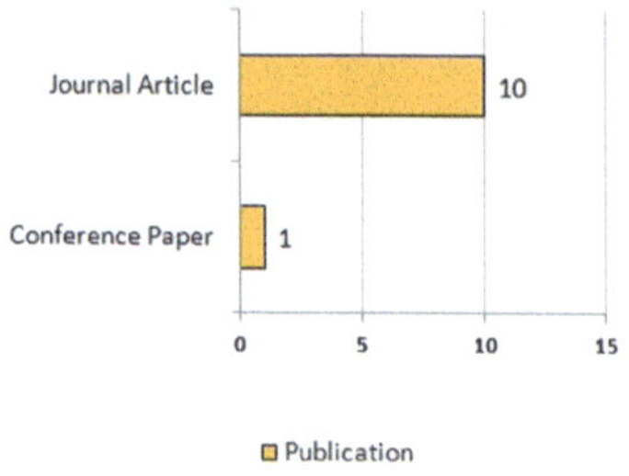

Fig. 4. Venue Types.

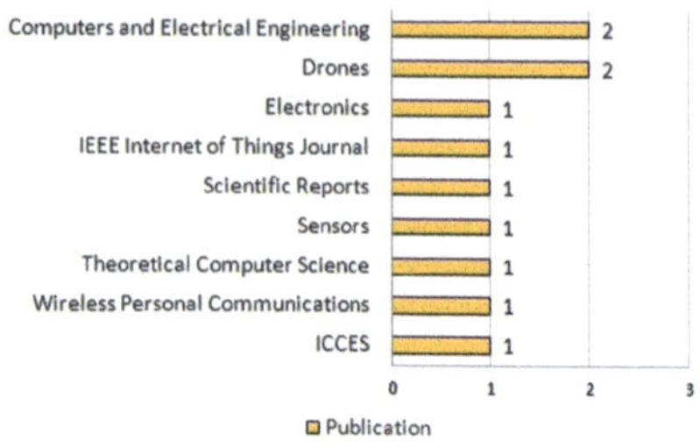

Fig. 5. Publications by Journal/conference.

Evaluation Metrics: Common evaluation metrics included accuracy, precision, recall, F1-Score, ROC-AUC, and false positive rate (FPR). These metrics provide insights into the effectiveness and reliability of the IDS models (see Fig. 7).

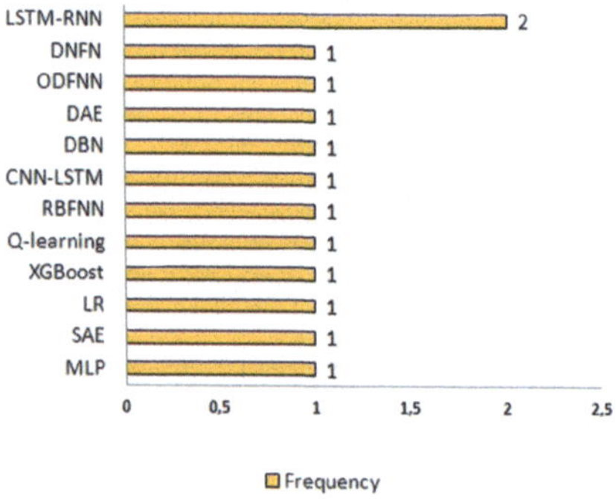

Fig. 6. Frequency of Machine Learning models.

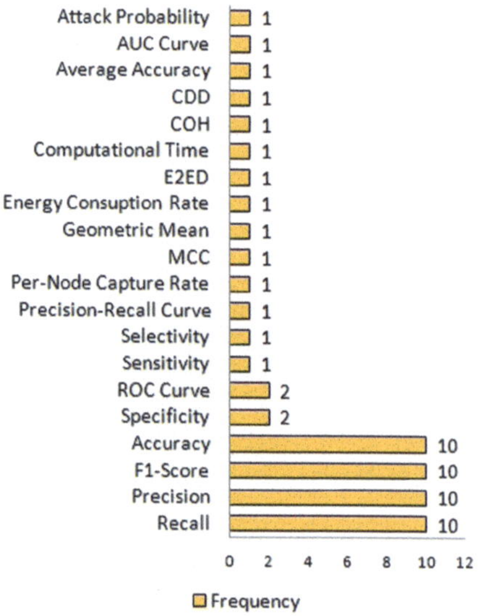

Fig. 7. Frequency of Evaluation Metrics Usage.

3.1 Synthesis and Key Findings:

RQ1, Machine Learning Models: The study identifies several machine learning algorithms employed to enhance security and detect intrusions effectively (see Fig. 6). The algorithms include LSTM-RNN (Long Short Term Memory-

Recurrent Neural Network) [4,14], which appears most frequently to reflect its robustness in handling sequential data and temporal dependencies; thus, it is essential for dynamic drone environments. Other algorithms, like Deep Neuro-Fuzzy Network (DNFN) [11], Optimal Deep Feed-forward Neural Network (ODFNN) [1], Radial Basis Function Neural Network (RBFNN) [8], and CNN-LSTM [10], combine Neural Networks with other techniques to leverage their strengths for improved detection accuracy. Researchers utilize unsupervised learning methods such as Deep AutoEncoder (DAE) [2], Deep Belief Networks (DBN) [6], and Sparse AutoEncoder (SAE) [3] for their capability to learn intricate patterns and identify anomalies without labeled data. Additionally, researchers employ Q-learning [16], a reinforcement learning method, to enable adaptive and responsive intrusion detection. The presence of traditional algorithms like Linear Regression (LR) and Multi-Layer Perceptron (MLP) [15], and ensemble methods like XGBoost [12] highlight the diversity of approaches, ensuring comprehensive security measures. The frequency and variety of these algorithms underscore their critical role in addressing the complex challenges of intrusion detection on the Internet of Drones.

The authors employed meta-heuristic algorithms in two main techniques for model construction: feature selection and parameter fine-tuning. For feature selection, which involves selecting a subset of relevant features, they used algorithms such as Opposition Poor and Rich Optimization (OPR) [1], Modified Deer Hunting Optimization (MDHO) [2], Sea Turtle Foraging Algorithm (STFA) [6], Oppositional Aquila Optimizer (OAO) [12], and Adaptive Mongoose Optimizer Algorithm (AMOA) [3]. For parameter fine-tuning, aimed at optimizing the model's performance, they utilized the Improved Mayfly Optimization (IMFO) [1], Crystal Structure Optimization (CSO) [2], Sparrow Search Optimization (SSO) [6], Coyote Optimization Algorithm (COA) [12], and Harris Hawks Optimizer (HHO) [3].

This study highlights the diverse range of machine learning algorithms employed to enhance security and effectively detect intrusions on the Internet of Drones. The integration of neural networks with other techniques underscores their importance in dynamic drone environments. The adoption of unsupervised learning methods and reinforcement learning, alongside traditional and ensemble algorithms, demonstrates a comprehensive approach to intrusion detection. Additionally, the use of meta-heuristic algorithms for feature selection and parameter fine-tuning further enhances model performance, addressing the complex challenges of securing the IoD.

RQ2, IDS Types: The IDS implementations were categorized as Network-based (NIDS), Host-based IDS (HIDS), or hybrid. Anomaly-based IDS were more prevalent, reflecting the need to detect unknown threats in the dynamic IoD environment (see Fig. 8).

The study reveals a predominant focus on Network-Based IDS, with eleven papers. In contrast, Host-based IDS is entirely absent, and Cooperative-Based IDS (CIDS) [16] is minimally represented with only one paper. This emphasis on NIDS reflects the importance of monitoring network traffic to secure drone communications. Still, it highlights a significant gap in research on HIDS, which could address individual drone security, and CIDS, which could enhance detection through collaborative strategies. The absence of HIDS suggests challenges related to resource constraints in drones. At the same time, the limited adoption of CIDS indicates its early research stage despite its potential for improving detection accuracy and robustness. These findings underscore the need for more balanced research efforts to develop comprehensive, multi-layered security solutions for the IoD, integrating the strengths of NIDS, HIDS, and CIDS to enhance overall system resilience.

RQ3, Datasets: Commonly used datasets include NSL-KDD, UNSW-NB15, CICIDS2017, Bot-IoT, and custom IoD datasets. These datasets vary in terms of the types of attacks they encompass and their applicability to the IoD context (see Fig. 9).

The study reveals a preference for comprehensive and realistic datasets, with NSL-KDD, KDDCup 99, CICIDS2017, and UNSW-NB15 being the most frequently used. NSL-KDD appears in five papers, reflecting its improvements over the original KDDCup 99, which is still widely utilized due to its historical significance. Modern datasets like CICIDS2017 and UNSW-NB15, appearing in four and three papers, respectively, highlight the shift towards capturing contemporary network behaviors and threat scenarios. Including IoT and drone-specific datasets, such as TON-IoT and UAV-IDS-2020, reveals an awareness of the unique security challenges in interconnected systems. The diverse range of datasets used emphasizes the field's efforts to develop robust and flexible IDS models capable of addressing various threat landscapes. This trend towards using realistic and varied datasets is crucial for ensuring IDS models' effectiveness in real-world applications.

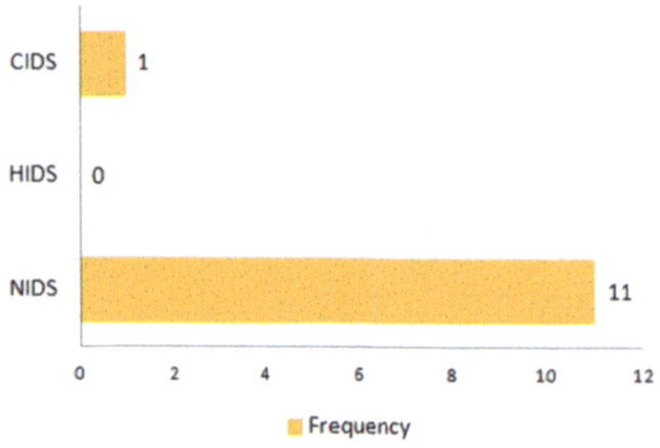

Fig. 8. Frequency of IDS Types Usage.

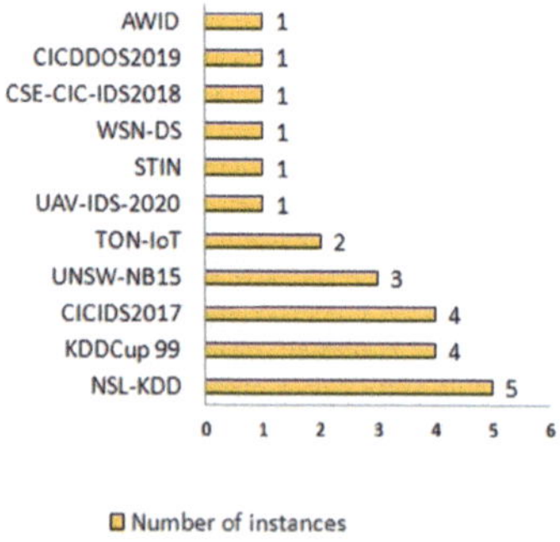

Fig. 9. Frequency of Dataset Usage.

RQ4, Software Environments: The study demonstrates an evident affinity for Python (see Fig. 11), which has been used in seven publications, mainly due to its flexibility and rich library support. Python libraries such as Keras, Scikit-Learn, and TensorFlow represent high-level machine-learning libraries. Their significant role lies in simplifying the implementation of intricate models, thereby enabling more effective and efficient research activities. Specialized software tools, including MATLAB, VSCode, Spark, OpenCV, Network Simulator 2 (NS2), NS3, OMNeT++, and PyTorch, are each used in one paper to meet specific research requirements, such as development environments, computer vision, and network simulation. Several studies [1,4,12] did not specify the software environment employed and, therefore, were excluded from consideration in this systematic mapping study. The use of various software environments underlines the need for flexible, scalable, and customized solutions to advance research on IDS and IoD.

RQ5, Attack Classification: Both binary and multi-class classifications were studied. Binary classification typically focused on distinguishing between normal and malicious activities, while multi-class classification aimed to identify specific types of attacks (see Fig. 10).

The study reveals a strong emphasis on multi-class classification in intrusion detection research, with eleven papers dedicated to it, and only one paper focusing on binary classification. This trend reflects the research community's devotion to tackling the complexity and specificity of different cyber threats, offering more refined detection and tailored protection techniques. Although multi-class classification requires much effort and resources, it offers valuable insights into developing tidy countermeasures. Although benefits exist, challenges such as class imbalance and computational requirements persist. Although less prevalent, the binary classification remains efficient because of its simplicity and ability to identify threats quickly. This highlights the importance of selecting the most appropriate approach that fits the application's requirements.

3.2 Limitations

The limitations of this study arise from its restricted scope, which exclusively analyzes intrusion detection systems based on machine learning in the domain of IoD. By focusing only on this specific field, the SMS risks excluding valuable insights and methodologies from associated fields, including drone or unmanned aerial vehicle networks, Flying ad-hoc networks (FANET), and the broader Internet of Things. These related areas provide valuable knowledge and techniques that could improve the comprehension and development of intrusion detection systems for the IoD. Not including any of these related domains in the SMS could limit the depth of its findings, and obstruct the discovery of novel ways and the applicability of its conclusions in broader contexts within the field of autonomous aerial systems.

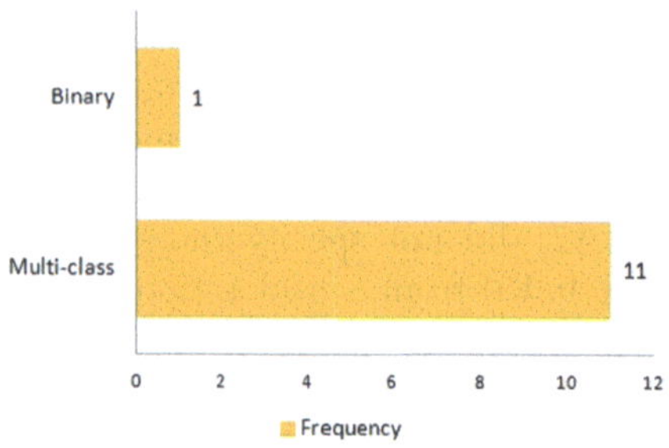

Fig. 10. Frequency of Attack Classification.

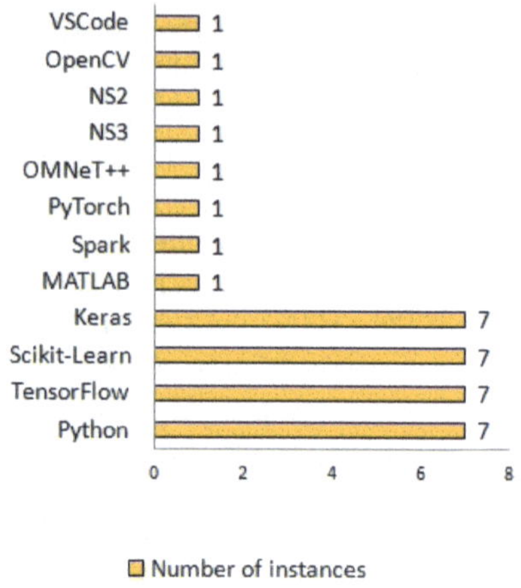

Fig. 11. Frequency of Software Environment Usage.

4 Conclusion

A systematic mapping study of machine learning-based intrusion detection systems for the Internet of Drones was conducted in this paper. The following section provides answers to the research questions posed in the methodology and discussed in the results and discussion section.

- **RQ1,** Machine Learning Models: The study reveals a diverse set of machine learning algorithms used for intrusion detection systems in the Internet of Drones. However, most researchers are increasingly leaning toward using hybrid models and deep learning techniques, along with employing meta-heuristic algorithms for feature selection and parameter fine-tuning, to improve model interpretability, reduce over-fitting, decrease training times, and enhance overall model performance.
- **RQ2,** IDS Types: The study predominantly focuses on network-based IDS, which monitors network traffic to secure Internet of drones network, but it does not cover attacks like GPS jamming and spoofing that affect drones during disconnection, highlighting the need for hybrid IDS that protect both network traffic and the drones themselves.
- **RQ3,** Datasets: The most used datasets are NSL-KDD, KDDCup 99, CICIDS2017, and UNSW-NB15; however, these datasets are considered outdated and have issues like imbalanced data, failing to fully cover IoD-specific scenarios, which highlights the need for recent, large, labeled, specialized datasets to train models.
- **RQ4,** Software environments: Most studies use Python and its libraries, as its flexibility and rich ecosystem make it the preferred choice for implementing machine learning models.

- **RQ5,** Attack Classification: The research predominantly focuses on multi-class classification, which aims to identify specific types of attacks but generally requires greater computational resources and longer processing times for both training and prediction.

Future research should address this study's limitations by expanding the scope to include related domains such as UAV networks, UAVs, FANET, and IoT, integrating insights and methodologies to enhance the understanding and development of an IDS for the IoD. Additionally, there should be a focus on developing lightweight, energy-efficient deep-learning models that can operate on resource-constrained drone hardware.

Acknowledgments. This work is fully funded by the General Direction of Scientific Research and Technological Development, Algerian Ministry of Higher Education and Scientific Research DGRSDT/MESRS.

Compliance with ethical standards

Disclosure of Interests. The authors declare that they have no competing interests.

Appendix

The table below presents the evaluation metrics used to assess the models discussed in Sect. 3 (see Table 5).

Table 5. Evaluation Metrics.

Metric	Meaning	Description
TPR	True Positive Rate, Sensitivity, Recall	$TPR = \frac{TP}{TP+FN}$
TNR	True Negative Rate, Specificity	$TNR = \frac{TN}{TN+FP}$
FPR	False Positive Rate	$FPR = \frac{FP}{FP+TN}$
FNR	False Negative Rate	$FNR = \frac{FN}{FN+TP}$
Accuracy		$Accuracy = \frac{TP+TN}{TP+TN+FP+FN}$
Precision		$Precision = \frac{TP}{TP+FP}$
F1-Score		$F1Score = 2 \times \frac{Precision \times Recall}{Precision+Recall}$
AUC Curve	Area Under Curve	$AUC = \sum_{i=1}^{n-1} \frac{TPR_i+TPR_{i+1}}{2} \times (FPR_{i+1} - FPR_i)$
		– n is the number of thresholds (distinct predicted probabilities).
ROC Curve	Receiver Operating Characteristic Curve	Plot TPR vs. FPR for each threshold to create the ROC curve.
CDD	Critical Distance Diagram	$CD = q_\alpha \sqrt{\frac{k(k+1)}{6N}}$
		– q_α is the critical value from the Studentized range statistic. – k is the number of classifiers. – N is the number of datasets.
MCC	Matthews Correlation Coefficient	$MCC = \frac{TP \cdot TN - FP \cdot FN}{\sqrt{(TP+FP)(TP+FN)(TN+FP)(TN+FN)}}$
COH	Communication Over-Head	$COH = \frac{TotalCommunicationCost - UsefulCommunicationCost}{UsefulCommunicationCost}$
PNCR	Per-Node Capture Rate	$PNCR = \frac{TotalPacketsCapturedbyaNode}{TotalTime}$
ECR	Energy Consumption Rate	$ECR = \frac{TotalEnergyConsumed}{TotalTime}$
GM	Geometric Mean	$GM = (\prod_{i=1}^{n} x_i)^{\frac{1}{n}}$
E2ED	End-to-End-Delay	$E2ED = N \cdot (d_{trans} + d_{prop} + d_{queue} + d_{proc})$

References

1. Al-Wesabi, F.N., et al.: Oppositional poor and rich optimization with deep learning enabled secure internet of drone communication system. Comput. Electr. Eng. **104**, 108368 (2022). https://doi.org/10.1016/j.compeleceng.2022.108368
2. Alissa, K.A., et al.: Crystal structure optimization with deep-autoencoder-based intrusion detection for secure internet of drones environment. Drones **6**, 297 (2022). https://doi.org/10.3390/drones6100297, https://www.mdpi.com/2504-446X/6/10/297
3. Alotaibi, S.S., Sayed, A., Elhameed, E., Alghushairy, O., Assiri, M., Ibrahim, S.S.: Enhancing security in IOT-assisted UAV networks using adaptive mongoose optimization algorithm with deep learning. IEEE Access **12**, 63768–63776 (2024). https://doi.org/10.1109/ACCESS.2024.3392618
4. Ashraf, S.N., et al.: IOT empowered smart cybersecurity framework for intrusion detection in internet of drones. Sci. Rep. **13**, 18422 (2023). https://doi.org/10.1038/s41598-023-45065-8
5. Derhab, A., et al.: Internet of drones security: taxonomies, open issues, and future directions. Veh. Commun. **39**, 100552 (2023). https://doi.org/10.1016/j.vehcom.2022.100552
6. Escorcia-Gutierrez, J., et al.: Sea turtle foraging algorithm with hybrid deep learning-based intrusion detection for the internet of drones environment. Comput. Electr. Eng. **108**, 108704 (2023). https://doi.org/10.1016/j.compeleceng.2023.108704
7. Gharibi, M., Boutaba, R., Waslander, S.L.: Internet of drones. IEEE Access **4**, 1148–1162 (2016). https://doi.org/10.1109/ACCESS.2016.2537208
8. Heidari, A., Navimipour, N.J., Unal, M.: A secure intrusion detection platform using blockchain and radial basis function neural networks for internet of drones. IEEE Internet Things J. **10**, 8445–8454 (2023). https://doi.org/10.1109/JIOT.2023.3237661
9. Khraisat, A., Gondal, I., Vamplew, P., Kamruzzaman, J.: Survey of intrusion detection systems: techniques, datasets and challenges. Cybersecurity **2**(1), 1–22 (2019). https://doi.org/10.1186/s42400-019-0038-7
10. Ntizikira, E., Lei, W., Alblehai, F., Saleem, K., Lodhi, M.A.: Secure and privacy-preserving intrusion detection and prevention in the internet of unmanned aerial vehicles. Sensors **23**, 8077 (2023). https://doi.org/10.3390/s23198077
11. Perumalla, S., Chatterjee, S., Kumar, A.P.S.: Block chain-based access control and intrusion detection system in IOD. In: 2021 6th International Conference on Communication and Electronics Systems (ICCES), pp. 511–518 (2021). https://doi.org/10.1109/ICCES51350.2021.9488948
12. Perumalla, S., Chatterjee, S., Kumar, A.: Modelling of oppositional Aquila optimizer with machine learning enabled secure access control in internet of drones environment. Theoret. Comput. Sci. **941**, 39–54 (2023). https://doi.org/10.1016/j.tcs.2022.08.019
13. Petersen, K., Vakkalanka, S., Kuzniarz, L.: Guidelines for conducting systematic mapping studies in software engineering: an update. Inf. Soft. Technol. **64**, 1–18 (2015). https://doi.org/10.1016/j.infsof.2015.03.007
14. Ramadan, R.A., Emara, A.H., Al-Sarem, M., Elhamahmy, M.: Internet of drones intrusion detection using deep learning. Electronics **10**, 2633 (2021). https://doi.org/10.3390/electronics10212633

15. Tandra, N., Babu, C.N., Dhanke, J., Sudhakar, A.V., Rao, M.K., Ravichandran, S.: Enhancing security and privacy in small drone networks using 6G-IoT driven cyber physical system. Wireless Pers. Commun. (2024). https://doi.org/10.1007/s11277-024-11138-8
16. Wu, M., Zhu, Z., Xia, Y., Yan, Z., Zhu, X., Ye, N.: A Q-learning-based two-layer cooperative intrusion detection for internet of drones system. Drones **7**, 502 (2023). https://doi.org/10.3390/drones7080502
17. Yahuza, M., et al.: Internet of drones security and privacy issues: taxonomy and open challenges. IEEE Access **9**, 57243–57270 (2021). https://doi.org/10.1109/ACCESS.2021.3072030
18. Yang, W., Wang, S., Yin, X., Wang, X., Hu, J.: A review on security issues and solutions of the internet of drones (2022). https://doi.org/10.1109/OJCS.2022.3183003

A Comprehensive Study on Artificial Intelligence of Things Optimization Approaches

Mohamed Amine Ghamri(✉), Badis Djamaa, Mohamed Akrem Benatia, and Takieddine Boumediri

Ecole Militaire Polytechnique, Bordj el Bahri, Algiers, Algeria
mohammedamine.ghamri@emp.mdn.dz
http://www.emp.mdn.dz/

Abstract. The convergence of artificial intelligence (AI) concepts with the Internet of Things (IoT) has ushered in the era of Artificial Intelligence of Things (AIoT), where interconnected devices collect and analyze data from the environment. This amalgamation presents a novel challenge of delivering rapid services, especially in sensitive domains requiring real-time responses. Edge computing emerges as a solution, bringing computational and storage capabilities closer to data sources. However, this demands efficient deep neural networks (DNNs) tailored for AIoT applications. This article explores optimization strategies for AIoT, including collaborative learning, model optimization techniques such as pruning, quantization and deployment considerations. We present a comprehensive study for developing and deploying DNNs optimized for AIoT inference, encompassing model design with autoML techniques, optimization through pruning and quantization, exploiting collaboration between different entities, and deployment onto edge devices. This study aims to address the challenges of AIoT by optimizing DNNs for efficient and scalable deployment on resource-constrained edge devices.

Keywords: Artificial Intelligence of Things · Neural Architectural Search · Model Pruning · Model Quantization · Split computing · Federated Learning

1 Introduction

The advancement of information and communication technology has ushered in a transformative era known as the Internet of Things (IoT), where devices are interconnected to collect data from the environment through on-device sensors. Concurrently, the evolution of artificial intelligence (AI), particularly deep neural networks (DNNs), has revolutionized various domains such as computer vision and speech recognition, enabling the delivery of higher-quality services by analyzing collected data. This convergence of AI concepts and IoT devices has given rise to the concept of Artificial Intelligence of Things (AIoT).

T. Marir et al. (Eds.): ICCSA 2024, ISEM 45, pp. 16–30, 2025.
https://doi.org/10.1007/978-3-031-90758-6_2

However, this amalgamation presents a novel challenge: the need for rapid delivery of services, especially in sensitive domains requiring real-time responses (e.g., healthcare or autonomous vehicles). One solution to this challenge is to bring computational and storage capabilities closer to the data collection point, known as the edge computing paradigm. Due to the resource constraints of edge devices, including computational complexity, energy consumption inherent in AI models, and fluctuating network connectivity, specialized optimization techniques are essential to adapt AI execution effectively in these environments.

This article explores optimization strategies designed for AIoT applications, encompassing a wide array of techniques found in current literature. Our aim is to offer a comprehensive review of these approaches, highlighting trends, challenges, and opportunities to assist researchers in identifying and addressing key issues in the field.

This paper is structured as follows:

- Sect. 2 delves into Context and motivation.
- In Sect. 3, we focus state of the art works according to the given taxonomy
- Sect. 4 is for discussion, trends and future opportunities.
- The final section pertains to the conclusion and future prospects of this study.

2 Context and Motivation

AIoT, encompassing diverse applications across numerous domains, is exemplified through various impactful use cases.

In [29], researchers focused on diabetic retinopathy detection, emphasizing deployment in rural areas with limited network connectivity. They employed EfficientNet-B4 for classifying fundus images using a MSE loss function to ensure accurate severity level classification. Hardware included the U2P Board, leveraging the Intel Distribution of OpenVINO toolkit for optimized model deployment in resource-constrained environments.

In another notable study [24], the authors proposed an AIoT system for intelligent industry monitoring. This system collects image data from IoT devices, using cameras for classifying defective products and estimating defect severity. The system uses model partitioning, deploying lower model layers on the fog layer and higher layers on cloud servers. An early exit layer in the fog layer models allows for quick classification and regression tasks locally. If the early exit is inefficient based on a predefined loss threshold, intermediate results are sent to the cloud for further processing. This approach balances quick inference with the accuracy benefits of cloud processing

In the domain of smart agriculture, authors in [31] developed an edge-based model using Tiny-YOLOv3 for automatic apple detection, crucial for improving supply chain management. The lightweight model was optimized for small object detection and deployed on edge devices like Jetson Nano and Raspberry Pi, accelerated by a Neural Computing Stick (NCS). Evaluation metrics, including frames per second (FPS) and energy efficiency, confirmed its effectiveness in delivering fast and cost-efficient apple detection.

In [43], authors proposed a smart monitoring system for students that estimates stress levels using multimodal data (facial expressions, voice signals, and vocal content). Data from cameras and voice recorders are collected at the IoT layer, with preprocessing tasks such as normalization, centering images, and NLP processing (removing stop words, tokenizing, and stemming) handled at the fog layer. DNN models, including VGG16 for facial expression, LSTM for voice pitch, and Naive Bayesian techniques for text, are deployed on the cloud layer for stress detection. The system automatically generates alerts to school authorities upon detecting stress, leveraging IoT to reduce preprocessing overhead and network bandwidth by processing data locally at the fog layer.

These studies span diverse application domains, underscoring the pervasive relevance of the Artificial Intelligence of Things (AIoT). By identifying commonalities across these works, we can formalize a definition for AIoT. An AIoT system is characterized by the triplet ¡E, M, R¿: **E** denotes participant entities like devices and cloud platforms, **M** includes AI models such as DNN, and **R** encompasses relationships intrinsic to interconnected systems, encompassing collaborations between entities and models.

Several surveys in the literature [28,32,45,57] have categorized AIoT optimization approaches and highlighted their challenges. For instance, the survey in [57] categorizes works based on their primary objectives: 'perceiving' involves data collection and processing, 'learning' focuses on dynamic adaptation to environmental changes, 'reasoning' centers on deriving insights or making decisions, and 'behaving' encompasses operational capabilities, from passive response to active exploration and control.

In our paper, we explicitly align these categories with the elements of AIoT, providing a clearer framework for researchers to contextualize their work. This approach strengthens the connection to AIoT and enables comprehensive optimization across all its defined components.

3 State of The Art

In this section, we present a taxonomy of AIoT optimization approaches and discuss notable works in the field.

3.1 Model Optimization Works

Works focus on acting directly on the DNN model, which corresponds to **M** element in the above definition, so it better fits AIoT requirements, through works like pruning, quantization and distillation [46]. the optimization operation can occur at many possible stages. We can consider that such works mainly follow the pipeline outlined in Fig. 1.

The process begins with the design of AI models tailored to specific tasks and optimized for deployment on particular devices. Research in this area focuses

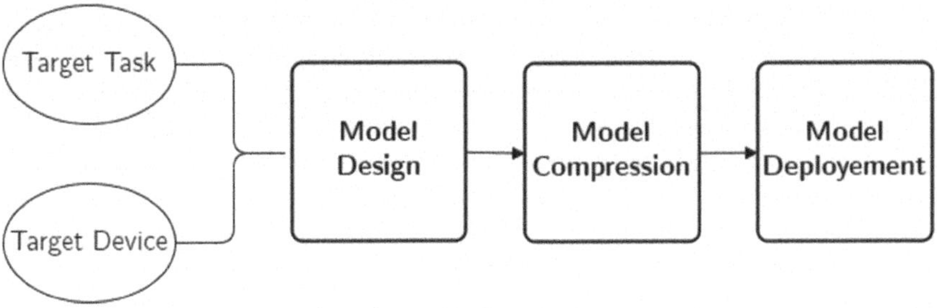

Fig. 1. The main model optimization pipeline

on developing and adapting DNN and other AI models to meet these requirements effectively. One well-known model family is MobileNet family, starting with MobileNetV1 [13], based on depth-wise separable convolutions, optimizing for mobile and embedded devices. Subsequent versions include MobileNetV2 [40], utilizing inverted residuals for improved accuracy. MobileNetV3 [12] integrates Neural Architecture Search for efficiency, with advanced non-linearities. The latest iteration, MobileNetV4 [36], introduces designs like the Universal Inverted bottleneck block and Mobile MQA for mobile acceleration, ensuring universal efficiency in architecture designs. Another notable model is Squeezenet [14], which employs strategic design approaches to minimize model parameters while preserving accuracy. This is achieved through innovations like the fire module, a fundamental component of the architecture. Previous works did what we can refer to as manual model design. Another line of research is automatic model design, which involves automatically selecting a set of models from a population, performing selections based on a search strategy, and applying automated modifications. This process is repeated iteratively until a satisfactory optimization target is achieved according to an evaluation function. This type of search is known as neural architecture search (NAS) [5]. Where works can be categorized according to the search strategy, With the fact that a model can be seen as a variable length string, authors in [58] developed NAS-RL a NAS search technique based on reinforcement learning (RL), this technique is built around a Recurrent neural network (RNN) based controller that generates the DNN description (string) from which the corresponding DNN architecture is built, this NAS process is then adjusted by RL algorithm in order to optimize the expected validation accuracy of the model. Another RL based search strategy is MetaQNN [1], where authors trained an agent with Q Learning algorithm, so it sequentially selects layers, forming a path that represents the DNN architecture. They employed epsilon-greedy policy with the state space being the layers parameters, and the action is the next layer decision. Similarly to [58], the reward function consists of the validation accuracy. Other works rather employ evolutionary algorithms, like in [38], where authors encoded each architecture as a graph, and referred to it as DNA. The population comprises trained architectures, and their fitness is evaluated based on validation accuracy. Offspring are generated through mutation operations such as adding or removing convolutional operations. The evolutionary process retains architectures with superior fitness while discarding others, continually generating offspring to improve model performance. This approach has shown promising results in optimizing classification models. the mentioned NAS works emphasize task-specific optimizations, like for image classification [1,38,58]. However, both architecture and hardware significantly influence latency as observed in [22]. As a result, researchers focuse on the more specialized hardware-aware NAS (HW-NAS) that prioritizes achieving a balance between accuracy and performance rather than just maximizing model accuracy [6]. This approach is crucial in AIoT, as it directly addresses the constraints of interconnected hardware networks. An example of such works is the study [44], where authors introduced an evolutionary NAS method focused on maximizing similarity to a baseline model using Representation Mutual Information (RMI). They constrained their search by hardware limitations, ensuring that computational costs like FLOPs remain within specified thresholds. Genetic algorithms were employed to optimize models across various edge devices, demonstrating efficient search and promising performance improvements.

The initial design of the model completed, research then shifts towards compressing these models through various learning techniques. One significant area of focus is applying optimization techniques to existing DNN models to enhance their time or energy efficiency while maintaining accuracy. This stage corresponds to the second phase of the pipeline illustrated in Fig. 1. Among the various compression techniques, model pruning stands out.

Model pruning involves selectively removing connections, neurons, or weights from a pre-trained neural network, aiming to create a more compact and streamlined model while preserving or even improving its performance. This technique is founded on the "lottery ticket hypothesis" [9], which suggests that each DNN model contains a submodel, or sparse model, that, when retrained, can achieve comparable accuracy to the original model. This hypothesis presents a valuable opportunity in the AIoT context, as it indicates the potential for creating sparse models, which require lower computational and storage capacities.

Based on the objective of removing elements, pruning techniques can be categorized into unstructured and structured pruning approaches. Unstructured pruning involves removing weights based on heuristics like magnitude [11], whereas methods like Single Shot Network Pruning (SNIP) [21] gauge the impact on the loss function when deciding whether to retain or discard a weight. In their study [18], authors explored both heuristics by evaluating how pretraining (using large models) versus training from scratch affects parameter changes. They introduced a single-shot pruning method that utilizes a metric combining loss gradients and weight magnitudes. In contrast to typical approaches that concentrate on pruning feedforward networks, the authors of [23] introduce a magnitude-based pruning technique tailored for recurrent neural networks (RNNs). They categorize connections into three types based on their interactions with recurrent neurons and apply an optimization-based pruning method to compress each category of connections individually.

On the other hand, structured pruning involves removing grouped sets of weights, such as entire neurons or channels, instead of individual weights. Like in the work of [34], where authors proposed an iterative method for pruning, that iterate over pruning and fine tuning, the pruning criteria is based on the Taylor approximation of the loss function. To enhance efficiency in IoT scenarios, authors in [35] propose a novel pruning approach for DNNs. Their method involves two main phases: structure learning and weight learning. Channel significance is evaluated using the Taylor criterion and a long-term feature-saliency metric derived from batch-wise evaluations. Channels are reallocated based on layer saliency, enhancing efficiency without sacrificing performance. Furthermore, the authors optimize the distributed learning process for IoT applications. Instead of training large DNNs solely on cloud servers, they retrain and prune cloud-trained models to create compact networks suited for edge nodes. This strategy preserves model knowledge, leading to faster convergence and improved efficiency tailored to specific IoT tasks.

Besides, within the network compression techniques illustrated in Fig. 1, research on network quantization [10] is of significant interest. This technique involves converting high-precision parameter values (typically floating-point types) into lower-precision integer values, while ensuring minimal loss in accuracy.

The authors in [54] introduce two innovative network quantization methods: single-level network quantization (SLQ) for high-bit quantization and multi-level network quantization (MLQ) for extremely low-bit quantization (ternary). This study pioneers in addressing network quantization from both width and depth perspectives. At the width level, parameters are divided into two groups, one undergoing quantization and the other being re-trained to mitigate quantization loss, with SLQ enhancing width-level performance through parameter distribution. At the depth level, incremental layer compensation is used to iteratively quantize layers, effectively reducing quantization loss with each iteration. While authors in [55] proposed a novel approach of quantization by reformulating the quantization operation as combination of binary quantization that can be considered as scaled unit step functions.

3.2 Hardware Optimization Works

As edge computing becomes increasingly integral to AIOT applications, researches focus on accelerating technologies to optimize computing time, particularly in DNN applications, in other words, optimizing at the **E** element of the AIoT definition. Such works include acceleration using Graphical Processing Units (GPUs). An example is the work in [41], where authors developed a robust hand-held device utilizing AI and Chest X-ray (CXR) imaging for COVID-19 detection. The device, powered by an NVIDIA Jetson Nano GPU module, employs a ResNet-50 CNN model to classify CXR images in real-time for COVID-19, Tuberculosis, and Normal cases. This high-performance device shows promise in enhancing diagnostic accuracy and efficiency, potentially aiding healthcare systems during outbreaks like COVID-19.

In [3], researchers explored GPU virtualization via frameworks to enhance edge device computing resources without significantly increasing power consumption. They utilized rCuda middleware [42] for transparent GPU computing, where edge nodes dynamically request GPU computation via CUDA function calls pipelined to rCuda servers. The approach demonstrated improved computational efficiency and substantial energy savings by offloading GPU workloads to backend servers, maintaining transparency to programmers.

Beyond GPUs, other works explore hardware platforms such as Field Programmable Gate Arrays (FPGAs). Such works include [17], where authors developed an FPGA system-on-chip architecture using a Zynq 7030 device to accelerate Machine Learning (ML) algorithms in edge environments. This architecture supports dynamic deployment of ML functions locally or remotely, demonstrating competitive performance with reasonable resource usage. It includes a software component with an OpenStack worker for task deployment and a hardware component featuring an AXI interconnection module for shared memory access between CPU and programmable logic. In another significant contribution to agriculture, authors in [26] developed 'LiteCNN,' a lightweight CNN model for plant disease identification. Deployed on a ZYNQ Z7-Lite 7020 FPGA board, the model was optimized using techniques like loop unrolling and pipelining. The study highlights the development of a low-power and high-accuracy terminal for real-time plant disease identification in agricultural settings.

In addition to research focusing on computation acceleration, another line of inquiry emphasizes advancements in communication facilities. The authors in [47] focused on optimizing IEEE 802.15.4 Time-Slotted Channel Hopping (TSCH) within the IPv6 stack for Industrial IoT (IIoT). They addressed challenges such as traffic variability and burstiness by enhancing the Minimal Scheduling Function (MSF) with their Hybrid Proactive Scheduler with Adaptive Channel Selection (HPS). HPS dynamically calculates cell requirements based on resource utilization, Expected Transmission Count (ETX), and buffer allocation. It incorporates autonomous scheduling for burst traffic and a whitelisting mechanism for optimal channel selection, implemented in Contiki-ng. Evaluation in simulated and real-world settings showed significant improvements in reliability, latency, and power consumption, highlighting HPS's effectiveness in enhancing network performance and overcoming existing limitations.

The authors in [20] investigated IP multicast protocols for IoT communication, focusing on the Multicast Protocol for Low-Power and Lossy Networks (MPL). They identified MPL's drawbacks and proposed enhancements to mitigate resource consumption, including memory, bandwidth, and energy. Through simulations and experiments, their improved MPL protocol showed reduced resource usage, improved end-to-end delay performance, and maintained data packet delivery reliability, showcasing significant optimizations for multicast routing in resource-constrained LLNs.

3.3 Relationship Optimization Works

Optimizing AIOT systems involves leveraging collaboration to maximize efficiency, thus, the better exploitation of the **R** element of the AIoT definition. Notably, powerful yet costly models highlight the potential for improved device connectivity. Vertical collaboration, such as computation offloading between different models, and horizontal collaboration, through distributed AI or multi-agent RL [2], play significant roles. One innovative concept is split computing, depicted in Fig. 2, where models can be divided into sub-models by selecting cut layers (L_i). This approach allows an entity to perform partial inferences up to the selected cut layer and transmit results to another model for further computation [30].

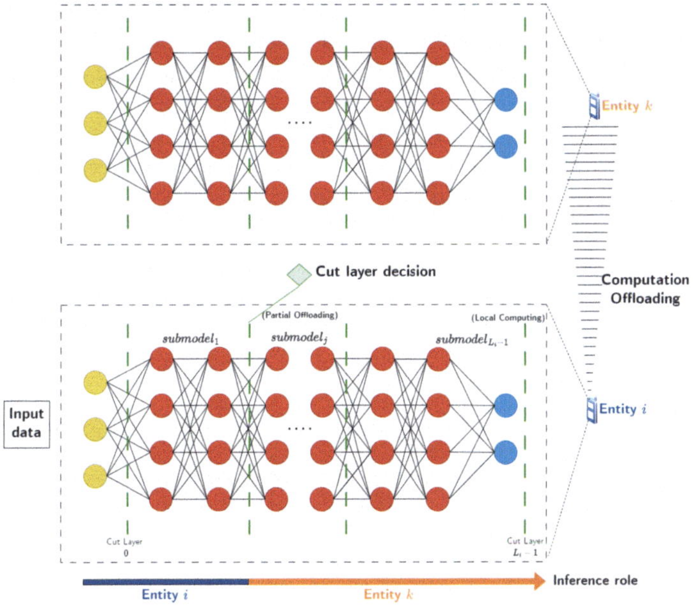

Fig. 2. Split computing process

This property is relevant In the AIOT context, where connectivity plays a crucial role, influencing dynamic decisions on inference cut layers based on criteria like energy consumption and time constraints, as explored in [16]. Their approach involves selecting cut layers dynamically using a multi-objective function considering energy, time, and accuracy. Recent research trends favor sequential decision-making with RL for runtime computation offloading decisions, ideal for AIOT environments with dynamic temporal variations.

In [37], researchers introduced a Deep Reinforcement Learning (DRL) algorithm for delivering high-quality service under strict time constraints and achieving a predefined confidence level. The system utilized 'fast' models on edge devices and 'full' models on edge servers, with task offloading decisions based on local caching, cumulative confidence levels, and time constraints. They integrated a resource allocation subroutine using second-order cone programming (SOCP) to optimize shared resource allocation like server CPU and network bandwidth.

In their study [53], Yilin et al. focused on optimizing offloading decisions, particularly selecting inference cut layers and offloading targets to minimize latency and energy consumption. They employed a Multi-Agent Reinforcement Learning (MARL) approach with Q-learning and Deep Actor-Critic networks, incorporating a policy exchange mechanism for improved training. State variables considered included data input size, network conditions, and historical latency and energy consumption.

In [8], researchers developed an optimal offloading strategy for Large Language Models (LLMs), focusing on GPT-J-6B [51] deployed on edge and cloud servers. Devices selected offloading servers based on available resources, considering server workload, network bandwidth, and distance. The approach used a Partially Observed Markov Decision Process (POMDP) formulated to learn a generative model capturing state probabilities from partial observations, guided by a reward mechanism optimizing response latency and quality. Unlike previous studies, this work integrated resource allocation directly into the reinforcement learning framework. Where the previously mentioned works focused on partitioning the DNN computation between entities, other works rather work on partitioning the data between the entities, such works perform what we refer to as the federated learning [56].

4 Discussion

We provide in this section a summary and discussion of state-of-the-art works, highlighting challenges, trends, research opportunities, and future directions.

4.1 Model Based

Model-based optimization directly targets AI models, especially DNNs, to enhance efficiency. Manual design techniques incorporate architectural elements to improve accuracy or reduce model complexity, making them suitable for constrained devices. These methods provide flexibility and greater explainability but require advanced knowledge and thorough mathematical analysis. On the other hand, automatic design techniques are simpler and do not need advanced understanding, though they offer less explainability. They focus on developing efficient search methods with specific target objectives, with the main challenge being the modeling aspects, such as designing objectives and selecting search algorithms.

A prevalent trend in model design combines automatic and manual techniques. Automatic design searches through models, while manual design incorporates specific architectural updates. Examples include MobileNetV3 [12],

MobileNetV4 [36], and the EfficientNet family [48]. EfficientNet began with a neural architecture search for EfficientNet-B0, followed by compound scaling for EfficientNet-B1 to B7, achieving competitive results in classification and object detection, leading to the EfficientDet family [49].

In NAS, encoding is crucial for achieving good results and shaping the search strategy. The main challenge is effectively modeling the search problem to capture all DNN characteristics. NAS methods are computationally demanding, leading efforts to optimize both the problem modeling and the search procedure. A comprehensive survey [39] identified these challenges and proposed a related taxonomy. HW-NAS operates like classical NAS but includes hardware-related constraints, making it more suitable for AIoT applications. These constraints address hardware limitations and guide the search procedure by reducing the search space or using penalty mechanisms, ensuring the resulting models are efficient and deployable on constrained devices.

Pruning plays a crucial role in model deployment pipelines by reducing parameters, thereby enhancing computation and storage efficiency through neural network sparsity. Unstructured pruning provides flexibility in model size reduction, whereas structured pruning accelerates computations using hardware accelerators like GPUs. For example, leveraging 2:4 sparsity with Nvidia's TensorCores has been explored [33]. Challenges include mitigating high training times, which are addressed through flexible training schedules and distributed learning architectures tailored for IoT networks, as demonstrated in recent studies [35].

Quantization not only reduces network size but also enhances performance by leveraging faster integer operations over floating-point equivalents. In their study, authors in [19] extensively analyzed both pruning and quantization techniques, highlighting that while quantization generally outperforms pruning, there are specific scenarios where each technique excels. Therefore, an effective approach in research involves integrating both methods to maximize their synergistic benefits.

All works emphasize training with high computational resources before deployment, which, while effective, poses challenges in AIoT environments. These environments generate large volumes of data, potentially requiring model retraining due to novel distributions. Novel approaches like continual learning [52] focus on integrating new data while preserving prior knowledge. On-device learning, as opposed to the time-consuming pipeline in Fig. 1, conducts both inference and training processes on constrained devices, as explored in [32].

4.2 Hardware Based

Studies focusing on hardware-based optimization often employ hardware acceleration for computationally intensive operations, leading to substantial time savings. These efforts also investigate enhancements in specialized communication technologies.

Critical to edge computing system development is the creation of GPU algorithms optimized for parallel computing, particularly for DNNs. This field benefits from robust hardware and supporting software tools, including deployment platforms like OpenVINO[1], format exchange frameworks such as ONNX[2], and high-level computation frameworks enabling automatic compilation. Hardware acceleration provides substantial time savings but raises energy consumption challenges, necessitating careful management of the trade-off between speed and energy efficiency. Hardware virtualization technology shows promise by enabling lightweight edge devices to offload computation to more powerful edge servers, acting as clients that request resources. Other essential hardware includes system-on-chip (SoC) solutions and microcontrollers like FPGA and Raspberry Pi, crucial for efficient edge computing.

In communication, protocols like Bluetooth Low Energy (BLE) [25] are crucial for energy-efficient IoT nodes. Optimizing these protocols can enhance efficiency, emphasizing the integration of communication advancements with hardware and algorithmic improvements in edge computing. Intelligent Reflecting Surfaces (IRS)[15] utilize backscatter technologies to dynamically adjust signal reflections, improving signal strength and coverage in environments with obstacles or interference. This innovation enhances communication efficiency and reliability in IoT networks.

4.3 Relationships Based

A growing research trend in the AIoT field involves leveraging collaboration between entities, capitalizing on its inherent potential. In the realm of RL-based methodologies, which is emerging as a trend, split computing works can be categorized into two main stages: problem design involving state, action, reward, etc., and RL algorithmic enhancements. Effective design must encompass the dynamic characteristics of IoT networks, influencing decision-making processes. Algorithms must also be efficient, considering deployment scenarios such as constrained devices where lightweight algorithms like Q-learning (tabular and simple) are preferable. In contrast, more complex state spaces necessitate DRL variants.

Due to security challenges, federated learning (FL) appeared as a new strategy yo maintain the devices'data confidentiality [27]. A basic FL setup allows edge devices to collaboratively learn an AI model without exchanging their data. It starts from a server which send the initial configuration to these devices, they, in their turn, train local models and send them to the server for aggregation. This process continues until convergence of the model.

Federated learning when applied in AIoT suffer from many communication challenges. The bandwidth of participating devices may lead to a problem of not uploading the model. A solution proposed by [4] describes the collaborative federated learning, in which, a device with limited bandwidth uploads its local model to a neighbor with high capacites of communication. An other challenge

[1] https://docs.openvino.ai/2024/index.html.
[2] https://onnx.ai/.

is the storage capacity of connected things. To tackle this problem, split learning may be integrated with federated learning [50].

Hence, to fully harness collaboration among entities in AIoT systems, researchers are exploring the integration of split computing and federated learning into a unified approach known as Split Federated Learning (SL-FL) [7]. This hybrid approach aims to capitalize on the strengths of both methodologies, enhancing the overall effectiveness of collaborative AIoT systems.Enhancing split computing performance can also involve implementing early exits [30], which helps improve latency and reduce computational costs.

5 Conclusion

In conclusion, the field of AIoT optimization is advancing rapidly with a diverse range of strategies and techniques aimed at enhancing performance and efficiency in edge computing environments. Through our comprehensive review, we have highlighted the evolution from traditional AI models to specialized optimizations tailored for AIoT, addressing challenges such as computational complexity, energy efficiency, and network variability. By categorizing and analyzing these approaches, we have identified promising trends like federated learning, split computing, and hardware-aware NAS, each offering unique solutions to optimize AI deployment in IoT ecosystems. Moving forward, continued research efforts should focus on refining these techniques, integrating new advancements in AI and IoT technologies, and addressing emerging challenges to realize the full potential of AIoT in various application domains.

References

1. Baker, B., Gupta, O., Naik, N., Raskar, R.: Designing neural network architectures using reinforcement learning. arXiv preprint arXiv:1611.02167 (2016)
2. Busoniu, L., Babuska, R., De Schutter, B.: A comprehensive survey of multiagent reinforcement learning. IEEE Trans. Syst. Man Cybern. Part C (Appl. Rev.) **38**(2), 156–172 (2008)
3. Cecilia, J.M., Morales-García, J., Imbernón, B., Prades, J., Cano, J.C., Silla, F.: Using remote GPU virtualization techniques to enhance edge computing devices. Fut. Gen. Comput. Syst. **142**, 14–24 (2023). https://doi.org/10.1016/j.future.2022.12.038, https://www.sciencedirect.com/science/article/pii/S0167739X22004423
4. Chen, M., Poor, H.V., Saad, W., Cui, S.: Wireless communications for collaborative federated learning. IEEE Commun. Mag. **58**(12), 48–54 (2020)
5. Chitty-Venkata, K.T., Emani, M., Vishwanath, V., Somani, A.K.: Neural architecture search benchmarks: insights and survey. IEEE Access **11**, 25217–25236 (2023). https://doi.org/10.1109/ACCESS.2023.3253818
6. Chitty-Venkata, K.T., Somani, A.K.: Neural architecture search survey: a hardware perspective. ACM Comput. Surv. **55**(4), 1–36 (2022)
7. Duan, Q., Hu, S., Deng, R., Lu, Z.: Combined federated and split learning in edge computing for ubiquitous intelligence in internet of things: state-of-the-art and future directions. Sensors **22**(16), 5983 (2022)

8. Fang, J., He, Y., Yu, F.R., Li, J., Leung, V.C.: Large language models (LLMs) inference offloading and resource allocation in cloud-edge networks: an active inference approach. In: 2023 IEEE 98th Vehicular Technology Conference (VTC2023-Fall), pp. 1–5. IEEE (2023)
9. Frankle, J., Carbin, M.: The lottery ticket hypothesis: finding sparse, trainable neural networks. arXiv preprint arXiv:1803.03635 (2018)
10. Guo, Y.: A survey on methods and theories of quantized neural networks (2018). https://arxiv.org/abs/1808.04752
11. Han, S., Pool, J., Tran, J., Dally, W.: Learning both weights and connections for efficient neural network. Adv. Neural Inf. Process. Syst. **28** (2015)
12. Howard, A., et al.: Searching for mobileNetV3 (2019). https://arxiv.org/abs/1905.02244
13. Howard, A.G., et al.: MobileNets: efficient convolutional neural networks for mobile vision applications (2017). https://arxiv.org/abs/1704.04861
14. Iandola, F.N., Han, S., Moskewicz, M.W., Ashraf, K., Dally, W.J., Keutzer, K.: SqueezeNet: alexnet-level accuracy with 50X fewer parameters and <0.5 MB model size (2016). https://arxiv.org/abs/1602.07360
15. Jalali, J., Khalili, A., Berkvens, R., Famaey, J.: Joint offloading policy and resource allocation in IRS-aided MEC for IOT users with short packet transmission. In: 2023 IEEE 98th Vehicular Technology Conference (VTC2023-Fall), pp. 1–7. IEEE (2023)
16. Kang, Y., et al.: NeuroSurgeon: collaborative intelligence between the cloud and mobile edge. ACM SIGARCH Comput. Architect. News **45**(1), 615–629 (2017)
17. Karras, k, et al.: A hardware acceleration platform for AI-based inference at the edge. Circ. Syst. Signal Process. **39**(2), 1059–1070 (2020)
18. Kohama, H., Minoura, H., Hirakawa, T., Yamashita, T., Fujiyoshi, H.: Single-shot pruning for pre-trained models: rethinking the importance of magnitude pruning. In: Proceedings of the IEEE/CVF International Conference on Computer Vision, pp. 1433–1442 (2023)
19. Kuzmin, A., Nagel, M., Van Baalen, M., Behboodi, A., Blankevoort, T.: Pruning vs quantization: which is better? Adv. Neural Inf. Process. Syst. **36** (2024)
20. Lakhlef, I.E., Djamaa, B., Senouci, M.R., Bradai, A.: Enhanced multicast protocol for low-power and lossy IoT networks. IEEE Sens. J. **24**(9), 15393–15408 (2024). https://doi.org/10.1109/JSEN.2024.3375797
21. Lee, N., Ajanthan, T., Torr, P.H.: SNIP: single-shot network pruning based on connection sensitivity. arXiv preprint arXiv:1810.02340 (2018)
22. Li, C., et al.: HW-NAS-Bench: hardware-aware neural architecture search benchmark (2021). https://arxiv.org/abs/2103.10584
23. Li, G., Yang, P., Qian, C., Hong, R., Tang, K.: Stage-wise magnitude-based pruning for recurrent neural networks. IEEE Trans. Neural Netw. Learn. Syst. **35**(2), 1666–1680 (2022)
24. Li, L., Ota, K., Dong, M.: Deep learning for smart industry: efficient manufacture inspection system with fog computing. IEEE Trans. Industr. Inf. **14**(10), 4665–4673 (2018)
25. Liu, C., Zhang, Y., Zhou, H.: A comprehensive study of Bluetooth low energy. In: Journal of Physics: Conference Series. vol. 2093, p. 012021. IOP Publishing (2021)
26. Luo, Y., Cai, X., Qi, J., Guo, D., Che, W.: FPGA-accelerated CNN for real-time plant disease identification. Comput. Electron. Agric. **207**, 107715 (2023). https://doi.org/10.1016/j.compag.2023.107715, https://www.sciencedirect.com/science/article/pii/S0168169923001035

27. Mammen, P.M.: Federated learning: opportunities and challenges. arXiv preprint arXiv:2101.05428 (2021)
28. Marchisio, A., et al.: Deep learning for edge computing: current trends, cross-layer optimizations, and open research challenges. In: 2019 IEEE Computer Society Annual Symposium on VLSI (ISVLSI), pp. 553–559. IEEE (2019)
29. Mathew, G., Ramachandran, S.S., Suchithra, V.: EdgeAI: diabetic retinopathy detection in intel architecture. In: 2020 IEEE/ITU International Conference on Artificial Intelligence for Good (AI4G), pp. 75–80. IEEE (2020)
30. Matsubara, Y., Levorato, M., Restuccia, F.: Split computing and early exiting for deep learning applications: survey and research challenges. ACM Comput. Surv. **55**(5), 1–30 (2022)
31. Mazzia, V., Khaliq, A., Salvetti, F., Chiaberge, M.: Real-time apple detection system using embedded systems with hardware accelerators: an edge AI application. IEEE Access **8**, 9102–9114 (2020)
32. Merenda, M., Porcaro, C., Iero, D.: Edge machine learning for AI-enabled IoT devices: a review. Sensors **20**(9), 2533 (2020)
33. Mishra, A., et al.: Accelerating sparse deep neural networks. arXiv preprint arXiv:2104.08378 (2021)
34. Molchanov, P., Tyree, S., Karras, T., Aila, T., Kautz, J.: Pruning convolutional neural networks for resource efficient inference. arXiv preprint arXiv:1611.06440 (2016)
35. Qi, C., et al.: An efficient pruning scheme of deep neural networks for internet of things applications. EURASIP J. Adv. Signal Process. **2021**(1), 31 (2021)
36. Qin, D., et al.: MobileNetV4 – universal models for the mobile ecosystem (2024). https://arxiv.org/abs/2404.10518
37. Qu, K., Zhuang, W., Wu, W., Li, M., Shen, X., Li, X., Shi, W.: Stochastic cumulative DNN inference with RL-aided adaptive IoT device-edge collaboration. IEEE Internet Things J. **10**(20) (2023)
38. Real, E., et al.: Large-scale evolution of image classifiers. In: Precup, D., Teh, Y.W. (eds.) Proceedings of the 34th International Conference on Machine Learning. Proceedings of Machine Learning Research, vol. 70, pp. 2902–2911. PMLR (2017). https://proceedings.mlr.press/v70/real17a.html
39. Ren, P., et al.: A comprehensive survey of neural architecture search: challenges and solutions. ACM Comput. Surv. **54**(4) (may 2021). https://doi.org/10.1145/3447582
40. Sandler, M., Howard, A., Zhu, M., Zhmoginov, A., Chen, L.C.: MobileNetV2: inverted residuals and linear bottlenecks (2019). https://arxiv.org/abs/1801.04381
41. Satheeshkumar, K., Arunachalam, V., Deepika, S.: Hand-held GPU accelerated device for multiclass classification of X-ray images using CNN model. Microprocess. Microsyst. **106**, 105046 (2024). https://doi.org/10.1016/j.micpro.2024.105046, https://www.sciencedirect.com/science/article/pii/S0141933124000413
42. Silla, F., Iserte, S., Reaño, C., Prades, J.: On the benefits of the remote GPU virtualization mechanism: the rCUDA case. Concurrency Comput. Prac. Exper. **29**(13), e4072 (2017)
43. Singh, M., et al.: A facial and vocal expression based comprehensive framework for real-time student stress monitoring in an IoT-fog-cloud environment. IEEE Access **10**, 63177–63188 (2022)
44. Sinha, N., El Rahman Shabayek, A., Kacem, A., Rostami, P., Shneider, C., Aouada, D.: Hardware aware evolutionary neural architecture search using representation similarity metric. In: Proceedings of the IEEE/CVF Winter Conference on Applications of Computer Vision, pp. 2628–2637 (2024)

45. Surianarayanan, C., Lawrence, J.J., Chelliah, P.R., Prakash, E., Hewage, C.: A survey on optimization techniques for edge artificial intelligence (AI). Sensors **23**(3), 1279 (2023)
46. Surianarayanan, C., Lawrence, J.J., Chelliah, P.R., Prakash, E., Hewage, C.: A survey on optimization techniques for edge artificial intelligence (AI). Sensors **23**(3), 1279 (2023)
47. Tabouche, A., Djamaa, B., Senouci, M.R.: HPS: a hybrid proactive scheduler with adaptive channel selection for industrial 6TiSCH networks. Ad Hoc Netw. **161**, 103527 (2024). https://doi.org/10.1016/j.adhoc.2024.103527, https://www.sciencedirect.com/science/article/pii/S1570870524001380
48. Tan, M., Le, Q.: EfficientNet: rethinking model scaling for convolutional neural networks. In: International Conference on Machine Learning., pp. 6105–6114. PMLR (2019)
49. Tan, M., Pang, R., Le, Q.V.: EfficientDet: scalable and efficient object detection (2020). https://arxiv.org/abs/1911.09070
50. Thapa, C., Arachchige, P.C.M., Camtepe, S., Sun, L.: SplitFed: when federated learning meets split learning. In: Proceedings of the AAAI Conference on Artificial Intelligence, vol. 36, pp. 8485–8493 (2022)
51. Wang, B., Komatsuzaki, A.: GPT-J-6B: a 6 billion parameter autoregressive language model (2021)
52. Wang, L., Zhang, X., Su, H., Zhu, J.: A comprehensive survey of continual learning: theory, method and application. IEEE Trans. Pattern Anal. Mach. Intell. **46**(8) (2024)
53. Xiao, Y., et al.: Reinforcement learning based energy-efficient collaborative inference for mobile edge computing. IEEE Trans. Commun. **71**(2), 864–876 (2022)
54. Xu, Y., Wang, Y., Zhou, A., Lin, W., Xiong, H.: Deep neural network compression with single and multiple level quantization. In: Proceedings of the AAAI Conference on Artificial Intelligence, vol. 32 (2018)
55. Yang, J., et al.: Quantization networks. In: Proceedings of the IEEE/CVF Conference on Computer Vision and Pattern Recognition., pp. 7308–7316 (2019)
56. Zhang, C., Xie, Y., Bai, H., Yu, B., Li, W., Gao, Y.: A survey on federated learning. Knowl. Based Syst. **216**, 106775 (2021)
57. Zhang, J., Tao, D.: Empowering things with intelligence: a survey of the progress, challenges, and opportunities in artificial intelligence of things. IEEE Internet Things J. **8**(10), 7789–7817 (2020)
58. Zoph, B., Le, Q.V.: Neural architecture search with reinforcement learning. arXiv preprint arXiv:1611.01578 (2016)

Fault Prediction in IoT Applications Using Federated Learning Based on Local Differential Privacy

Nabila Azeri[1(✉)], Ouided Hioual[2], Djamel Benmerzoug[3], and Ouassila Hioual[2,3]

[1] ICOSI Laboratory, Abbes Laghrour University, BP 1252 Road of Batna, El Hamma, 40004 Khenchela, Algeria
azeri.nabila@univ-khenchela.dz

[2] Mathematics and Informatics Department, Abbes Laghrour University, BP 1252 Road of Batna, El Hamma, 40004 Khenchela, Algeria
{hioual.ouided,hioual_ouassila}@univ-khenchela.dz

[3] LIRE Laboratory, Constantine 2 University, BP67A, Ali Mendjeli, 25000 Constantine, Algeria
djamel.benmerzoug@univ-constantine2.dz

Abstract. The Internet of Things (IoT) has grown rapidly, affecting many industries. However, it also raises serious concerns about data privacy and system security. Ensuring accurate fault prediction in IoT environments is essential to maintaining system performance and minimizing unexpected downtime. Unlike the usual centralized fault finding methods, which can expose sensitive data because they collect everything in one place, this study uses a Federated Machine Learning (FML) framework mixed with local differential privacy (LDP). The idea is to distribute the learning process across the Edge devices and use local differential privacy, which helps keep privacy secure and in the same time keep a good accuracy in predicting faults. The results from the experiments show that this model performs well in keeping sensitive data secure without compromising performance.

Keywords: IoT · Fault prediction · Federated Machine Learning · Local Differential Privacy

1 Introduction

The fast growth of the Internet of Things (IoT) has linked many different devices from common smart home gadgets to big industrial machines. This creates new chances to make operations better using data insights [8,13]. This connected network makes a huge amount of data that can lead to big improvements in how efficiently things run and how reliable systems are, especially using Machine Learning (ML) methods [14]. One important use of ML here is fault prediction.

T. Marir et al. (Eds.): ICCSA 2024, ISEM 45, pp. 31–43, 2025.
https://doi.org/10.1007/978-3-031-90758-6_3

This is about finding and fixing problems in IoT systems before they cause expensive downtime or disruptions [1,6].

In the last few years, ML has become a very important tool for doing good fault prediction in IoT settings [5,7]. Traditional centralized ML systems usually depend on collecting data from many IoT devices and sending it to one central server, where the models get trained on this big dataset. Even though this method provides strong insights, it also comes with many big problems. Centralized systems often create serious privacy issues because lots of sensitive data has to go to a central server, which makes it easy for attacks to happen. Plus, putting all the data in one place raises the chances of having single points of failure.

Federated Machine Learning (FML) is a different option because it lets model training happen on many devices, which helps keep data private and reduces communication costs [9,10]. However, FML faces several challenges, particularly in ensuring data security.

Based on our previous work for enhancing privacy and security in self-adaptive systems [2–4], this paper introduces a novel approach combining Local Differential Privacy (LDP) with Federated Machine Learning, termed MDP-FML. This hybridisation permits to enhance both privacy and security in fault prediction systems for IoT applications. LDP ensures that data privacy is maintained by adding noise to the data, making it difficult for external entities to discern any individual information, while FML facilitates collaborative model training without compromising the confidentiality of the data.

The rest of this paper is organised as follows. Section 2 reviews some related work, highlighting current approaches and their limitations in the context of fault prediction using FML in IoT systems. Section 3 introduces the proposed LDP-FML based architecture. In Sect. 4, we present a case study on fault prediction in industrial IoT environments, demonstrating the practical application of our approach. Section 5 conducts a comparative analysis, evaluating the performance and effectiveness of the LDP-FML approach against our previous work. Section 6 concludes the paper and outlines future directions.

2 Related Work

A novel approach to collaborative learning without data exchange is federated learning. Because devices may learn from one another without providing all of their data to a central server, this is very helpful for IoT-based sustems. The most significant studies that have looked into the use of FML in IoT contexts are presented in this section.

Zhang and colleagues' research [17] investigate how FML might enhance cybersecurity in IoT networks, particularly for troubleshooting linked devices. They developed the FedIoT platform, which detects anomalies using the Fed-Detect algorithm. The study demonstrates that FML is highly effective in identifying various cyberattack types on a range of IoT devices. The outcomes also demonstrate the platform's efficiency, particularly in light of the fact that it requires less memory and training time, making it a suitable choice for low-resource devices.

FML is investigated by Rey et al. [15] for malware detection in IoT devices. Through a performance comparison between federated models and traditional centralized models, the authors shows that FML can accomplish comparable results while maintaining data privacy. Also, they point out serious flaws in the federated approach, especially with regard to hostile actors' ability to poison the model. They assess several model aggregation functions as countermeasures and observe that more development is required to strengthen FML's resilience in hostile envirments.

In [18], an optimization framework is introduced to improve FML within distributed Industrial IoT environments. The proposed solution uses a multi-layer FML architecture to enable distributed training of deep neural networks, where local models trained at industrial gateways are aggregated at edge or cloud servers. To enhance the efficiency of device selection and resource management, the authors present a stochastic optimization approach, along with a novel reinforcement learning-based strategy. Their experimental results demonstrate that this method outperforms existing solutions in minimizing both loss and energy consumption.

Discussion: While the aforementioned works contribute valuable insights into the application of FML in IoT, they largely overlook the critical aspects of privacy and security, particularly in the context of collaborative learning. Although some studies address privacy concerns to a certain extent, the robustness of FML against malicious attacks and the integration of advanced privacy-preserving techniques such as Local Differential Privacy remain underexplored. Our work aims to fill this gap by proposing a fault prediction architecture that not only leverages FML for collaborative learning but also enhances privacy and security through the integration of LDP, providing a more secure and reliable solution for IoT environments.

3 The Proposed LDP-FML Based Architecture

In our previous work, we proposed an architecture for self-adaptive systems using ML [2]. This architecture encompassed essential software modules designed to support self-adaptive functionalities, including resource and data management, process planning, QoS analysis, process optimization, predictive maintenance, and fault detection. Implemented with centralized ML [4], our approach demonstrated commendable accuracy and precision. However, it faced constraints related to data security, privacy issues, and communication overhead, revealing the need for an alternative strategy that could maintain predictive accuracy while mitigating the drawbacks of centralized data handling and computation.

In response to these limitations, our subsequent work [3] integrated FML into the architecture. This FML-based approach shifts processing and intelligence to the network's edge, enabling local data analysis and decision-making. By using

FML, our architecture facilitates collaboration among edge devices to train ML models without centralizing data collection. This decentralized model not only preserves data privacy and reduces communication overhead but also enhances system resilience, all while retaining the high accuracy and precision of the original centralized approach. In this paper, we further enhance our architecture by integrating LDP, addressing additional privacy concerns and providing a more robust solution to secure collaborative learning in IoT environments.

3.1 Architecture Overview

As shown in Fig. 1, the LDP-FML based architecture consists of three main layers: physical layer, Edge aggregators and a Cloud layer.

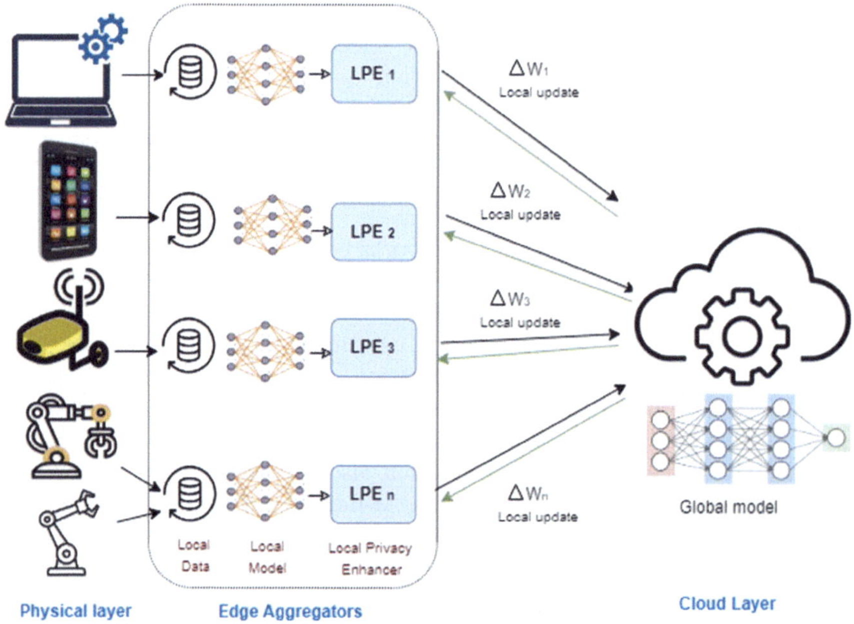

Fig. 1. The proposed LDP-FML based architecture

Physical layer: This layer is responsible for gathering and managing data from local devices. It includes a variety of devices such as industrial machinery, individual smartphones, IoT devices, sensors, and other endpoints.

Edge Aggregators: Aggregating data from multiple devices in physical layer within a specific geographical or organizational boundary. Edge Aggregators contribute to global model training by forwarding aggregated updates to the Cloud Layer. They implement LDP mechanisms to safeguard sensitive information before forwarding perturbed data to Edge Aggregators.

Cloud Layer: This central component orchestrates the federated learning process, managing global model updates, and coordinating communication between Edge Aggregators.

3.2 Integrating LDP Into Our Architecture

To enhance the privacy of the federated learning process, we introduce the Local Privacy Enhancer (LPE) module. This module is integrated into the Edge Aggregators to safeguard sensitive user data. The LPE employs differential privacy techniques, specifically the Laplace mechanism, to obfuscate raw data before it undergoes further processing.

Given a function f mapping datasets to real-valued outputs and a privacy parameter ϵ, the Laplace Mechanism adds Laplace noise to the output:

$$M(D) = f(D) + Lap(b) \tag{1}$$

where:

- $M(D)$ is the perturbed output
- $f(D)$ is the original function output
- $Lap(b)$ is Laplace noise with scale parameter $b = \frac{\Delta f}{\epsilon}$
- Δf is the sensitivity of the function f

In our architecture, the Client Side, represented by the Edge Aggregator, plays a crucial role in local model training while preserving data privacy. The algorithm for the Edge Layer with LDP is explained in Algorithm 1, and it work like this:

Algorithm 1. Client Side with LDP

Require: M_0 : initial global model, D : local dataset, ϵ: privacy budget
Ensure: Perturbed gradients
1: $M_{local} \leftarrow M_0$
2: $M_{local} \leftarrow TrainLocalModel(M_{local}, D)$
3: $gradientsLocal \leftarrow ComputeGradients(M_{local}, D)$
4: $sensitivity \leftarrow computeSensitivity(gradientsLocal)$
5: $scale \leftarrow sensitivity/\epsilon$
6: $noise \leftarrow Laplace(0, scale, size = gradientsLocal.shape)$
7: $perturbedGradients \leftarrow gradientsLocal + noise$
8: $SendToServer(perturbedGradients)$

Initialization: The initial global model M_0 is generated by the Cloud layer then it will be sent to Edge Aggregators. These latter ones start setting up their local models M_{local}, using the initial global model M_0.

Local Model Training: The Edge Aggregator uses the local dataset D to train the local model M_{local}. This process involves adjusting the model's parameters to reduce the local loss function based on the data it has. The result is an updated local model that reflects the local data characteristics.

Gradient Computation: After training, the Edge Aggregator computes the gradients $gradientsLocal$ of the local model. These gradients indicate the necessary adjustments to the model parameters based on the local training process.

Sensitivity Calculation: The algorithm calculates the sensitivity $sensitivity$ of the computed gradients. Sensitivity measures the maximum impact that a single data point can have on the gradient, which is crucial for determining the appropriate level of noise to add for privacy.

Noise Generation: To ensure privacy, the algorithm generates Laplace noise with a scale parameter derived from the sensitivity and the privacy budget ϵ. The scale $scale$ is calculated as $sensitivity/\epsilon$. This noise is added to the gradients to obscure the exact details of the local data.

Perturbation of Gradients: The local gradients $gradientsLocal$ are perturbed by adding the generated Laplace noise, resulting in the perturbed gradients $perturbedGradients$. This perturbation ensures that sensitive information is obfuscated while still allowing for useful model updates.

Transmission to Server: Finally, the Edge Aggregator sends the perturbed gradients to the Cloud Layer. This aggregated and perturbed data will be used by the server to update the global model, preserving the privacy of the individual client's data throughout the collaborative learning process.

Since the LDP mechanism is integrated into the Edge aggregators, there is no change required in the Cloud layer algorithm. The Cloud layer simply processes the perturbed gradients as it would with unperturbed gradients, ensuring the integrity of the federated learning process while maintaining data privacy. In this layer, the server is responsible for managing and updating the global model by aggregating gradients received from participating Edge aggregators (clients). The behavior of the Cloud layer remains consistent with our previous architecture [3].

4 Case Study: Fault Prediction in Industrial IoT

The primary objective of the Fault Prediction module in our architecture is to determine whether a device within the Industrial IoT network is likely to experience a fault. Additionally, the module aims to identify the type of fault when a device is deemed faulty. To achieve these goals, the module leverages both supervised and unsupervised learning techniques.

As presented in Fig. 2, the Fault Prediction can be broken down into three key stages: Classification, Clustering, and Prediction. These stages work together to classify the operational status of devices, identify the nature of faults, and predict future device behavior.

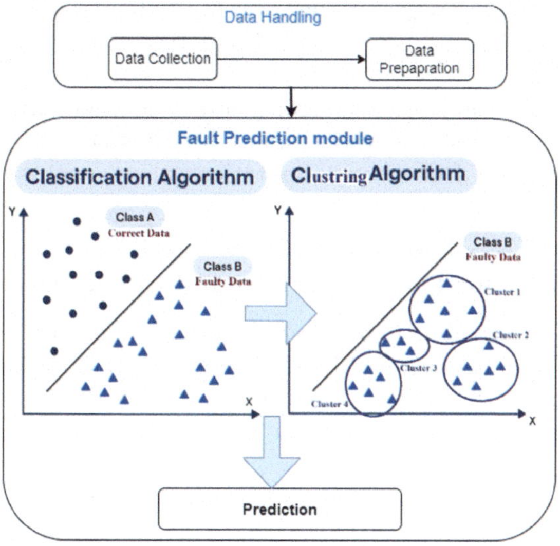

Fig. 2. Fault Prediction structure

Before deploying models within the Fault Prediction module, it is crucial to perform thorough data preparation. Properly preprocessed data is vital for the model's ability to accurately predict faults and classify potential issues. In this study, we used the same dataset as in our previous work to ensure consistency and facilitate a meaningful comparison of results.

4.1 Data Collection and Preparation

The dataset, obtained from Kaggle [11], simulates real-world predictive maintenance scenarios. The dataset includes the following key features:

- Product ID: A unique code for each product, which includes a letter showing its quality (L for low, M for medium, H for high) and a number specific to that quality.
- Air Temperature [K]: The temperature is created by a random process, staying close to an average of 300 K with a variation of about 2 K.
- Process Temperature [K]: This is the air temperature plus 10 K, adjusted to vary within 1 K using a random process.
- Rotational Speed [rpm]: Based on a power of 2860 W, with some random noise added.
- Torque [Nm]: Centered around 40 Nm with some variation (about 10 Nm), but it never goes below zero.

- Tool Wear [min]: The time the tool wears out, depending on product quality: 5 min for high, 3 min for medium, and 2 min for low.

Additionally, the dataset includes a 'Target' label indicating machine failure at a specific data point. When the process fails, this label is set to 1, which serves as a crucial indicator for the Fault Prediction module's classification and prediction processes.

For data preparation, steps such as feature extraction, data balancing, and other preprocessing techniques were employed to enhance the dataset's quality and model performance. Due to space limitations, these details are not presented here.

4.2 Methodology for Fault Detection and Categorization

The Fault Prediction module in our architecture is designed to detect and categorize faults in the data generated by devices within an Industrial IoT environment. The process is divided into three main steps: Classification, Clustering, and Prediction, as outlined in Algorithm 2.

Algorithm 2. Fault Prediction Algorithm

Require: X: dataset, k: number of clusters
Ensure: Predicted fault type for new data x_{new}

Step 1: Classification
Train classifier f on dataset X
$y \leftarrow f(X)$ ▷ y contains labels where 0 = correct, 1 = faulty
Step 2: Clustering
Extract faulty data $X_f \leftarrow x_i \mid f(x_i) = 1$
Apply clustering algorithm on X_f to get $C = C_1, C_2, \ldots, C_k$
Step 3: Prediction
$y_{new} \leftarrow f(x_{new})$
if $y_{new} = 0$ **then**
    Return "Correct Data"
else
    Assign x_{new} to closest cluster C_j
    Return "Faulty Data of type C_j"
end if

The process starts with a classification step, where a classifier model is trained on the dataset X. This model assign for each data point an output label. If data is correct then it assign the value 0, else it assign the value 1.

In the next step which is Clustering, we only take the Faulty data (data with the value 1). Faulty data, denoted as X_f, then grouped into k distinct clusters. Each cluster C_j represents a different type of fault. As a last step, we evaluate new data x_{new}. If the classifier finds that x_{new} is correct (label 0), it is marked as "Correct Data". But if it is labeled as faulty (label 1), it gets assigned to the nearest fault cluster C_j, which helps in identifying the type of fault.

Classification: The classification stage is about assigning each data point to one of two predefined categories : *Correct Data* and *Faulty Data.* This is done using supervised learning algorithms, where the classifier is trained on labeled data to differentiate between normal and abnormal behavior.

Let $X = \{x_1, x_2, \ldots, x_n\}$ represent the dataset, with each x_i being a feature vector corresponding to the operational data of a device. The classifier learns a function $f : X \rightarrow \{0, 1\}$, where $f(x_i) = 0$ indicates correct data, and $f(x_i) = 1$ indicates faulty data.

Clustering: Once the faulty data is identified, the next step involves using clustering algorithms to group the faulty data into different clusters based on the type of fault. Clustering helps in understanding the nature of the faults by bringing similar faulty data points together.

For the subset of faulty data $X_f = \{x_i \mid f(x_i) = 1\}$, the aim is to partition X_f into k clusters $C = \{C_1, C_2, \ldots, C_k\}$, where each cluster C_j represents a specific type of fault. This can be expressed as minimizing the within-cluster variance:

$$\arg\min_C \sum_{j=1}^{k} \sum_{x \in C_j} \|x - \mu_j\|^2 \tag{2}$$

where μ_j is the centroid of cluster C_j.

Prediction: In the prediction stage, the trained model is used to predict the likelihood of new data points being faulty and to identify the type of fault. For a new data point x_{new}, the classifier first predicts whether it is correct or faulty. If faulty, it is assigned to one of the fault clusters identified in the clustering stage.

The prediction can be represented as:

$$f(x_{\text{new}}) = \begin{cases} \text{Correct Data,} & \text{if } f(x_{\text{new}}) = 0 \\ \text{Faulty Data,} & \text{if } f(x_{\text{new}}) = 1, \text{ and assign } x_{\text{new}} \text{ to } C_j \end{cases}$$

4.3 Implementation of the LDP-FML Architecture

In this section, we explain how we built our FML system combined with LDP. The aim is to create a secure, privacy-focused way to enable collaborative learning for fault prediction in Industrial IoT applications.

Experimental Setup: Our architecture leverages the TensorFlow Federated (TFF) framework [16] to handle the parallel training of client models. The federated learning process is based on the Federated Averaging (FedAvg) algorithm [12], which aggregates model updates from multiple clients (Edge aggregators) to form a global model. The training data is divided into batches and distributed among individual clients, with each client training its local model on its subset of data.

Global Model Initialization: The training process begins with the initialization of the global model. This step requires designing the model architecture by determining key elements such as the number of layers, the specific types of layers, the activation functions used in each layer. Then we define the global model using TensorFlow's Keras API. The *tff.learning.from_keras_model* function is used to convert the Keras model into a format compatible with TFF's framework.

Client Selection and Local Training: We utilize TFF's built-in functionalities to manage the selection of clients for each round of training. In our case, the *@tff.federated_computation* decorator is used to define the client selection function.

The local training process occurs on each selected client, where the global model is downloaded and trained on the client's local dataset. We define a local training function that takes the global model and a batch of local data as input. The function then uses TFF's gradient tape mechanism to compute the loss and update the model parameters locally.

Privacy-Preserving Model Updates with LDP: LDP is applied to ensure that the gradients shared by each client do not reveal sensitive information. As presented in the code below, this is achieved by adding noise to the gradients before sending them to the server. The noise is generated using the Laplace mechanism, where the noise scale is determined by the sensitivity of the data and the privacy budget ϵ.

```
def apply_ldp(gradients, epsilon):
    sensitivity = compute_sensitivity(gradients)
    scale = sensitivity / epsilon
    noise = tf.random.laplace(gradients.shape, scale)
    perturbed_gradients = gradients + noise
    return perturbed_gradients
```

Model Aggregation: The server aggregates the perturbed gradients received from the clients using the FedAvg algorithm. The aggregation process computes the weighted average of the gradients to update the global model.

```
@tff.federated_computation
def model_aggregation(global_model, local_models):
    weighted_local_models = [
        (client_weight, local_model) for client_weight,
            local_model in zip(client_weights,
            local_models)
    ]
    averaged_gradients = tff.federated_mean(
        weighted_local_models)
```

```
    updated_model = update_global_model(global_model,
        averaged_gradients)
    return updated_model
```

After updating the global model, it is broadcasted to all clients for the next round of training, ensuring synchronization across the entire system.

5 Comparative Analysis

In this section, we evaluate the performance of our current approach, which integrates FML with LDP, against our previous work that employed Centralized Machine Learning [4] and Basic Federated Machine Learning [3]. To ensure a fair comparison, we have used the same dataset and maintained consistent data distribution across clients in both FML-based approaches.

The comparative results are summarized in Table 1. The table highlights the key performance metrics-Recall, Accuracy, and Precision-for each approach.

Table 1. Comparison of Performance Metrics

	Recall	Accuracy	Precision
MDP - FML	0.9519	0.9503	0.9404
Basic FML	0.9584	0.9544	0.9420
Centralized ML	0.9580	0.9785	0.9695

The results indicate that while the Centralized ML approach achieves the highest accuracy and precision, the introduction of LDP in the FML setup ensures a competitive performance with slightly lower accuracy and precision but with the added benefit of enhanced privacy preservation. The Basic Federated ML without LDP also shows competitive results, demonstrating that even without privacy enhancements, FML can perform well in distributed environments. However, the privacy guarantees provided by LDP make the slight trade-off in performance worthwhile for scenarios where data privacy is paramount.

6 Conclusion

This paper presented a novel approach for fault prediction in IoT applications using LDP with FML. We introduced an architecture designed to enhance privacy while maintaining high prediction accuracy. The integration of LDP within the federated learning framework ensures that individual data privacy is preserved, addressing critical concerns in data-sensitive applications.

Through extensive experimentation, we demonstrated that our proposed method not only provides robust privacy guarantees but also achieves competitive performance in fault prediction compared to traditional centralized and basic federated learning methods. The results highlight the effectiveness of combining LDP with FML to strike a balance between privacy and model performance.

Future work will explore further optimizations of the proposed architecture and its application to other domains. Additionally, investigating the integration of other privacy-preserving techniques could provide additional insights into enhancing both privacy and accuracy in collaborative learning scenarios.

References

1. Atassi, R., Alhosban, F.: Predictive maintenance in IoT: early fault detection and failure prediction in industrial equipment. J. Intell. Syst. Internet Things **9**(2) (2023)
2. Azeri, N., Hioual, O., Hioual, O.: Towards an approach for modeling and architecting of self-adaptive cyber-physical systems. In: 2022 4th International Conference on Pattern Analysis and Intelligent Systems (PAIS), pp. 1–7. IEEE (2022)
3. Azeri, N., Hioual, O., Hioual, O.: Enhancing self-adaptive cyber-physical systems using federated machine learning. In: TACC 2023 : 3rd Tunisian-Algerian Joint Conference on Applied Computing., pp. 108–119. ceur-ws.org (2023)
4. Azeri, N., Zouikri, Z., Rezgui, M., Hioual, O., Hioual, O.: Fault prediction using supervised and unsupervised learning algorithms in cyber physical systems. In: 2022 2nd International Conference on New Technologies of Information and Communication (NTIC), pp. 1–6. IEEE (2022)
5. Firouzi, F., Farahani, B., Ye, F., Barzegari, M.: Machine learning for IoT. Intelligent Internet of Things: From Device to Fog and Cloud pp. 243–313 (2020)
6. Goodarzi, P., Schütze, A., Schneider, T.: Comparison of different ML methods concerning prediction quality, domain adaptation and robustness. tm-Tech. Mess. **89**(4), 224–239 (2022)
7. Hussain, F., Hussain, R., Hassan, S.A., Hossain, E.: Machine learning in IoT security: current solutions and future challenges. IEEE Commun. Surv. Tutorials **22**(3), 1686–1721 (2020)
8. Laghari, A.A., Wu, K., Laghari, R.A., Ali, M., Khan, A.A.: A review and state of art of internet of things (IoT). Archives of Computational Methods in Engineering, pp. 1–19 (2021)
9. Li, T., Sahu, A.K., Zaheer, M., Sanjabi, M., Talwalkar, A., Smith, V.: Federated optimization in heterogeneous networks. Proc. Mach. learn. syst. **2**, 429–450 (2020)
10. Mammen, P.M.: Federated learning: Opportunities and challenges. arXiv preprint arXiv:2101.05428 (2021)
11. Matzka, S.: Explainable artificial intelligence for predictive maintenance applications. In: 2020 Third International Conference on Artificial Intelligence for Industries (AI4I), pp. 69–74. IEEE (2020)
12. McMahan, B., Moore, E., Ramage, D., Hampson, S., y Arcas, B.A.: Communication-efficient learning of deep networks from decentralized data. In: Artificial intelligence and statistics, pp. 1273–1282. PMLR (2017)
13. Mouha, R., et al.: Internet of things (IoT). J. Data Anal. Inf. Process. **9**(02), 77 (2021)
14. Ouadah, A., Zemmouchi-Ghomari, L., Salhi, N.: Selecting an appropriate supervised machine learning algorithm for predictive maintenance. Int. J. Adv. Manuf. Technol. **119**(7), 4277–4301 (2022)

15. Rey, V., Sánchez, P., Celdrán, A.H., Bovet, G.: Federated learning for malware detection in IoT devices. Comput. Netw. **204**, 108693 (2022)
16. TFF: TensorFlow Federated: machine learning on decentralized data (2023). https://www.tensorflow.org/federated
17. Zhang, T., He, C., Ma, T., Gao, L., Ma, M., Avestimehr, S.: Federated learning for internet of things. In: Proceedings of the 19th ACM Conference on Embedded Networked Sensor Systems, pp. 413–419 (2021)
18. Zhang, W., et al.: Optimizing federated learning in distributed industrial IoT: a multi-agent approach. IEEE J. Sel. Areas Commun. **39**(12), 3688–3703 (2021)

Deep Reinforcement Learning-Based Resource Allocation for the Internet of Vehicles: A Systematic Mapping Study

Ramzi Bouzoubia[1(✉)], Sofiane Zaidi[1], Lazhar Khamer[1,2], Abdelhabib Bourouis[1], and Mostafa Ogab[1]

[1] Department of Mathematics and Computer Science, Research Laboratory on Computer Science's Complex Systems (ReLa(CS)2), University of Oum El Bouaghi, Bouaghi, Algeria
{ramzi.bouzoubia,zaidi.sofiane,bourouis.abdelhabib, mostafa.ogab}@univ-oeb.dz, l.khamer@univ-soukahras.dz

[2] Department of Mathematics and Computer Science, University of Souk Ahras, Souk Ahras, Algeria

Abstract. The Internet of Vehicles (IoV) integrates vehicle communication and IoT technologies to enhance driving safety with low-latency, and support compute-intensive applications such as autonomous driving and traffic management. Mobile Edge Computing (MEC) processes tasks near data sources to reduce latency and power consumption, but the dynamic nature of vehicular networks complicates the resource allocation required by data sources. Deep Reinforcement Learning (DRL) offers a promising solution, managing complex state spaces effectively. This systematic mapping study reviews 16 studies published between 2013 and 2024 on DRL-based resource allocation in IoV. It categorizes the algorithms, evaluation criteria, and software environments used, highlighting the advantages of DRL over traditional methods. Techniques like Deep Q-Learning (DQL) and Deep Deterministic Policy Gradient (DDPG) have shown strong performance in various IoV scenarios. The study recommends future research in areas such as developing sophisticated algorithms, integrating with 5G networks, and exploring federated learning to support decentralized training and preserve data privacy. This comprehensive overview of DRL applications in IoV resource allocation provides valuable insights for future advancements and practical implementations in this emerging field, aiming to enhance the efficiency and performance of IoV systems.

Keywords: Internet of Vehicles · IoV · Resource Allocation · Resource Management · V2X Communication · Mobile Edge Computing · MEC

1 Introduction

The Internet of Vehicles (IoV) is an emerging communication network that integrates vehicles and Internet of Things (IoT) technologies[1]. It promises

T. Marir et al. (Eds.): ICCSA 2024, ISEM 45, pp. 44–59, 2025.
https://doi.org/10.1007/978-3-031-90758-6_4

to significantly improve the driving and safety experience by enabling vehicle-to-everything (V2X) communication[18]. IoV supports numerous applications and services with compute-intensive, and low-latency requirements, including autonomous driving, urban traffic management, and in-vehicle infotainment services [9]. With the rapid increase in the demands for intelligent devices and compute-intensive computing, efficient management of computing resources becomes crucial.

Mobile Edge Computing (MEC) offers a potential solution by enabling computational tasks to be processed close to data sources, thereby reducing latency and energy consumption [20]. However,resource allocation is a challenging task due to the heterogeneous and mobile nature of vehicular networks. Traditional optimization techniques rely on mathematical models, but struggle with NP-hard problems [3,9]. Therefore, deep reinforcement learning (DRL) is proposed to address this problem.DRL is emerging as a promising approach that leverages neural networks to handle vast state spaces. It improves the learning process and enhances the effectiveness of the proposed solutions.

For instance, algorithms like Deep Q-Learning (DQL) and Proximal Policy Optimization (PPO) have shown strong performance in several IoV network situations, such as determining network slices, managing broadcast coverage, allocating resource blocks, and controlling power[21].

This systematic mapping study aims to provide a comprehensive and up-to-date overview of research on resource allocation based on DRL for IoV. By examining existing work, this research offers a detailed mapping of the various DRL algorithms used for resource allocation, evaluation metrics and optimization criteria. Moreover, it reviews the contributions of various studies and the software environments used in this field. The goal is to provide perspectives for future research and practical implementation.

The paper is structured as follows: Sect. 2 describes the review protocol used to conduct this systematic mapping study. Section 3 presents the selected studies, answers the research questions, and discusses the results. Finally, Sect. 4 offers conclusions and suggests future research directions.

2 Methodology

This systematic mapping study aims to provide an overview of DRL-based resource allocation research in IoV as illustrated in Fig. 1. We have conducted this study according to the methodology described by Petersen et al. [10], which provides a structured framework to organize and schedule systematic mapping studies. The essential steps of our systematic mapping study, as illustrated in Fig. 2, are the definition of the research questions, the search and selection of relevant articles, the analysis and discussion of the results, and the final synthesis. The main purpose is to ensure a rigorous and comprehensive review.

2.1 Research Questions

The main objective of this first stage is to formulate precise and relevant research questions to orient the scope of the study towards the purpose to be achieved. The research questions (RQ) have been designed to address key aspects of the research area. Table 1 groups the research questions, focusing on the type of resources to be optimized, the deep reinforcement algorithms used in resource optimization, the advantages of using deep reinforcement learning in resource allocation, and the software environments used in the evaluation experiments.

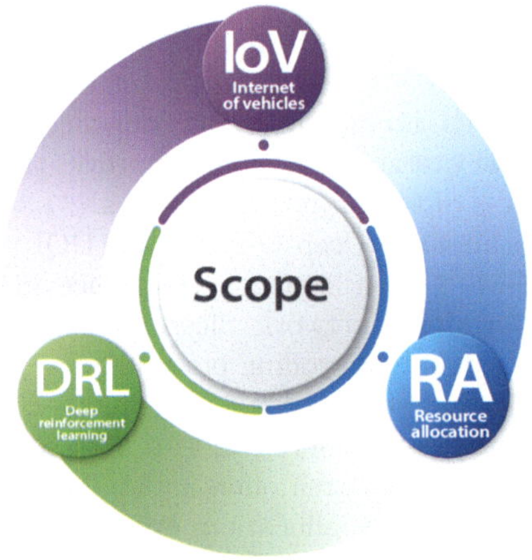

Fig. 1. Research area.

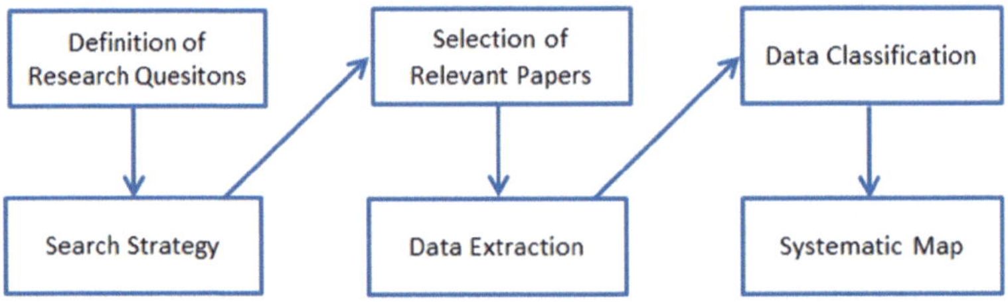

Fig. 2. Systematic Mapping Study Process.

Table 1. Research Questions.

RQ	Research Question
RQ1	Which specific DRL algorithms do IoV networks use for resource allocation and optimization?
RQ2	Which software environments do these experiments use?
RQ3	How can deep reinforcement learning optimize resource allocation in vehicle networks?
RQ4	What are the comparative advantages of using deep reinforcement learning over traditional model-based optimization techniques in IoV networks?
RQ5	What performance metrics are applicable for evaluating the effectiveness of DRL algorithms in resource allocation within IoV networks?

2.2 Search Strategy:

Our research strategy includes the following steps. First, we extracted keywords from the search questions. Next, we have used the derived keywords to create a search query, referred to as protocol. Finally, this query was applied to several databases to ensure complete coverage of relevant studies. The results of the search protocol application are shown in Table 2.

Databases: ScienceDirect, Scopus, ACM Digital Library, IEEE Xplore, SpringerLink, MDPI.

Protocol: ("Internet of vehicles" OR IoV) AND ("Resource Allocation" OR "Resource management") AND ("Deep Reinforcement Learning" OR "DRL").

Table 2. Number of studies per database.

Database	Search results
Scopus	96
IEEE Xplore	34
ScienceDirect	8
SpringerLink	52
ACM Digital Library	3
MDPI	8

2.3 Study Selection

The procedure for selecting the papers studied was twofold: an initial selection based on titles, abstracts, and keywords, followed by an in-depth examination of the full text. Inclusion and exclusion criteria were established to determine which articles or documents should be included in the study as illustrated in Table 3.

Table 3. Inclusion and Exclusion Criteria.

Inc Exc	Criteria
Inclusion	Peer-reviewed journal articles, conference papers, and review articles. Studies have been carried out specifically on the use of DRL for resource allocation in IOV Articles published between 2013 and 2024 Articles published in English.
Exclusion	Articles duplicated. Articles without available full texts. Non-peer-reviewed articles (e.g., editorials, opinion pieces). Articles that are not related to computer science.

2.4 Data Extraction

The data extraction process involved capturing relevant information from each included study to provide a detailed overview of research on resource allocation in DRL-based IoV networks. The following data were extracted: reference, year of publication, AI model used, set of resources optimized and allocated, software environment, and contribution of each paper.

3 Results and Discussion

Statistics show that the distribution of publications across various databases indicates significant contributions from IEEE Xplore with seven publications, followed by SpringerLink with five, Scopus with three, ScienceDirect with one, and MDPI with one publication as illustrated in Fig. 4. The temporal analysis reveals an increasing interest in recent years, starting in 2020 as illustrated in Fig. 3. The study includes several journals, which reveals the value of disseminating research findings through DRL-based resource allocation for the Internet of Vehicles as illustrated in Fig. 5.

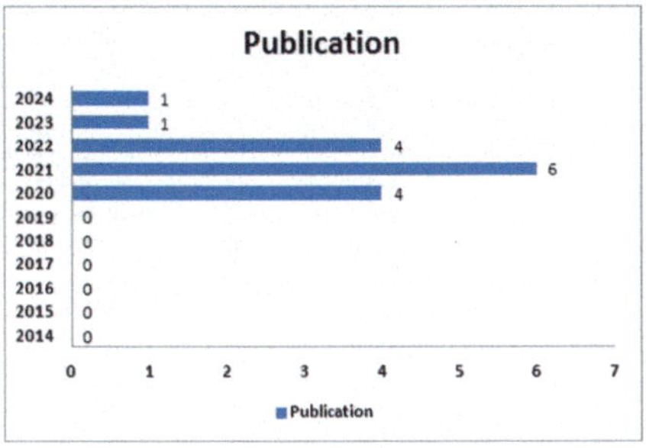

Fig. 3. Publications by Year.

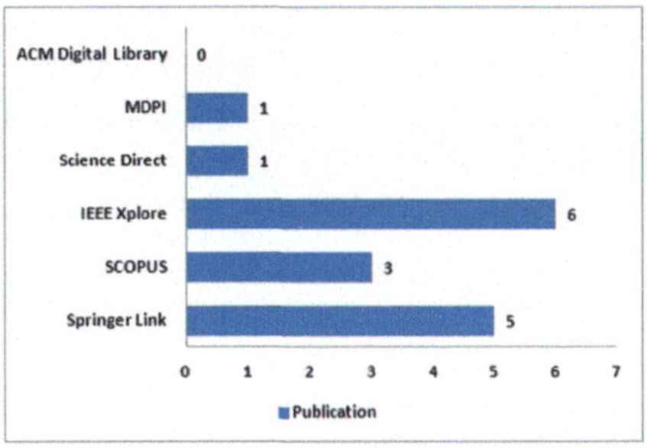

Fig. 4. Publications by Database.

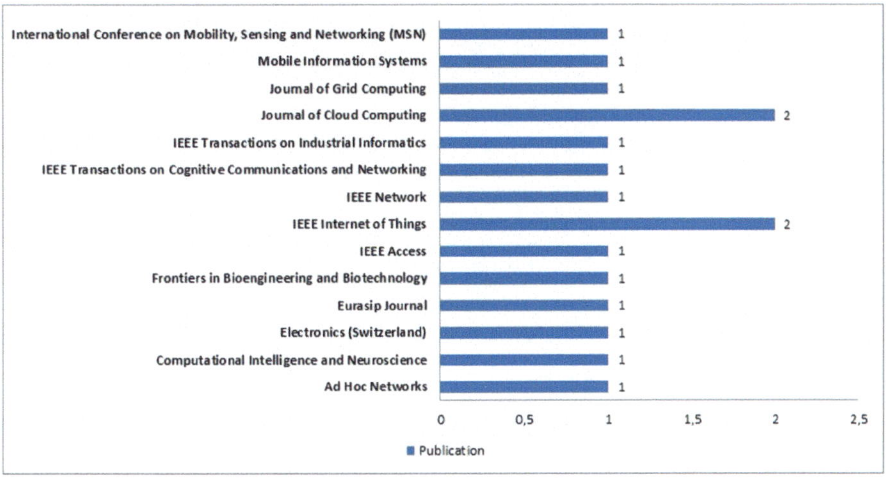

Fig. 5. Publications by Journal.

3.1 Synthesis and Key Findings:

Tables 4 and 5 present a comparison of different studies on the use of DRL for optimizing resource allocation and task execution in MEC systems and vehicular networks , citing the evaluation criteria and a summary of the contribution, as well as the different DRL algorithms used in each paper studied. This table provides a solid basis for the future development of DRL solutions, offering avenues for continuous performance improvements.

RQ1, DRL Algorithms: Document analysis has enabled us to identify the various Deep Reinforcement Learning algorithms used in the contributions. It is important to note that the proposed solutions often employ combinations

Table 4. Analysis of the chosen studies.

Ref	Year	DRL Algorithm	Platform	Evaluation Criteria	Contribution
Li et al.[6]	2020	AC, DDPG, DQN	Veins simulator with OpenAI Gym	• Service delay. • Transmission time. • Task completion rate.	Introduced a collaborative edge caching scheme using DRL to optimize caching decisions in vehicular networks, resulting in reduced service delay and transmission time.
Zhang et al.[19]	2020	DDPG, DNN	Python, TensorFlow	• Convergence speed. • System cost. • Resource utilization.	Developed a joint optimization scheme for content caching and computation offloading, improving system cost and resource utilization.
Xiao et al.[14]	2020	PPO	SUMo, Flow	• Dynamic traffic flow. • Resource consumption. • Total Loss. • Mean Reward.	Development of a framework that combines blockchain and DRL for optimal resource allocation in an IoV, using the PPO algorithm to adapt to traffic dynamics while guaranteeing network security and efficiency.
Fu et al.[2]	2020	SAC-DRL	Python, TensorFlow	• Video Quality (Bitrate). • Latency. • Learning Speed and Stability.	A new transcoding and live video broadcasting scheme that maximises video throughput and reduces delays and throughput variations in the IoV thanks to vehicular fog computing and joint optimisation of vehicle programming.
Mlika et al.[9]	2021	DQN	Julia programming language using Flux.jl	• Packet reception ratio (PRR). • Service Delay. • Task Completion Rate.	Introduced a model-free DRL framework for resource allocation in MEC-enabled IoV networks using network slicing and NOMA, showing robust performance under different network conditions.
Wang et al.[11]	2021	MDP, Q-Learning	MATLAB	• Latency. • Energy consumption. • Task Completion Rate.	Proposed a reliable computing offloading strategy to reduce latency and energy consumption in MEC systems.
Li et al.[7]	2021	Q-Learning, DNN	MATLAB	• Latency. • Energy consumption. • Computational cost • System security. • QoS.	Developed a resource allocation scheme using DRL to optimize computing offloading in MEC systems.
Zhang et al.[20]	2021	Q-Learning, DQN	MATLAB	• Latency. • Energy Consumption. • Resource Utilization. • Efficiency.	Proposed a resource allocation scheme for IoV to improve delay and energy management.
Xu et al.[15]	2021	DQN, DNN	TensorFlow with OpenAI Gym	• Latency. • Energy Consumption. • Task Completion Rate.	Developed a resource allocation strategy using DRL to optimize task scheduling in MEC systems.

Table 5. Analysis of the chosen studies.

Ref	Year	DRL Algorithm	Platform	Evaluation Criteria	Contribution
Yu et al.[16]	2021	DQN, DNN	Python	• Task offloading •Latency. •Energy Consumption. •Efficiency.	Proposed a task offloading and resource allocation strategy based on DQN, enhancing task offloading efficiency and reducing latency and energy consumption.
Kong et al.[5]	2022	AC-DDPG	Python, TensorFlow	• Average Reward. • Task Completion Rate. • Energy Consumption. • Task Delay.	Proposed an energy-efficient edge computing framework for IoV using DRL to balance energy consumption and task performance.
Wu et al.[12]	2022	DQN, DNN	MATLAB	• Resource Utilization. • Task Completion Rate. • Energy Consumption. • Task Delay.	Proposed a hybrid task offloading scheme (HyTOS) using DRL to achieve vehicle-to-edge (V2E) and vehicle-to-vehicle (V2V) offloading, significantly reducing task delay and energy consumption.
Zhang et al.[17]	2022	MAS, DNN	MATLAB	• Resource Utilization. • Throughput. • Latency.	Proposed a multi-factor multi-agent reinforcement learning approach for resource allocation, improving throughput and reducing latency in vehicular networks.
Li et al.[8]	2023	DQN, DNN	Python, TensorFlow	• Latency. • Resource Utilization. • Energy Consumption.	Developed a resource allocation strategy for IoV using DQN, improving resource utilization and reducing latency and energy consumption.
Jamil et al.[4]	2023	PPO, A2C, DQN	Veins Simulator, OMNet++, SUMO, OpenAI Gym	• Percentage of completed tasks. • Task waiting time. • Task delay. • Packet loss. • System overhead.	Developed an intelligent priority and deadline-aware resource allocation and task scheduling algorithm, demonstrating better performance in task completion and reduced packet loss compared to other algorithms.
wu et al.[13]	2024	DNN	Python, TensorFlow	• System Cost. • Resource Utilization. • Load Balance.	Introduced a TD3-based task offloading scheme to improve resource utilization and load balancing in vehicular edge computing.

of DRL algorithms, sometimes in association with other mechanisms such as Markov Decision Processes (MDP) or game-theoretic algorithms. Figure 6 shows the different types of DRL used in the articles selected for this study. It is worth mentioning that of all the articles selected for study, Deep Deterministic Policy Gradient (DDPG) and Deep Q-Network (DQN) are used more than other types of DRL.

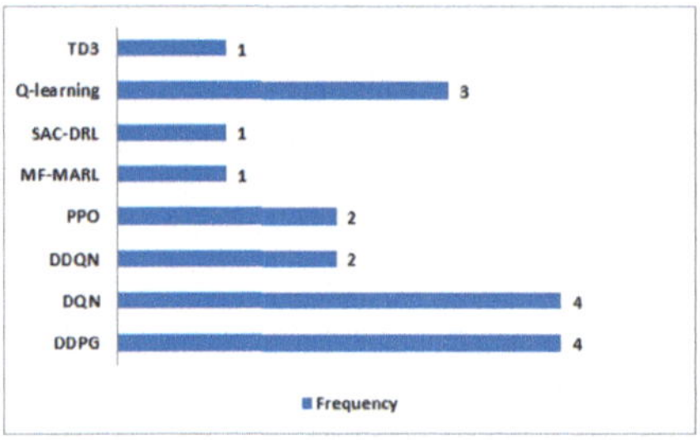

Fig. 6. Frequency of DRL Algorithms.

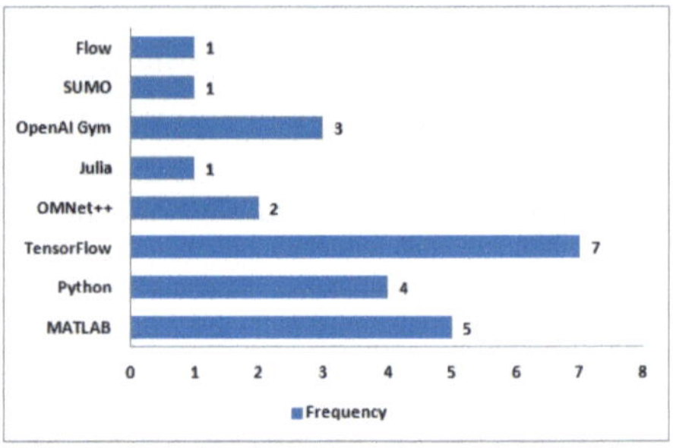

Fig. 7. Frequency of Used Software.

RQ2, Software Environments: Among the main environments used to evaluate the effectiveness of DRL algorithms in IoV network resource allocation are the Veins simulator with OpenAI Gym for collaborative caching simulations, Python with TensorFlow for most experiments, notably in joint optimization and video transcoding, the Julia language with Flux.jl for a model-free framework in MEC, MATLAB for computational offloading strategies and resource management, as well as OMNet++, SUMO, and OpenAI Gym for integrated simulations. These tools enable realistic scenarios to be created and DRL algorithms to be tested under varied and complex conditions, demonstrating their ability to optimize IoV network performance. The comparison of the frequency of used software is illustrated in Fig. 7.

RQ3, Optimizing Resource Allocation in IoVs Using DRL: To optimize resource allocation in IoVs, DRL combines deep learning capabilities and real-time decision-making. In [11], the authors model the problem as a Markov decision process, with resources serving as states and resource allocation decisions as actions. They store experiments and randomly sample them to stabilize learning, subsequently updating the Q-network by minimizing the error between the estimated and target values. By continuously interacting with the environment and learning optimal strategies through trial and error, DRLeffectively optimizes resource allocation in MEC environments. DRL optimizes resource allocation in IoV networks by learning efficient allocation policies from experience, dynamically adapting to changes in the network, and maximizing overall system performance.

RQ4, Advantages of DRL Algorithms : Based on the comparison Table 6 of the DRL algorithms used in the contributions, the comparison algorithms used in the different DRLs clearly outperform traditional methods when it comes to resource allocation in connected vehicle (IoV) networks. Here are a few key points that make the difference:

1. Traditional optimization problems: Traditional optimization methods require complex calculations and iterations to achieve optimal results, making them

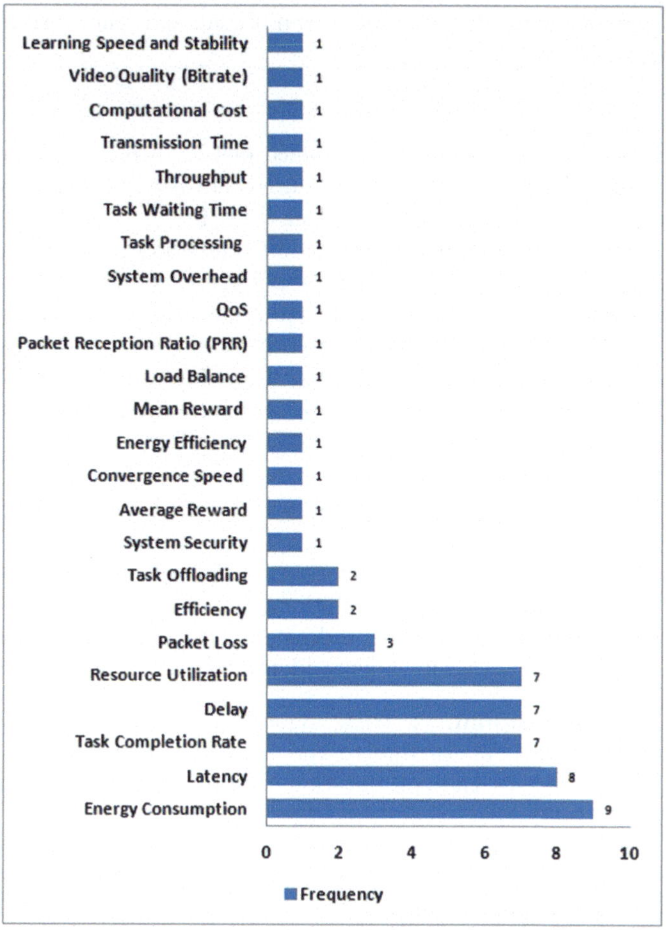

Fig. 8. Frequency of Evaluation Metrics.

unsuitable for real-time MEC systems. In contrast, the Reinforcement Learning (RL) algorithm can create an experience to learn and complete the optimization objective through a trial-and-error feedback mechanism, which is different from traditional optimization algorithms.

2. Advantages of DRL algorithms: DRL algorithms combine the perceptive capabilities of deep learning with the decision-making capabilities of Reinforcement Learning. This enables DRL to optimize decisions in real-time according to the current state of the IoV system, unlike dynamic programming methods, which are computationally expensive.
3. Superior performance: Simulation results shown in Table 6 show that DRL-based strategies outperform comparison methods in terms of time and energy consumption. In conclusion, DRL methods are particularly effective in dynamic and uncertain environments, such as vehicle networks, where tra-

ditional methods struggle to adapt to rapidly changing network and vehicle conditions.

RQ5, Performance Metrics: Performance metrics are crucial for assessing the effectiveness of DRL algorithms for resource allocation. The 16 papers under study encompass a variety of 24 types of metrics. IoV networks frequently use latency, energy consumption, delay, and Task Completion Rate in resource allocation due to their crucial importance for performance and energy efficiency. Latency and delay are essential for real-time applications such as autonomous driving and traffic management, which require rapid responses to ensure safety and efficiency. Energy consumption is also critical for extending vehicle autonomy, and task completion rates are essential for measuring algorithm efficiency and optimizing system performance as illustrated in Fig. 8.

3.2 Limitations and Challenges

The application of Deep Reinforcement Learning in resource allocation for the Internet of Vehicles offers promising solutions, but also presents several limitations and challenges that need to be addressed. The following sections provide a detailed analysis of these limitations as identified in various studies.

- **Curse of Dimensionality:** DRL algorithms suffer from the curse of dimensionality, which arises due to the exponential growth of states and actions. This increases the computational and memory complexity significantly [5,8,9,11,17].
- **High Computational Complexity:** The high computational complexity of DRL algorithms poses significant challenges, especially in dynamic and resource-constrained environments like IoV [8,9,13,15,20].
- **Slow Convergence:** DRL methods often exhibit slow convergence, making them less suitable for highly dynamic environments where quick adaptation is necessary [5,11,13,15,20].
- **Energy Consumption:** Performing computation offloading and resource allocation with DRL can lead to high energy consumption, which is a significant challenge in IoV systems where energy efficiency is crucial [5,7,12,16,19,20].
- **Real-Time Decision Making:** Achieving real-time decision-making capabilities is difficult with DRL due to its inherent need for extensive exploration and exploitation of policies [2,5,11,12].
- **Scalability Issues:** DRL algorithms can face significant scalability issues due to the exponential growth of states and actions, which complicates the learning process and increases computational requirements [2,5,8,9,16,19].
- **Stability and Convergence:** Ensuring stable and efficient convergence in DRL is challenging. The learning process can be unstable due to the nonlinear function approximators (like deep neural networks) used in DRL [8,9,15,16,20].

Table 6. Comparison of the DRL Algorithms.

Ref	DRL Algorithm	Comparison Algorithms	Delay	Energy Consumption
[6]	AC, DDPG, DQN	• Non-cooperative strategy • Random cache strategy	-	-
[14]	PPO	• not explicitly compared to other algorithms	-	-
[19]	DDPG, DNN	• Non-cooperative strategy • V2V-assisted random cache strategy	-	-
[2]	SAC-DRL	• TACA • SPGRA	• SAC-DRL :0.45 ms • SAC-DRL :0.45 ms TACA:0.52 ms • SAC-DRL :0.45 ms SPGRA: 0.48 ms	• SAC-DRL: 0.42 J • TACA: 0.54 J • SPGRA: 0.50 J
[9]	DQN	• OMA-MP • NOMA-MP • NOMA-RP	• DRL (DQL): 0.35 s • Rules-based allocation: 0.48 s	• DRL (DQL): 0.30 J • Rules-based allocation: 0.38 J
[11]	MDP, Q-Learning	• PEACE. • DTO. • GCO.	• Q-learning 0.442 s • PEACE: 1.161 s • 0.993 s • 0.876 s	-
[7]	Q-Learning, DNN	• COS • MS • CCC	• DQL :0,442 s • COS: 0,647 s • MS: 0,397 s • CCC: 0,192 s	-
[20]	Q-Learning, DQN	• GA	• DRL(DQN): 0.32 s • GA: 0.45 s	• DRL(DQN): 0.28 J • GA: 0.34 J
[15]	DQN, DNN	• EF • RANDOM	• DRL (DDPG et P-DQN): 0.32 s • EF: 0.40 s • RANDOM:0.43 s	• DRL(DDPG et P-DQN): 0.28J • EF: 0.33 J • RANDOM: 0.35 J
[16]	DQN, DNN	• Greedy Algorithm	• DRL(DDPG): 0.29 s • Greedy Algorithm: 0.40 s	• DRL(DDPG): 0.25 J • Greedy Algorithm: 0.33 J
[5]	AC-DDPG	• HA	• AC-DDPG : 0.30 s • HA : 0.42 s	• AC-DDPG : 0.26 J • HA : 0.34 J
[12]	DQN	• LC. • USS. • GT-hybrid .	• DQN: 0.442 s • LC: 0.876 s • USS: 0.993 s • GT-hybrid:1.161 s	-
Li et al.[8]	DDQN	• EHA. • MRL.	• DDQN:0,22 s. • EHA:0,3 s. • MRL:0,33 s.	-
[4]	IRATS	• A2C. • Random.	• IRATS:0.25 s • A2C:0.40 s. • Random:0.45 s.	-
[13]	TOLB	• TD3-TR. • DDPG-TR. • AEO.	• TOLB0:24 s. • TD3-TR:0.26 s. • DDPG-TR:0.32 s. • AEO0:44 s.	• TOLB:0.35 J. • TD3-TR:0.40 J. • DDPG-TR:0.48 J. • AEO:0.59 J.
[17]	MF-MARL	• CG-MARL. • MARL.	-	-

While DRL offers robust solutions for resource allocation and task offloading in IoV, its limitations must be addressed to enhance its applicability. Improving sample efficiency, stability, and adaptability to dynamic environments, along with optimizing energy consumption and computational resource usage, are critical areas for future research. Additionally, hybrid models should be streamlined to reduce complexity and improve real-time decision-making capabilities Table 7.

Table 7. Nomenclature.

Acronym	Description	Acronym	Description
A2C	Advantage Actor-Critic	MARL	Multi-Agent Reinforcement Learning
AEO	All Edge Offloading	LC	Local unloading algorithm
AC	Actor-Critic	MF-MARL	Mean-Field Multi-Agent Reinforcement Learning
CCC	Cooperative Content Caching in 5G Networks	MAS	Multi Agent System
COS	Computation Offloading Scheme to Improve QoE in Vehicular Networks	MRL	Meta-Reinforcement Learning Algorithm
CG-MARL	Cooperative Game-based Multi-Agent Reinforcement Learning	MS	Mobility Support for Vehicular Cloud Radio-Access-Networks
DDPG	Deep Deterministic Policy Gradient	NOMA-MP	Non-Orthogonal Multiple Access - Maximum
DDPG-TR	DDPG-based Task Offloading and Resource Allocation	NOMA-RP	Non-Orthogonal Multiple Access - Random Power
DDQN	Double Deep Q-Network	OMA-MP	Orthogonal Multiple Access - Maximum Power
DTO	Dynamic Task Offloading and Resource Allocation for Mobile Edge Computing	PEACE	Privacy-Preserving and Cost-Efficient
DQN	Deep Q-Network	PPO	Proximal Policy Optimization
EHA	Enhanced Heuristic Algorithm	SAC	Soft Actor-Critic
EF	Edge First	SPGRA	Standard Policy Gradient Resource Allocation Algorithm
GA	Genetic Algorithm	TACA	Traditional Actor-Critic Algorithm
GCO	Game-Based Computation Offloading Method in Vehicular	TD3	Twin Delayed Deep Deterministic Policy Gradient
GT-hybrid	Hybrid unloading algorithm based on Game Theory	TD3-TR	TD3-based Task Offloading and Resource Allocation
HA	Heuristic Algorithm	TOLB	Task Offloading and Load Balancing
IoV	Internet of Vehicles	USS	Unloading algorithm for a single scenario
IRATS	Intelligent Resource Allocation and Task Scheduling		

4 Conclusion

This systematic mapping study explores the use of Deep Reinforcement Learning for resource allocation in IoV. DRL algorithms, such as DQL and DDPG, have shown promising performance in various IoV scenarios, offering advantages like reduced latency and improved energy efficiency. The study underscores the potential of DRL to optimize real-time decisions in dynamic environments, outperforming traditional methods in many cases.

However, despite these promising results, several challenges and limitations must be acknowledged. DRL algorithms often suffer from the curse of dimensionality, leading to high computational and memory complexities. The high computational complexity and slow convergence rates of DRL methods pose significant hurdles, especially in dynamic and resource-constrained environments like IoV. Additionally, the energy consumption associated with DRL-based resource allocation can be substantial, which is a critical concern for IoV systems that prioritize energy efficiency. Achieving real-time decision-making capabilities remains difficult due to the extensive exploration and exploitation required by DRL algorithms.

Scalability issues are another significant challenge, as the exponential growth of states and actions complicates the learning process and increases computational requirements. Ensuring stability and convergence in DRL is also challenging, given the use of nonlinear function approximators like deep neural networks. These factors can lead to instability and inefficiencies in the learning process.

This study of DRL-based resource optimization in IoV networks has enabled us to demonstrate DRL algorithms' superiority over traditional resource allocation methods. However, it limits the performance comparison between different DRL algorithms and neglects the possibility of comparing them with other sophisticated deep learning techniques.

Future research should focus on addressing these challenges by improving sample efficiency, enhancing stability, and developing more sophisticated algorithms. Integrating DRL with 5G networks, exploring federated learning for decentralized training, and leveraging advanced techniques such as transfer learning and continuous learning could further enhance the applicability and performance of DRL in IoV systems.

In conclusion, while DRL offers a promising approach for resource allocation in IoV, it is crucial to address its inherent challenges and limitations to fully realize its potential. This balanced perspective provides a comprehensive overview and valuable insights for future advancements and practical implementations in this emerging field.

Acknowledgments. This work is fully funded by the General Direction of Scientific Research and Technological Development, Algerian Ministry of Higher Education and Scientific Research DGRSDT/MERS.

Disclosure of Interests. The authors declare that they have no competing interests.

References

1. Elfatih, N.M., et al.: Internet of vehicle's resource management in 5G networks using AI technologies: current status and trends. IET Commun. **16**(5), 400–420 (2022)
2. Fu, F., Kang, Y., Zhang, Z., Yu, F.R., Wu, T.: Soft actor-critic DRL for live transcoding and streaming in vehicular fog-computing-enabled IOV. IEEE Internet Things J. **8**, 1308–1321 (2021). https://doi.org/10.1109/JIOT.2020.3003398
3. Gao, D.: Computing resource allocation strategy based on mobile edge computing in internet of vehicles environment. Mobile Inf. Syst. **2022** (2022). https://doi.org/10.1155/2022/5781366

4. Jamil, B., Ijaz, H., Shojafar, M., Munir, K.: IRATS: a DRL-based intelligent priority and deadline-aware online resource allocation and task scheduling algorithm in a vehicular fog network. Ad Hoc Netw. **141** (3 2023). https://doi.org/10.1016/j.adhoc.2023.103090
5. Kong, X., et al.: Deep reinforcement learning-based energy-efficient edge computing for internet of vehicles. IEEE Trans. Ind. Inf. **18**, 6308–6316 (2022). https://doi.org/10.1109/TII.2022.3155162
6. Li, M., Gao, J., Zhao, L., Shen, X.: Deep reinforcement learning for collaborative edge computing in vehicular networks. IEEE Trans. Cogn. Commun. Netw. **6**, 1122–1135 (2020). https://doi.org/10.1109/TCCN.2020.3003036
7. Li, X.: A computing offloading resource allocation scheme using deep reinforcement learning in mobile edge computing systems. J. Grid Comput. **19**(3), 1–12 (2021). https://doi.org/10.1007/s10723-021-09568-w
8. Li, Y., Liu, Z., Tao, Q.: Resource allocation strategy of internet of vehicles using reinforcement learning in edge computing environment (2023). https://doi.org/10.21203/rs.3.rs-974516/v1, https://doi.org/10.21203/rs.3.rs-974516/v1
9. Mlika, Z., Cherkaoui, S.: Network slicing with MEC and deep reinforcement learning for the internet of vehicles. IEEE Netw. **35**, 132–138 (2021). https://doi.org/10.1109/MNET.011.2000591
10. Petersen, K., Vakkalanka, S., Kuzniarz, L.: Guidelines for conducting systematic mapping studies in software engineering: an update. Inf. Softw. Technol. **64**, 1–18 (8 2015). https://doi.org/10.1016/j.infsof.2015.03.007
11. Wang, K., Wang, X., Liu, X.: A high reliable computing offloading strategy using deep reinforcement learning for IoVs in edge computing. J. Grid Comput. **19**(2), 1–15 (2021). https://doi.org/10.1007/s10723-021-09542-6
12. Wu, C., Huang, Z., Zou, Y.: Delay constrained hybrid task offloading of internet of vehicle: a deep reinforcement learning method. IEEE Access **10**, 102778–102788 (2022). https://doi.org/10.1109/ACCESS.2022.3206359
13. Wu, Z., Jia, Z., Pang, X., Zhao, S.: Deep reinforcement learning-based task offloading and load balancing for vehicular edge computing. Electronics (Switzerland) **13** (2024). https://doi.org/10.3390/electronics13081511
14. Xiao, H., Qiu, C., Yang, Q., Huang, H., Wang, J., Su, C.: Deep reinforcement learning for optimal resource allocation in blockchain-based IoV secure systems. In: 2020 16th International Conference on Mobility, Sensing and Networking (MSN), pp. 137–144 (2020). https://doi.org/10.1109/MSN50589.2020.00036
15. Xu, J., Xu, Z., Shi, B.: Deep reinforcement learning based resource allocation strategy in cloud-edge computing system. Front. Bioeng. Biotechnol. **10** (2022). https://doi.org/10.3389/fbioe.2022.908056
16. Yu, Z., Xu, X., Zhou, W.: Task offloading and resource allocation strategy based on deep learning for mobile edge computing. Comput. Intell. Neurosci. **2022** (2022). https://doi.org/10.1155/2022/1427219
17. Zhang, H., et al.: Mean-field-aided multiagent reinforcement learning for resource allocation in vehicular networks. IEEE Internet Things J. **10**, 2667–2679 (2022). https://doi.org/10.1109/JIOT.2022.3214525
18. Zhang, K., Leng, S., Peng, X., Pan, L., Maharjan, S., Zhang, Y.: Artificial intelligence inspired transmission scheduling in cognitive vehicular communications and networks. IEEE Internet Things J. **6**(2), 1987–1997 (2018)
19. Zhang, M., Wang, S., Gao, Q.: A joint optimization scheme of content caching and resource allocation for internet of vehicles in mobile edge computing. J. Cloud Comput. **9**(1), 1–12 (2020). https://doi.org/10.1186/s13677-020-00182-x

20. Zhang, Y., Zhang, M., Fan, C., Li, F., Li, B.: Computing resource allocation scheme of IOV using deep reinforcement learning in edge computing environment. EURASIP J. Adv. Signal Process. **2021**(1), 1–19 (2021). https://doi.org/10.1186/s13634-021-00750-6
21. Zhang, Y., Luo, Z.: A review of research on spectrum sensing based on deep learning. Electronics **12**(21), 4514 (2023)

Clustering Using Markov Blanket Approximation to Analyze Agricultural Data in an IoT Environment

Karima Sid[1,2](✉) and Kenza Belhouchette[3]

[1] Department of Mathematics and Computer Science, Oum El Bouaghi University, Oum El Bouaghi, Algeria
sid.karima@univ-oeb.dz

[2] Laboratory of Data Science Computing and Artificial Intelligence (LISIA), Abdelhamid Mehri Constantine 2 University, Constantine, Algeria

[3] Department of Mathematics and Computer Science, Computer Science's Complex Systems Laboratory (RELACS2), Oum El Bouaghi University, Oum El Bouaghi, Algeria
kenza.belhouchette@univ-oeb.dz

Abstract. The integration of information and digital communication tools in the agricultural sector has led to the emergence of precision agriculture. This field now utilizes Internet of Things (IoT) technologies, geographical data, and both historical and real-time information, all of which have the potential to transform farming into a smart, technologically advanced enterprise. IoT and machine learning can synergize to create fully intelligent systems. In precision agriculture, clustering is often employed to extract knowledge for optimizing crop productivity, irrigation control, and food quality and safety monitoring. This paper proposes an unsupervised clustering algorithm based on the concept of Markov Blanket Approximation to group agricultural data collected in an IoT environment. The algorithm is evaluated using a dataset extracted from the Meteoblue platform, and the results demonstrate its effectiveness compared to the K-means algorithm.

Keywords: Internet of Things · Machine learning · Markov Blanket Approximation · Precision Agriculture · Clustering

1 Introduction

The Internet of Things (IoT) is rapidly evolving and gaining widespread adoption across various sectors, including agriculture. Currently, there are over 5 billion devices connected to the Internet, and this number is projected to exceed 50 billion by 2025 [1]. The IoT enables real-time connectivity between physical and virtual objects, facilitating seamless interaction across diverse environments and networks [2]. In agriculture, IoT has become a vital tool, particularly in Precision Agriculture, which leverages real-time data about the origin and characteristics of the region and product, to enhance crop management, optimize resource use, and improve overall productivity [3, 4]. Through the deployment of IoT sensors, farmers can monitor environmental conditions such as

T. Marir et al. (Eds.): ICCSA 2024, ISEM 45, pp. 60–72, 2025.
https://doi.org/10.1007/978-3-031-90758-6_5

temperature, humidity, and soil moisture, thereby making data-driven decisions that maximize yields and reduce labor costs [5]. A well-known illustration that proves the effectiveness of using IoT in precision agriculture is the deployment of sensors by farmers in different regions or smart greenhouses to collect meteorological data (e.g. humidity, temperature and other factors), and then use this data to extract relevant information that improves the irrigation process.

Recently, the combination of Artificial Intelligence (AI) and IoT has played a transformative role in agriculture. IoT provides the hardware for collecting vast amounts of data, while AI, particularly machine learning supplies the software for analyzing and interpreting this data to guide decision-making. Machine learning can be categorized into supervised learning, where labeled data is available and unsupervised learning, where the goal is to uncover patterns without predefined labels [6]. Clustering, a key unsupervised learning technique is widely used in precision agriculture to group similar data points into clusters, aiding in tasks such as crop monitoring and irrigation management. However, traditional clustering algorithms like K-means and fuzzy C-means have several limitations, including the need to specify the number of clusters in advance and a tendency to converge to local optima. Additionally, their high computational complexity makes them inefficient for large datasets [7, 8].

To overcome these shortcomings, this paper introduces a novel clustering algorithm based on the Markov Blanket approximation. Generally, in the machine learning field, the Markov Blanket concept is employed to select relevant features and eliminate redundant data in order to improve the predictive model accuracy [9]. In this context, the concept is used to cluster agricultural data collected through IoT sensors. By identifying dependencies between samples, the Markov Blanket Approximation enhances the clustering process, making it well-suited for precision agriculture applications, where selected data quality directly impacts decision-making. This approach addresses the limitations of traditional algorithms and provides a more robust framework for analyzing agricultural data in IoT environments.

The remainder of this paper is organized as follows: Sect. 2 presents a review of related work; Sect. 3 outlines the methods and concepts used in this research; Sect. 4 describes the proposed algorithm in detail; Sect. 5 discusses the evaluation and results; and Sect. 6 concludes the paper with final thoughts and future directions.

2 Related Work

The integration of IoT-based systems in agriculture has grown significantly in recent years due to their ability to enhance productivity, efficiency, and sustainability. These systems allow for precise decision-making in smart farming environments by providing real-time data on crops, livestock, and environmental conditions [10]. Many studies have demonstrated the benefits of IoT in precision agriculture, including improvements in real-time monitoring, food safety management, and irrigation optimization [5].

A noteworthy example is a study [11] that introduced an IoT-based system for monitoring food safety and quality using a mobile application. This system assessed food freshness by comparing images captured through a mobile camera with reference images. The results were communicated to supervisors and consumers, demonstrating how IoT enhances food safety management.

In the domain of irrigation, another study [12] proposed using wireless sensor networks to optimize agricultural water use. The system employed node sensors and smartphone applications to provide real-time feedback on soil moisture levels, resulting in better water management and improved crop yields. This underscores the potential of IoT for use resource optimization in agriculture.

Livestock monitoring has also benefited from IoT innovations. For instance, study [13] suggested a system that controls animal health using IoT sensors to track vital signs such as heart rate and body temperature. The collected data is transmitted to a cloud server, enabling farmers to respond quickly to health issues, thus reducing losses and improving animal care practices.

Artificial intelligence (AI), particularly machine learning, has further expanded the capabilities of IoT in agriculture. AI-driven systems can optimize irrigation schedules, monitor soil and crop health, and manage the distribution of agricultural inputs like fertilizers. One of the most common AI techniques applied in precision agriculture is clustering, which helps identify patterns in environmental data, such as temperature, humidity, and soil moisture that are crucial for effective resource management [14].

However, traditional clustering algorithms, such as K-means, have inherent limitations, including random selection for centers and K-value, which have an impact on the K-means outcomes. To address these challenges, a hybrid clustering method that combines hierarchical clustering with K-means was proposed in [15]. This approach utilized the Elbow, Silhouette and gap statistics methods to find the optimal number of clusters. The evaluation of the proposed hybridization demonstrated better performance compared to standard K-means.

Further advancements in clustering for agricultural data analysis include algorithms like DBSCAN, E-DBSCAN, and CLARA. These methods address the limitations of K-means by being more robust in the presence of noise and not requiring the number of clusters to be predefined. For example, in [16], E-DBSCAN was shown to outperform DBSCAN, CLARA, and K-means in terms of clustering accuracy when applied to a dataset of different crops. The study demonstrated the effectiveness of advanced clustering algorithms in improving the accuracy of agricultural predictions, such as determining optimal growing conditions for crops.

Another study [17] employed a combination of K-means clustering and ensemble learning to classify crops based on environmental conditions and predict which crops would thrive under specific circumstances. The main objectives were to group crops into clusters that have similar growth requirements and to forecast suitable crop types based on a variety of environmental factors. This research highlighted how the integration of clustering with advanced machine-learning models could improve decision-making in agriculture, particularly in identifying high-yield crops for specific regions.

In contrast, the Markov Blanket concept from Bayesian network field [9] has been increasingly recognized for its effectiveness in feature selection and dependency analysis. It allows for the identification of the most relevant features by analyzing the dependencies between them, making it particularly useful in complex datasets where feature redundancy can negatively affect clustering/classification performance. For instance, [18] explored the use of Markov Blanket-based feature selection in high-dimensional data, demonstrating its ability to improve both classification accuracy and computational efficiency by reducing the feature space without sacrificing critical information. Furthermore, in [19], the Markov Blanket Approximation was applied in supervised learning to enhance classification performance. The study showed that using the Markov Blanket to identify relevant variables improved the algorithm's ability to group similar data points while minimizing noise from irrelevant features. This has significant implications for precision agriculture, where datasets are often large and heterogeneous.

Building on these advancements, this paper presents a novel unsupervised clustering algorithm based on the Markov Blanket approximation, which addresses some of the known limitations of conventional clustering methods. By leveraging IoT technologies to collect data in smart farming environment, the proposed algorithm can dynamically determine relevant knowledge and filter out redundant information, resulting in more efficient and accurate clustering. This makes the approach particularly well-suited for precision agriculture, where real-time data analysis is critical for informed decision-making.

3 Methodology

This section outlines the key concepts and methods used to create and implement the proposed algorithm.

3.1 Internet of Things in Agriculture

IoT in agriculture refers to a network in which various virtual "objects" of the agricultural system, as well as physical components like plants and animals, environmental factors, production tools, and agricultural information perception equipment, are connected to the internet via specific protocols to simplify information exchange and communication. It aims to make agricultural objects and processes intelligently identifiable, positioned, tracked, monitored, and managed [20].

IoT sensors are hardware components (devices) used to collect data and identify environmental changes. IoT sensors can measure variables like pressure, temperature, and motion. They can also share data with a network if they are connected to one. In agriculture, IoT sensors are used to track and gather data for various agricultural parameters, such as air quality, temperature, humidity, wind direction, pH, nutrient levels, and soil moisture. After that, this data can be examined and analyzed using AI methods (see Fig. 1) to maximize crop growth, lower expenses, safeguard resources, and enhance overall agricultural productivity [20].

In agriculture, one of the most popular IoT applications is precision farming, also named Precision Agriculture. It enhances the precision and management of farming. Precision agriculture seeks to increase productivity and efficiency by analyzing and interpreting sensor-generated data, using artificial intelligence tools, to make quick and informed decisions in a variety of scenarios, including crops monitoring, livestock management, irrigation management, and many more [21].

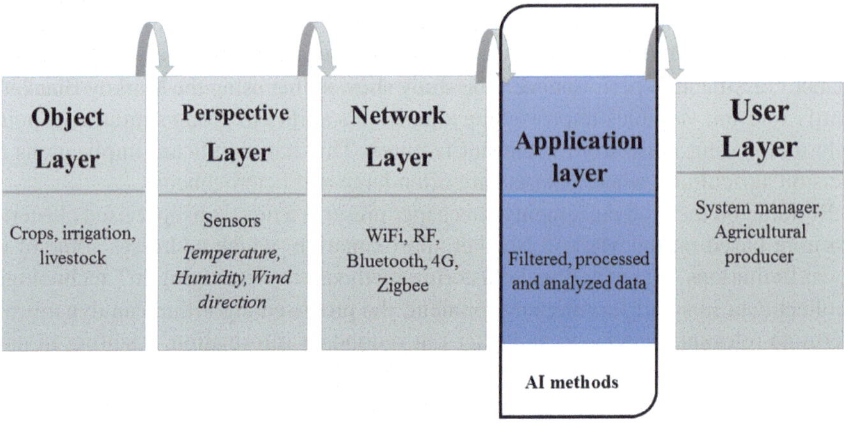

Fig. 1. Agricultural Internet of Things Workflow.

3.2 Unsupervised Learning and Data Clustering

Clustering, also known as data partitioning, is an unsupervised technique in machine learning, which involves a collection of learning algorithms with the aim of assembling unlabeled data (i.e., no desired outputs) with similar properties (see Fig. 2). The primary goals of clustering can be encapsulated in the subsequent four points [8]:

- Discover structures and patterns in large-scale data;
- Group data with similar patterns together;
- As a result, complexity is decreased and the relationships between data are simpler to understand;
- In some cases, partitioning is the intended outcome. For instance, in image segmentation process.

Similarity is a numerical measure of the degree of resemblance between two objects. Higher for a pair of objects that are more similar. Generally, a non-negative value $\in [0, 1]$. Assuming that Sim (X, Y) represents the degree of similarity between points X and Y, the main properties are [22]:

1. Sim (X, Y) = 1 only if X = Y. ($0 \leq \text{Sim} \leq 1$).
2. Sim(X, Y) = Sim (Y, X) for all X and Y.

There are several similarity measures in the literature, such as, Simple correspondence coefficient, Jaccard coefficient, cosine similarity and correlation [22].

The literature has proposed a wide range of clustering methods. There are three ways to classify clustering algorithms [23]:

- The type of input data used in the algorithm;
- The clustering criterion that determines the level of similarity between data points;
- The basic ideas and theory underlying clustering analysis methods.

Clustering algorithms can be broadly divided into two categories [23]:

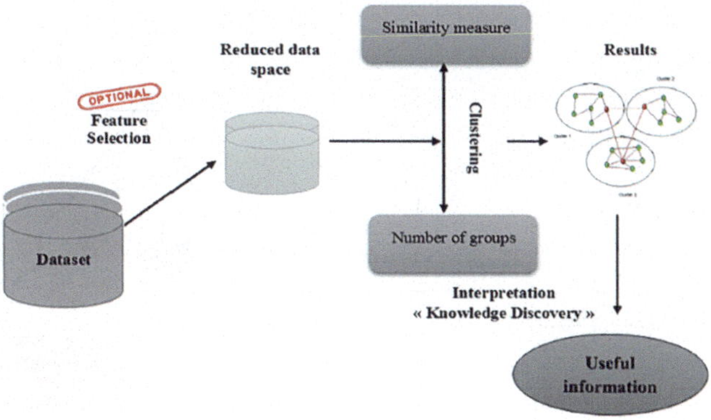

Fig. 2. General Clustering Process.

1. Partitionnal algorithms: generate different partitions then evaluate them according to certain criteria, such as, K-means algorithm and Fuzzy C-means. The primary limitations of these algorithms are:
   - Require the number of clusters in advance (limited);
   - Frequently ends at a local optimum;
   - Initialization is crucial.
2. Hierarchical algorithms: apply a specific criterion to create a hierarchical decomposition of all objects. The main shortcomings of this algorithms are:
   - They scale poorly: time complexity of at least $O(n^2)$, where n is the total number of objects;
   - Like any heuristic search algorithm, the local optimum problem.

4 Proposed Algorithm for Unsupervised Clustering

In precision agriculture, usually, the collected data using IoT technologies are heterogeneous and may include information that is redundant or irrelevant. In this context, we present in this section a novel unsupervised clustering algorithm, which is predicated on the idea of Markov Blanket approximation that employs the information theory metrics: mutual information and joint entropy.

The objective of clustering analysis in precision agriculture is to classify agricultural regions based on relevant information (similarity criterion), and then utilize the outcomes (clusters) to pinpoint areas in need of crop monitoring or irrigation management. An additional application of clustering results is the identification and the generation of labels for an unlabeled agricultural dataset, which can then be used as a training dataset for supervised or deep learning algorithms to produce accurate predictions. The aim of this work is to investigate optimal irrigation, whereby we pinpoint the areas that genuinely need irrigation (see Fig. 3).

The information theory metrics used in the calculation of the Markov Blanket Approximation are the entropy, the joint entropy, and the mutual information.

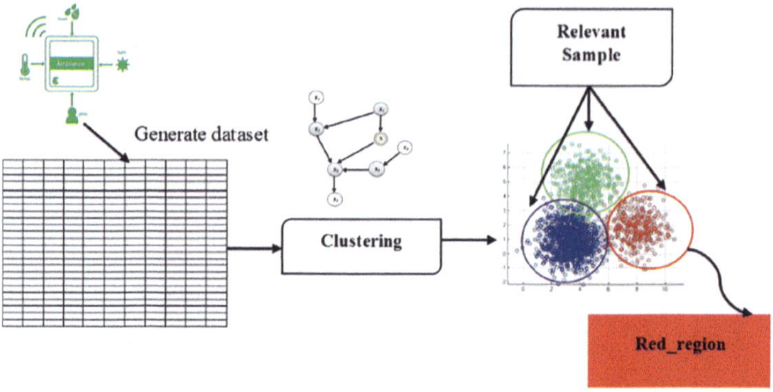

Fig. 3. General Workflow Including the Proposed Algorithm.

- The entropy **H (x)** (Eq. 1) permits us to measure the uncertainty in a random variable x with a marginal distribution probability **p (x)** as follows [24]:

$$H(x) = -\sum_{x} p(x) log p(x) \tag{1}$$

- The joint entropy **H(x,y)** (Eq. 2) allows measuring the information contained in a system of two random variables x and y, it is calculated using the joint probability **p(x,y)** as follows [24]:

$$H(x, y) = -\sum_{x}\sum_{y} p(x, y) log p(x, y) \tag{2}$$

- In information theory, the mutual information of two random variables is a quantity measuring the statistical dependence between these variables (according to Eq. 3).

$$MI(x, y) = H(x) + H(y) - H(x, y) \tag{3}$$

In a Bayesian network, the Markov blanket for a node X is the set of nodes ∂X composed of X's parents, its children and its children's parents. In a Markov network, the Markov coverage of a node is the set of its neighboring nodes, as shown in Fig. 4.

The Markov blanket approximation implemented in this work is based on a coefficient called the information correlation, defined as a normalization of the mutual information by the joint entropy as follows:

$$ICC(X, Y) = \frac{MI(X, Y)}{H(X, Y)} \tag{4}$$

Knowing that, MI(X, Y) is the mutual information which makes it possible to measure the dependence between the two variables (X and Y).

We have the following terms:

- **icc(X, C)** the information correlation coefficient, which represents the relevance between the sample X and the relevant sample C,
- **icc(Y, C)** the information correlation coefficient, which represents the relevance between the sample Y and the relevant sample C,

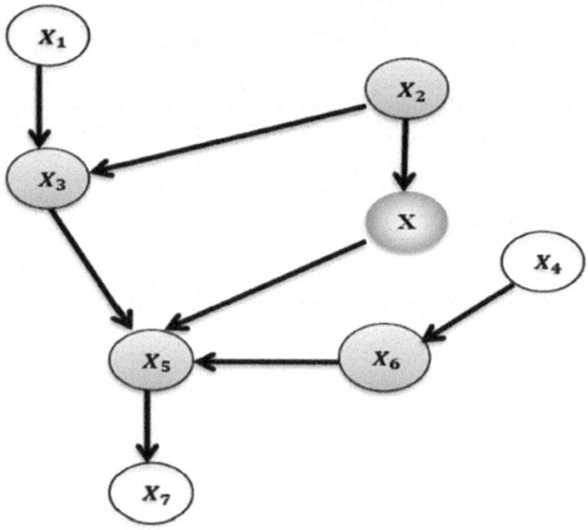

Fig. 4. Example of a Markov Blanket in a Bayesian network, The Markov Blanket of X, M(X) = {X_2, X_3, X_5, X_6} [25].

- **icc(X, Y)** the information correlation coefficient, which represents the relevance between the two samples X and Y.

X is a Markov Blanket Approximation of Y if and only if:

$$\begin{cases} icc(X, C) \geq icc(Y, C) \\ icc(X, Y) \geq icc(Y, C) \end{cases} \tag{5}$$

If X is a Markov Blanket Approximation of Y, then they have the same relevance to the problem, so we can group them into the same group.

Algorithm 1. Proposed algorithm for unsupervised Clustering

Inputs : n : number of samples, $X = (X_1; X_2; \ldots; X_n)$;
m: number of features, C: relevant sample $(C_1; C_2; \ldots; C_m)$;
Outputs : k: number of groups $\{ Gr_1, Gr_2, \ldots, Gr_k \}$;
Cr: vector of centers;
Begin
- Assign the first sample of X as center of the first group Gr_1,
$Cr_1 = X_1$;
- k=1;
// Generate clusters
For** i=2 to n* ***do
- for each sample X_i
If $\exists$ a group **Gr**, the center of this group is a Markov blanket approximation of this sample, according to equations Eq.4 and Eq.5, ***then***
- Group X_i to **Gr**;
else
-k=k+1;
-Assign X_i to a new group, $Cr_k = X_i$;
End if
Endfor
END
// Clustering results
- Generate the k groups $\{ \mathbf{Gr_1}, \mathbf{Gr_2}, \ldots, \mathbf{Gr_k} \}$ and vector Cr of k centers;

We use the Jaccard coefficient as a similarity metric in accordance with the following equation (Eq. 6) to identify the red region, which requires real-time irrigation.

$$Sim(Cr_i, C) = \frac{|Cr_i \cap C|}{|Cr_i \cup C|} \tag{6}$$

Algorithm 2. Generate the red_region

Begin
For** i=1 to k* ***do
- Calculate the similarity S (Cr_i, C)= / i=1, …, k according to Eq.6;
Endfor
- Sort the obtained values of S in a decreasing manner, we obtain the list of groups ordered according to S;
- Select the first group, as **red_region**;
END

5 Results and Discussion

Using the PyCharm IDE and Python as programming language, we implemented the suggested algorithm. The dataset used to assess our implementation extracted from the Meteoblue platform[1], which provides raw weather and climate data from different regions and under different conditions. The dataset includes 10825 samples that collected according to four parameters (characteristics): temperature (°C), humidity (%), wind speed (km/h) and wind direction (°).

We evaluate the performance of the proposed algorithm using a case study of tomato irrigation control. It can be inferred that the ideal conditions for irrigation are as follows:

- Temperature < 15 °C.[2]
- Humidity < 50%.[3]

Tomatoes grow best at 30–33 °C and good humidity. If the relative humidity drops below 50%, the crop becomes stressed and can no longer convert all the light it receives into additional production. In addition, at temperatures below 15 °C, the plant may exhibit symptoms of water and phosphorus deficiency. Due to a lack of information, the values for the other two characteristics—wind direction (°) and speed (km/h)—are chosen at random using irrigation logic.

According to this conclusion, we chose different relevant samples (C in the principal algorithm). The below table provides a summary of the clustering results that were obtained (see Table 1).

Table 1. Clustering results according to different relevant samples.

Relevant sample	Number of Groups	Red_region size
[14, 45, 20, 15]	23	123
[13, 45, 20, 15]	31	89
[14, 40, 20, 15]	25	101
[12, 40, 15, 10]	11	216
[9, 30, 6, 15]	19	212
[9, 30, 40, 25]	36	92

The above table shows the high sensitivity of our algorithm, i.e. even small changes in feature values have a noticeable impact on the results, which means that the similarity measure between samples for clustering is sensitive to changes in values, reflecting the quality of the similarity measure (Markov blanket approximation) for effective clustering.

[1] https://www.meteoblue.com.
[2] https://hortimedia.ma/tomate-dete-sous-serre-les-cles-de-la-reussite/.
[3] https://caldor.fr/cultures/tomates/.

Additionally, as demonstrated in the following figure (Fig. 5), we can observe a correlation between the size of the red region and the number of groups; that is, the size of the group representing the red_region decreases as the number of groups' increases.

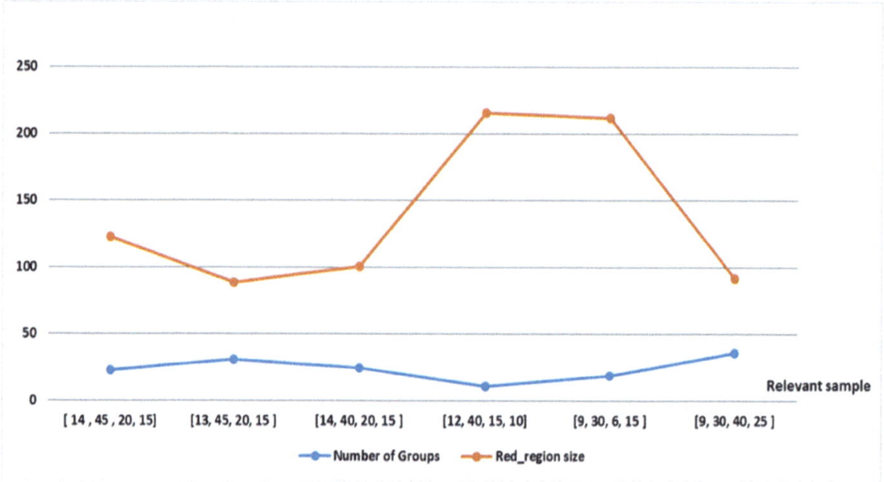

Fig. 5. Number of Groups Vs. Red_Region Size.

The principal of clustering algorithms is grouping data with maximum similarity within the same group and maximum dissimilarity between groups. In order to demonstrate the efficiency of the suggested algorithm, we compare with the K-means algorithm by measuring the similarity and dissimilarity within and between groups. To measure similarity, we use the Tanimoto index, according to the following equations:

$$sim(Cr_1, Cr_2) = \frac{|Cr_1 \cap Cr_2|}{(2 * |Cr_1 \cup Cr_2| - |Cr_1 \cap Cr_2|)} \tag{7}$$

$$sim(Gr_i) = \frac{\sum_{i,j} sim(X_i, X_j)}{group\_size} \tag{8}$$

Several values of **k** are used to evaluate the k-means algorithm. The following table recaps the results (Table 2) in terms of intra-group similarity, inter-group similarity and execution time expressed in seconds.

In general, the comparison results demonstrate that the suggested algorithm outperforms k-means and can be effectively applied to agricultural data analysis.

Table 2. Comparison between k-means and the proposed algorithm.

Algorithm		Intra-group Similarity $sim(Gr_i)$	Inter-group Similarity $sim(Cr_1, Cr_2)$	Execution Time (s)
K-means	K = 5	0.43	0.29	2.03
	K = 10	0.61	0.21	4.12
	K = 20	0.65	0.24	5.61
Proposed algorithm		0.91	0.07	5.98

6 Conclusion

Studies clearly indicate that clustering is a useful tool for knowledge extraction in precision agriculture. Traditional clustering algorithms, such as k-means, require the number of clusters as an initial parameter to distribute samples over these clusters according to a certain similarity measure, this has an impact on the performance of the algorithms. To address this issue, in this paper, we presented an algorithm for unsupervised clustering based on Markov Blanket approximation. We evaluated the workflow to optimize the tomato irrigation process. The goal of the algorithm is to locate the "red_region," or the area that genuinely requires irrigation. The efficiency of our algorithm is shown by comparing with the k-means algorithm. In the future, we intend to, first, implement the proposal using real data, and second, combine it with deep learning and supervised algorithms to create a fully integrated intelligent system that can enhance the prediction results in precision agriculture.

References

1. Verma, D.K., Mishra, A., Mishra, K.: Role of IOT in introducing smart agriculture. Int. Res. J. Eng. Technol. **9**, 883–887 (2022)
2. Talal, S.: Internet of Things-Iot: definition, architecture and applications. Egypt. J. of Appl. Sci. **34**(11), 81–95 (2019)
3. Muangprathub, J., et al.: IoT and agriculture data analysis for smart farm. Comput. Electron. Agric. **156**, 467–474 (2019)
4. Sethi, S.S., Priyanka, S.: New developments in the implementation of IoT in agriculture. SN Comput. Sci. **4**, 1503 (2023). https://doi.org/10.1007/s42979-023-01896-w
5. Tan, Y.S., et al.: IoT-based smart farming system. Int. J. Emerg. Multidisc. **3**(1), 1–14 (2024). https://doi.org/10.54938/ijemdcsai.2024.03.1.270
6. Sid, K., Batouche, M.: Distributed heterogeneous ensemble learning on Apache Spark for ligand-based virtual screening. Int. J. Data Min. Model. Manag. **13**, 11–2, 160–191 (2021). https://doi.org/10.1504/IJDMMM.2021.10035119
7. Aman, G.: Fuzzy C-Means Clustering (FCM) Algorithm (2021). https://medium.com/geekculture/fuzzy-c-means-clustering-fcmalgorithm-in-machine-learning-c2e51e586fff
8. Henry, L.: Clustering. 15-381 Artificial Intelligence. Carnegie Mellon School of Computer Science (2022)

9. Wang, H., Ling, Z., Yu, K., Wu, X.: Towards efficient and effective discovery of Markov blankets for feature selection. Inf. Sci. **509**, 227–242 (2020). https://doi.org/10.1016/j.ins.2019.09.010
10. Duguma, A.L., Bai. X.: Contribution of Internet of Things (IoT) in improving agricultural systems. Int. J. Environ. Sci. Technol. **121**, 2195–2208 (2024). https://doi.org/10.1007/s13762-023-05162-7
11. Witjaksono, G., et al.: IOT for agriculture: food quality and safety. IOP Conf. Ser. Mater. Sci. Eng. **343**, IOP Publishing (2018). https://doi.org/10.1088/1757-899X/343/1/012023
12. Jirapond, M., et al.: IoT and agriculture data analysis for smart farm. Comput. Electron. Agric. **156**, 467–474 (2019). https://doi.org/10.1016/j.compag.2018.12.011
13. Aparna, P., et al.: IoT based animal health monitoring system using raspberry Pi. Int. J. Adv. Res. Sci. Commun. Technol. **3**(17), 300–307 (2023). https://doi.org/10.48175/IJARSCT-9509
14. Nishant, K., et al.: Application of artificial intelligence (Ai) in agriculture: an Indian perspective. Soil Health Manag. Knowl. (2022)
15. Vandana, B., Sathish, K.S.: Hybrid k mean clustering algorithm for crop production analysis in agriculture. Int. J. Innov. Technol. Explor. Eng. (IJITEE) **9**, 12S (2019). https://doi.org/10.35940/ijitee.b1002.1292s19
16. Grace, K., Induja, L.M.: Enrichment of crop yield prophecy using machine learning algorithms. Intell. Autom. Soft Comput. **31**(1), 279–296 (2022). https://doi.org/10.32604/iasc.2022.019947
17. Swain, K.P., et al.: Empowering crop selection with ensemble learning and k-means clustering: a modern agricultural perspective. Open Agric. J. **18**, 1 (2024). https://doi.org/10.2174/0118743315291367240207093403
18. Tan, Y., Liu, Z.: Feature selection and prediction with a Markov blanket structure learning algorithm. BMC Bioinform. **14**(Suppl 17), A3 (2013). https://doi.org/10.1186/1471-2105-14-S17-A3
19. Hua, Z., Zhou, J., Hua, Y., Zhang, W.: Strong approximate Markov blanket and its application on filter-based feature selection. Appl. Soft Comput. **87**, 105957 (2020)
20. Jinyuan, X., Baoxing, G., Guangzhao, T.: Review of agricultural IoT technology. Artif. Intell. Agric. **6**, 10–22 (2022). https://doi.org/10.1016/j.aiia.2022.01.001
21. BizIntellia: IoT in Agriculture. https://www.biz4intellia.com/blog/5-applications-of-iot-in-agriculture/. Accessed 31 July 2024
22. Tarun, G.: Measures of Proximity in Data Mining & Machine Learning (2019). https://towardsdatascience.com/measures-of-proximity-in-data-mining-machine-learning-e9baaed1aafb
23. Halkidi, M., Batistakis, Y., Vazirgiannis, M.: On clustering validation techniques, **17**, 2–3 (2001). https://doi.org/10.1023/A:1012801612483
24. Marcos, L.d.P.: Information Theory Metrics. Hudson & Thames Quantitative Research, Cornell University (2019)
25. Visweswaran, S., Cooper, G.: Learning instance-specific predictive models. J. Mach. Learn. Res. **11**, 3333–3369 (2010)

GreenGrowth: Smart Irrigation and Data-driven Crop Management Framework for Sustainable Agriculture

Adil Chekati[1](✉), Skander Boutout[2], and Inas Benfetima[2]

[1] MISC Laboratory, University of Constantine 2 - Abdelhamid Mehri, Constantine, Algeria
adil.chekati@univ-constantine2.dz

[2] NTIC Faculty, University of Constantine 2 - Abdelhamid Mehri, Constantine, Algeria

Abstract. The rapid global population growth, combined with increasing environmental challenges such as drought, necessitates the adoption of sustainable agricultural practices. This paper introduces GreenGrowth, a comprehensive framework designed to optimize irrigation and crop management through the integration of smart technologies. Using Internet of Things (IoT) devices and machine learning algorithms, GreenGrowth gathers real-time meteorological and soil data to accurately predict irrigation needs and recommend optimal crop types. The framework's architecture includes a wireless sensor network, an ESP8266 microcontroller, and Firebase for data storage, enabling precise, data-driven decision-making. Key findings indicate that decision tree algorithms, outperform other models in predicting irrigation needs, while achieving a high accuracy rate in crop prediction. The implications of this research are significant for enhancing agricultural productivity, conserving water resources, and reducing manual intervention, thereby contributing to sustainable farming practices. This study not only demonstrates the feasibility of smart irrigation systems but also sets a foundation for future developments in data-driven agriculture.

Keywords: Internet of Things · Data-Driven Decision Making · Sustainable Agriculture

1 Introduction

According to estimates by the Food and Agriculture Organization of the United Nations (FAO), the world's population is set to reach around 9.6 billion by 2050 [17]. Thus, agriculture is the main source of human food, with the growing population and drought, the demand for quality and quantity of food production will increase considerably.

Faced with this challenge, There is a need to adopt sustainable agricultural practices in terms of resource use. Water is an absolutely vital resource for agricultural production. However, traditional irrigation methods have often been

T. Marir et al. (Eds.): ICCSA 2024, ISEM 45, pp. 73–85, 2025.
https://doi.org/10.1007/978-3-031-90758-6_6

inefficient, resulting in considerable water wastage. Faced with this situation, intelligent agriculture offers numerous opportunities for optimizing farming practices, particularly when it comes to crop irrigation. Irrigation automation enables the simplification of tasks, reduce waste and use water more precisely. A transition is underway towards systems based on the Internet of Things (IoT) and machine learning, adopted in the agro-industrial sector to automate and manage irrigation systems intelligently, providing essential data for more efficient and sustainable agriculture.

Agriculture is the largest consumer of freshwater, with 69% of the freshwater withdrawals in 2024 being allocated to agricultural activities, according to FAO data [6]. However, a significant portion of water and energy is wasted, with losses of approximately 40%. The core issue addressed by this paper is the inefficient irrigation of crops. The water needs of plants vary depending on their growth stage, weather conditions, and soil characteristics. Over-irrigation or poorly calibrated irrigation systems lead to substantial waste of this vital resource and reduce agricultural yields. Farmers alone cannot effectively address this problem, and traditional agricultural methods are ill-equipped to meet this challenge, necessitating reconsideration of these methods. The objective is to move toward an optimized and precise automation of the irrigation process

Currently, numerous research initiatives are focusing on the agricultural sector, increasingly emphasizing the integration of artificial intelligence (AI) techniques and the Internet of Things (IoT). In this study, GreenGrowth is introduced, a framework that takes into account weather data and other information collected by a network of soil sensors. This data will be transferred in real-time to Firebase. GreenGrowth uses highly accurate machine learning algorithms to process the collected data and accurately predict water requirements for the irrigation phase. Additionally, it proposes a predictive model that recommends the most suitable crops according to the soil type. This approach will synchronize the data and assist farmers in remotely managing and monitoring their farms, leading to optimized resource consumption, maximized yields, and minimized human intervention.

The Internet of Things (IoT) links billions of Internet-connected devices to share information with each other and with their environment. In 2025, the global IoT market is expected to reach $1,100 billion (excluding hardware) with 25.2 billion connections, compared with 6.3 billion in 2016 [8]. The integration of this technology in the agricultural sector has given rise to smart farming. Due to cutting-edge technologies, this revolutionary advance is set to change traditional farming methods.

"Smart agriculture" or "Smart Farming" different terms, but one basic concept: The intelligent use and preservation of the environment through digital solutions. Smart agriculture is a new approach that combines Information and Communication Technologies (ICT), including the Internet of Things (IoT), data analysis, robotics, drones and artificial intelligence in agricultural practices [12]. Its main aim is to optimize agricultural processes, maximize productivity and minimize resource consumption. Thanks to this method, farmers can collect and study information from sensors and other connected devices in real time, enabling them to make accurate decisions. The main aims of smart farming are to optimize production costs, improve farm profitability, reduce workloads and promote sustainable, environmentally-friendly agriculture [16].

Furthermore, the advent of intelligent tools, such as Data-driven decision making and Machine Learning based approaches, are providing unprecedented opportunities for optimizing crop management and sustainability. These innovations are not without challenges, but their potential to enhance agricultural productivity and resilience is immense. The following Table 1 presents a comprehensive review of the literature on Intelligent Agriculture, summarizing the advantages and disadvantages of various IoT applications in this field. This review highlights the critical role that these technologies play in the development of sustainable agricultural practices.

Table 1. Summary of IoT technology applications for agriculture

Type of technology	Advantages	Disadvantages
Intelligent drone deployment for agriculture [1,4,15]	- Plant recognition and classification using drones - Creation of yield maps - Monitoring agricultural areas and plant health	- Threat to privacy - Drone regulations - item Security
Greenhouse automation (weather monitoring)[5]	- Increased yields - Improved resilience to drought, pests and disease - Improved quality	- Very costly technology - Configuration and maintenance of complex systems - Any failure can lead to damage
Use of virtual reality by farmers [11,14]	- VR for disease management - Grape selection - Increasing cow productivity	- Any failure could disrupt cow productivity - Risk of injury or physical or material damage - Initial cost of equipment purchase

2 Background and Related Work

By leveraging the power of data-driven decision-making, Machine Learning (ML) algorithms offer a sophisticated means to enhance agricultural practices, leading to more sustainable and efficient outcomes. The integration of various machine learning approaches, such as Decision Trees (DT), K-Nearest Neighbors (KNN), Fuzzy Logic, and Convolutional Neural Networks (CNN), has enabled farmers to achieve high accuracy in irrigation systems, disease detection, and water flow prediction. These techniques, when combined with the appropriate tools and sensors, provide actionable insights that can significantly improve agricultural productivity and resilience against environmental challenges.

The Table 2 below presents a detailed summary of the literature on machine learning applications in agriculture. It outlines various projects that have utilized these advanced algorithms, highlighting the approaches taken, the metrics

Table 2. Summary of the literature review on Machine Learning applied to agriculture.

Works	ML Algorithms	Metrics	Results	Tools Tools
Intelligent irrigation system [2,3,7]	– DT	– Accuracy – Precision – Recall – F1-Score	– 89,68% – 91,11% – 82% – 86,31%	– ESP8266 – Arduino Uno – DH11 sensor – Soil Moisture Sensor
	– KNN	– Accuracy	– 94%	– ZigBee – WSN – HPT 675 – FC-28 – HTM2500 – THERM200 – Inclinometer
	– Fuzzy logic	– Optimization	– 80%	– nRF24L01 – WSN – Node-RED server – Sensors
Detect and identify infections and diseases, monitor and maintain favourable conditions [10]	– KNN	– Accuracy	– 96%	– GSM (SIM800) – ESP8266 – Video Camera – Arduino Uno – Sensors
Real-time water flow prediction system for irrigation system [9]	– CNN	– Accuracy – Précision – Recall – F1-Score	– 97,10% – 85,50% – 86,80% – 85,80%	– Moblile Camera – IMX789 sensor

used to evaluate performance, the results obtained, and the tools and technologies that supported these systems. This comprehensive review underscores the critical role of machine learning in driving the future of smart agriculture, where data-driven methodologies are becoming integral to maintaining sustainable agricultural ecosystems.

3 GreenGrowth Framework Architecture

3.1 Overview of the System Architecture

GreenGrowth framework implements an intelligent irrigation system using precise meteorological data from the farm's location, combined with soil moisture information collected by sensors connected to an ESP8266 microcontroller. This data is transmitted in real time to the Firebase Realtime Databas.

A trained machine learning model continuously analyzes streamed data froms the wirless sensor network, each time a change is detected, it determines whether or not irrigation is required.

The ESP8266 microcontroller retrieves the model's result: If irrigation is required, it activates the water pump to water the crops. Otherwise, the pump will remain deactivated to save water resources and electricity.

3.2 Dataset and Preprocessing

The success of the GreenGrowth framework's intelligent irrigation system relies heavily on the quality and relevance of the data used for training the machine learning models. To ensure accurate predictions, we have carefully selected two datasets that cover crucial parameters for both irrigation needs and crop prediction (Fig. 1).

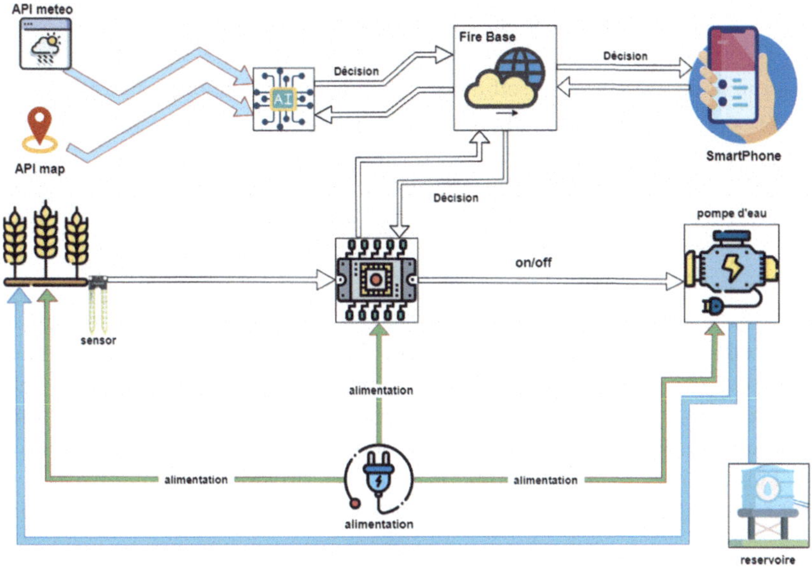

Fig. 1. GreenGrowth Framework Architecture

Irrigation Dataset. For the irrigation prediction model, we utilized the Crop Irrigation Scheduling Dataset from Kaggle [View Dataset 1]. This dataset provides detailed information on soil moisture, temperature, humidity, and crop growth stages, which are essential factors in determining irrigation requirements. The dataset contains over 381 records covering a wide range of crops common in the region, with features such as:

- Soil Moisture: Measured in percentage, indicating the water content in the soil.
- Temperature: Ambient temperature in degrees Celsius.
- Humidity: Air humidity as a percentage.
- Temperature: Ambient temperature measured in degrees Celsius.
- Crop Days: Number of days since the crop was planted.
- Irrigation: A binary variable indicating whether irrigation is required (1) or not (0).

We performed preprocessing on this dataset to ensure data uniformity and removed any irrelevant entries such as exotic crops not grown locally (e.g., coconut) using a Python script. The 'Crop Type' column is renamed 'Crop Num' which serves as a unique identifier assigned to each crop in place of their names, to ensure uniformity of data types. The cleaned dataset was then split into training (70%) and testing (30%) sets, ensuring that both sets contained a diverse range of samples (Table 3).

Table 3. Sample dataset for irrigation

crop num	crop days	soil moisture	temperature	humidity	irrigation
[1 - 210]	[120 - 990]	[14 - 236]	[11 - 85]	[0 - 100]	[0 | 1]
1	28	700	37	32	1
4	76	239	25	40	0

Crop Prediction Dataset. The second dataset used is the Crop Prediction Data, also sourced from Kaggle [View Dataset 2]. It includes variables related to soil nutrients (nitrogen, phosphorus, potassium), pH, temperature, humidity, and rainfall, all of which are critical for determining the next suitable crop to plant. This dataset contains 1,100 records and provides insight into the ideal crops for varying soil and weather conditions. The main features in this dataset are:

- N-soil (Nitrogen Content): Measured in parts per million (ppm), nitrogen is essential for plant growth and development.
- P-soil (Phosphorus Content): Measured in ppm, phosphorus is vital for root growth and flower/fruit development.
- K-soil (Potassium Content): Measured in ppm, potassium aids in disease resistance and plant strength.
- Temperature: Ambient temperature in degrees Celsius.
- Humidity: Relative humidity of the air as a percentage.
- pH: Soil pH, indicating the acidity or alkalinity of the soil, on a scale from 0 to 14. This is important for nutrient availability.
- Rainfall: The amount of rainfall during a specific period, measured in millimeters (mm).
- Crop: The target variable indicating the crop to be planted based on the current conditions.

Preprocessing steps included removing irrelevant columns such as price and location (state) using another Python script, as this information was not required for real-time predictions. Similar to the irrigation dataset, this data was split into training and testing sets (Table 4).

Table 4. Sample dataset for the next crop to be planted

N-soil	P-soil	K-soil	Temperature	Humidity	pH	Rainfall	Crop
[0_140]	[5_145]	[5_205]	[8.83_43.7]	[14.3_100]	[3.5_9.94]	[20.2_299]	/
90	42	43	20.8797	82.0027	6.5029	202.9355	RICE
71	54	16	22.6135	63.6907	5.7499	87.7595	MAIZE

Dataset Split. The split of the dataset is an important step. This split has been carried out randomly to ensure that the training and test sets contain a variety of samples (Table 5).

Table 5. Dataset split

Irrigation	Percentage	Number of lines according to percentage of irrigation dataset	Number of lines according to percentage of cycle dataset
Test	30%	115	330
Training	70%	266	770
Total	100%	381	1100

3.3 Irrigation Prediction Model

The learning process for the machine learning models in GreenGrowth involves using the cleaned datasets to train various algorithms for both irrigation and crop prediction. During training, the model learns the relationships between input features (e.g., soil moisture, temperature) and the target output (irrigation need or crop type).

Decision Tree. The decision tree (DT) is one of the best-known symbolic supervised machine learning methods in the field of artificial intelligence. It is one of the oldest machine learning classification techniques. Their modeling is based on a graphical representation of a set of rules, making them easy to understand, fast, simple and often yielding suitable results [13] Fig. 2.

The decision tree achieved an excellent accuracy of 93.91% on the test data, standing out as the best-performing model. Out of 67 non-irrigation cases, it correctly predicted 65 and misclassified only 2.
Similarly, out of 48 cases requiring irrigation, 43 were correctly predicted and only 5 were incorrectly predicted. Decision trees are known for their ability to handle binary classification problems, which explains the remarkable performance in this case of irrigation prediction.

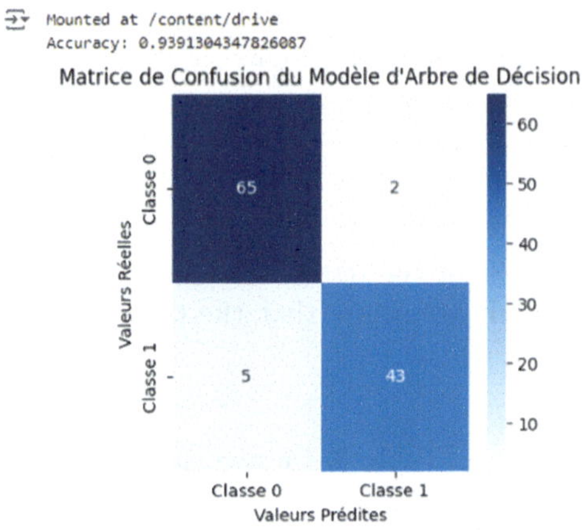

Fig. 2. Decision Tree Confusion Matrix for Irrigation Classification

K-Nearest Neighbors. The K-Nearest Neighbors (K-NN) method, also defined by Cover and Hart in 1967, is a supervised classification method based on the calculation of similarity between the instances to be classified and the instances in the training database. The K-NN method is used in data mining for its simplicity, efficiency and robustness Fig. 3.

Among the 67 cases where irrigation was not required, the KNN model correctly predicted 50, but misclassified the remaining 17 cases. Among the 48

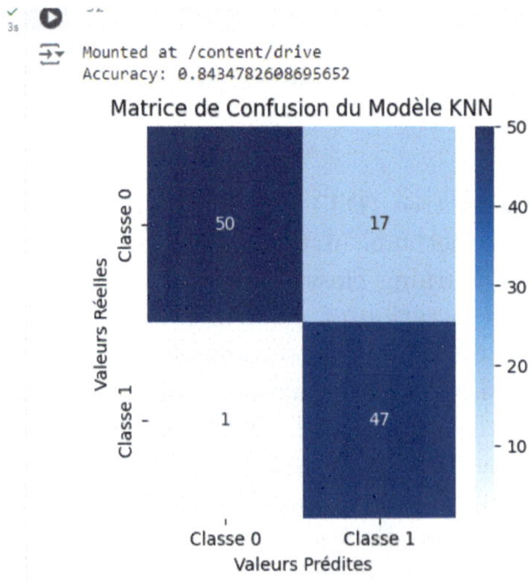

Fig. 3. K-nearest Neighbors Confusion Matrix for Irrigation Classification

cases requiring irrigation, the model correctly identified 47, while misclassifying 1. Overall, these results translate into an accuracy of 84.34% for the KNN model.

Support Vector Machines. Defined by Vapnik in 1995, SVM is a method based on supervised learning techniques introduced by Cortes and Vapnik. The idea of this method is to separate the training data, looking for the separator that maximizes the margin, in other words, to find the margin-maximizing criterion that gives it excellent generalization power (Fig. 4).

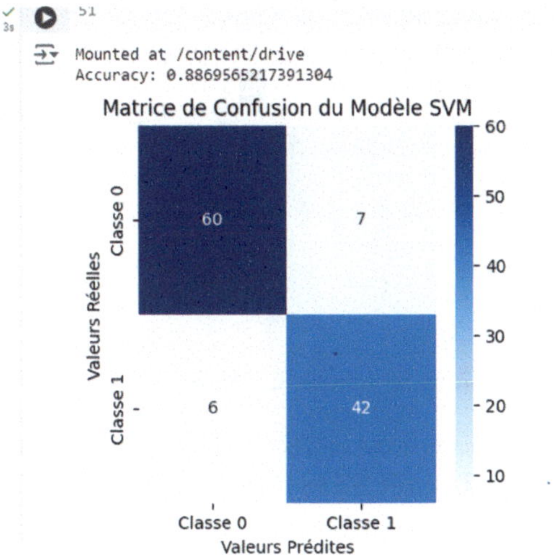

Fig. 4. SVM confusion matrix for intelligent irrigation classification

For the SVM model, it correctly predicted 60 of the 67 cases where irrigation was not required and misclassified 7. For the 48 cases requiring irrigation, it correctly classified 42 cases and misclassified 6 others. This translates into an accuracy of 88.69%.

Results: Having analyzed the results for each model, we can conclude that the Decision Tree (DT) has the best accuracy of all the models tested. Thanks to its ease and speed of interpretation and learning, it is perfect for intelligent irrigation needs in our situation.

3.4 Next Crop to Plant Prediction Model

In order to predict the next crop to be planted, we implemented two different algorithms: decision tree (DT) and linear regression (LR). Here are the results:

Linear Regression. Linear regression is defined as an algorithm that provides a linear relationship between an independent variable and a dependent variable to predict the outcome of future events.

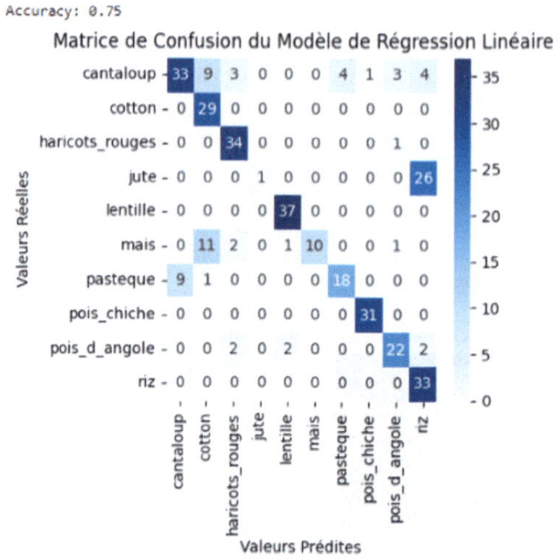

Fig. 5. Linear regression confusion matrix for Next crop to plant prediction

LR produced an low accuracy rate of 75%, which means that 248 cases out of 330 were correctly classified, while 82 cases were misclassified (Fig. 5).

Decision Tree. In prediction phase, DT achieved 98% accuracy: only 5 cases were misclassified, while the remaining 325 cases were correctly classified (Fig. 6).

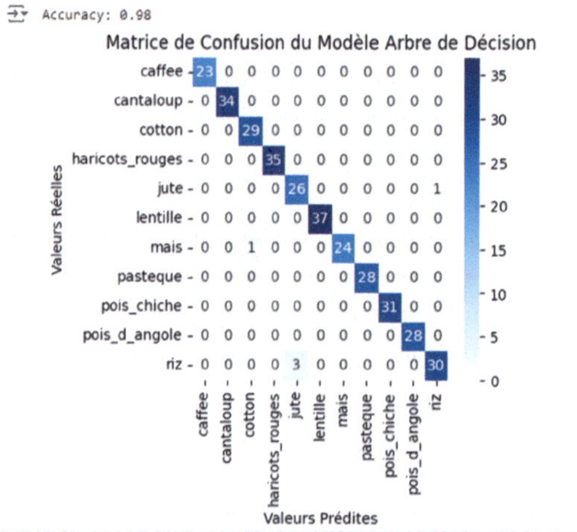

Fig. 6. Decision Tree confusion matrix for Next crop to plant prediction

Results: In conclusion, the decision tree algorithm appears to be the optimal choice. With an accuracy of 98%, it performs significantly better than linear regression. This indicates that the decision tree is more effective in predicting crops to be planted in the near future, which is important for efficient farmland management based on the main nutrients available in the soil.

3.5 Real-Time Learning and Adaptation

In real-time, the system continuously monitors environmental data collected by IoT sensors deployed across the farm. These sensors feed updated readings of soil moisture, temperature, and humidity to the system, allowing the trained machine learning model to make instantaneous predictions about whether irrigation is required.

The model adapts in real-time by leveraging the Firebase Realtime Database, which stores the incoming sensor data and sends it to the machine learning engine for analysis. Each time a change in environmental data is detected, the system re-evaluates the irrigation need or crop suitability based on the learned patterns. For instance, if there is a sudden increase in temperature or a drop in soil moisture, the model will trigger irrigation to optimize water usage.

Additionally, periodic retraining is planned to allow the system to learn from new data collected over time. This ensures that the model remains accurate as environmental conditions change or as new crops are introduced to the system. By continuously updating the training data with recent observations, GreenGrowth can evolve and improve its decision-making process, offering more precise recommendations for irrigation and crop selection in the future.

4 Conclusion

The GreenGrowth framework demonstrates the potential of integrating smart technologies and data analytics in modern agriculture to address critical challenges such as water scarcity and the need for sustainable farming practices. By leveraging IoT devices, machine learning algorithms, and real-time data collection, GreenGrowth not only optimizes irrigation but also enhances crop management decisions. The framework's deployment revealed that decision tree algorithms are particularly effective, achieving high accuracy rates in both irrigation and crop prediction, which underscores the value of advanced computational techniques in agricultural applications.

The successful implementation of GreenGrowth suggests that similar frameworks could be scaled and adapted to various agricultural contexts, thereby promoting more efficient resource use and improving crop yields. However, further research is needed to refine the system's adaptability to diverse environmental conditions and to explore the integration of additional data sources, such as satellite imagery and advanced weather forecasting models. Moreover, addressing the challenges of system accessibility and cost-effectiveness will be crucial in ensuring that the benefits of smart agriculture are widely distributed, particularly in developing regions. Overall, GreenGrowth represents a significant step towards the realization of sustainable, data-driven agriculture, with promising implications for global food security and environmental conservation.

References

1. Ait Issad, H.: Déploiement intelligent de drones pour une agriculture du futur. Ph.D. thesis, Université Mouloud Mammeri de Tizi-Ouzou (2020)
2. Benyezza, H., Bouhedda, M., Rebouh, S.: Zoning irrigation smart system based on fuzzy control technology and IoT for water and energy saving. J. Clean. Prod. **302**, 127001 (2021). https://doi.org/10.1016/j.jclepro.2021.127001, https://www.sciencedirect.com/science/article/pii/S0959652621012208
3. Blessy, A., et al.: Sustainable irrigation requirement prediction using internet of things and transfer learning. Sustainability **15**, 8260 (2023). https://doi.org/10.3390/su15108260, academic Editor: Juan Miguel Navarro
4. BlogduKosova: Les avantages de l'utilisation de drones dans l'agriculture (2023). https://blogdukosova.com/les-avantages-de-lutilisation-de-drones-dans-lagriculture/. Accessed 25 Aug 2024
5. DJEGHAIBEL, A., MAARADJI, S.: Automatisation intelligente d'une serre basée sur un système d'internet des objets en utilisant arduino (2020)
6. FAO: ntegrated agriculture water management and one health (2024). https://www.fao.org/one-health/areas-of-work/water/en#:~:text=Agriculture%20accounts%20for%2070%20percent,food%20and%20agriculture%20remains%20critical, accessed on August 25th, 2024
7. el Ghamry, A., et al.: Smart irrigation system using IoT and machine learning methods. In: Conference Paper (2023). https://doi.org/10.1109/NILES59815.2023.10296736, https://www.researchgate.net/publication/375212442
8. GSM-Association: Transformative IoT: market opportunities (2021). https://www.gsma.com/solutions-and-impact/technologies/internet-of-things/beyond-connectivity-iot-market-opportunities/. Accessed 25 Aug 2024
9. Kethavath, A., Dhruw, L.K., Choudhury, B.U., et al.: Optimization of irrigation timing for sprinkler irrigation system using convolutional neural network-based mobile application for sustainable agriculture. Smart Agr . Technol. (2023). https://doi.org/10.1016/j.atech.2023.100305
10. Kumar, S., Mishra, S., Khanna, P.: Precision sugarcane monitoring using SVM classifier. Proc. Comput. Sci. **122**, 881–887 (2017). https://doi.org/10.1016/j.procs.2017.11.492
11. Max_zero: Une nouvelle réalité pour l'agriculture : les entreprises qui exploitent apple vision pro & xr, vr et ar. Article de blog (février 2024)
12. Mirror Review: how smart farming is renovating traditional farming methods and tools (2023)
13. Nemouchi, S., Farah, N.: Reconnaissance de l'Écriture Arabe par Systèmes Flous. Master's thesis, Université Badji Mokhtar, Annaba, Algérie (2010)
14. Polge, A.: En turquie, des vaches équipées de casques de réalité virtuelle pour apaiser leur confinement en hiver. Article en ligne (2022)
15. Quora: Quels sont les avantages et les inconvénients de l'utilisation de drones ? (2023), https://fr.quora.com/Quels-sont-les-avantages-et-les-inconv%C3%A9nients-de-lutilisation-de-drones, . Accessed 25 Aug 2024

16. SMAG: Smart farming qu'est-ce que l'agriculture intelligente ? (2021). https://smag.tech/blog/smart-farming-a-z/ . Accessed 25 Aug 2024
17. UnitedNations: World population prospects: The 2017 revision (2017). https://www.un.org/en/desa/world-population-projected-reach-98-billion-2050-and-112-billion-2100 . Accessed 25 Aug 2024

Automated Weed Detection in Agriculture: A Performance Evaluation of YOLOv8 and InceptionV3 Models

Abdeljalil Gattal[1](✉), Mohamed-Cherif Nait-Hamoud[1,2], Mohamed-Faouzi Gattal[3], Hadjer Hadji[3], and Mohammed Cheriet[2]

[1] Laboratoire de Vision Et d'Intelligence Artificielle (LAVIA), Université Larbi Tébessi, Tébessa, Algérie
{Abdeljalil.Gattal,mohamed-cherif.nait-hamoud}@univ-tebessa.dz
[2] National School of Cyber Security, Sidi Abdellah, Algiers, Algeria
[3] Department of Computer Science, Echahid Cheikh Larbi Tebessi University, Tébessa, Algeria

Abstract. Weeds present a substantial challenge to agriculture, causing significant yield losses and elevating management costs. This study explores the automation of weed detection in crops through the deployment of advanced deep learning models. Two prominent models, YOLOv8 and InceptionV3, were extensively evaluated for their performance in accurately identifying weeds within agricultural imagery. The YOLOv8 model has proved to be the most effective method by achieving absolute accuracy in the differentiation of crops from weeds. The findings of the study highlighted the efficiency of weed management systems based on the YOLOv8 and its potential towards the sustainable development of agriculture through better weed control systems and then overall crop protection and resource optimization.

Keywords: Weed detection · InceptionV3 · YOLOv8 · Deep learning · Sustainable agriculture

1 Introduction

Weeds are a threat to worldwide agriculture since they reduce crop productivity and lead to complicated management strategies. In modern agriculture, detection of weeds within field crops is an important facet. As effective as it is now, imagining the increasing pressure on crops by weeds deepens the set on the need for effective detection and control measures.

Such non-significant and subfield surveys are not conventional, such as ML methods and computer vision, which provide practical alternatives to this problem, making considerable progress in weed control. These technologies allow for the efficient exploitation of imaging technology at an agricultural scale to analyze the crop vegetation cover for the identification and quantification of weeds. This method employs fewer herbicides and minimizes crop damage, thus making weed control more effective and feasible economically.

T. Marir et al. (Eds.): ICCSA 2024, ISEM 45, pp. 86–100, 2025.
https://doi.org/10.1007/978-3-031-90758-6_7

In weed detection, an emphasis on the use of AI has become common, and its importance in this particular area has grown. As a case in this paper, a few attempts to use A.I. methods to tackle the problem of weeds were described, focusing on the efficiency, accuracy, and application of the method to different types of agriculture. Additionally, some studies available in the reviews were undertaken to uncover new perspectives, present issues, and future opportunities of the weed detectors and controllers in modern agriculture.

In the research literature, studies have been found to identify increasing interest of the scientific community on this subject and increasing number of publications on the modern agricultural research problems. Various methodologies and strategies were explored in this study, synthesizing findings and insights organized within thematic frameworks. These later encompass a range of machine learning (ML) [1–4] and deep learning (DL) [5–10] algorithms and methodologies, including Convolutional Neural Networks (CNNs), AlexNet, VGGNet, and Deep Neural Networks (DNNs), as well as approaches that combine image processing and machine learning. The selected studies were categorized into two primary AI-classification based on the methodologies employed. Each classification includes a variety of algorithms that can be combined in different ways. The subsequent sections examine each method in details, outlining fundamental principles, the used datasets, and addressing the limitations identified in the studies providing a comprehensive overview of the used methods for weed detection based mainly on classical ML and deep learning approaches.

In this paper, we have also investigated the ability of two main deep learning (DL) models, YOLOv8 and InceptionV3, to address the challenge of weed detection in agricultural fields. The motivation is due to real-time object detection capabilities of YOLOv8 and InceptionV3, a convolutional neural network architecture, that demonstrated promising results in accurately identifying and classifying weeds. Our findings revealed that YOLOv8 outperforms InceptionV3 in terms of precision.

The rest of this paper is organized as follows: Sect. 2 explores AI-based weed detection techniques, examining various classical machine learning and deep learning methods in terms of performance. Section 3 presents a deep learning-based methodology for accurate weed detection, leveraging YOLOv8 and InceptionV3 models. Section 4 presents the experimental results, and finally, Sect. 5 concludes this manuscript.

2 Related Works

Several studies investigated various machine learning techniques for weed detection, including neural networks, dimensionality reduction methods, classification algorithms, including random forest technique. In 2021, Mahajan et al. [1] proposed a plant detection method combining morphological feature extraction with transfer algorithms, Support Vector Machines (SVM), and AdaBoost. This approach achieved 95.85% accuracy in agricultural and environmental protection applications. Bakhshipour et al. [2] (2018) examined weed detection using shape features, combining multiple features to establish patterns for each plant type. Ranjan et al. used SVM and ANN, reaching an overall accuracy of 95.0% and 92.92%, respectively, in the sugar beet fields. Likewise, Naresh et al. (2016) [3] also used modified local binary patterns for classifying medicinal plants and obtained a classification accuracy of over 97%. In an even more recent work, Yang et al. (2021) [4] reported a method combining shape and texture features using Multiscale Triangle Descriptor MTD and Local Binary Pattern Histogram Fourier, LBP-HF, for plant leaf identification with an accuracy of 99.1%.

Particularly, convolutional neural networks CNN have incorporated Alpha and Beta weed detection as advanced imaging features. Many approaches, AGROAVNET, YMufT, and Xception, have been utilized in them and proved to be successful. Chavan et al. [5] (2018) developed a hybrid model named AgroAVNET for weed detection, combining the characteristics of AlexNet and VGGNet with CNN, and achieved an accuracy of 98.23%. Trong et al. [6] (2012) developed the ymufT configuration for training deep neural architectures on imbalanced data sets and accomplished 97.18% of weed classification accuracy. Xu et al. (2021) [7] have exploited Xception based on transfer learning and reached a 99.63% of weed recognition accuracy. Besides, Olsen et al. [8] (2019) worked on the ResNet-50 and InceptionV3 and achieved 95.7% and 95.1% generalization tasks' accuracy, respectively. In addition, Ferreira et al. [9] (2019) tried JULE and DeepCluster, respectively, for image clustering and unsupervised picture analysis tasks, recording a 95% accuracy rate. Hu, B. et al. [10] (2020) proposed GWN, a deep neural network proposed for weed image recognition reaching an accuracy of 98.1%. Not too long ago, Dang et al. [11] (2023) evaluated several YOLO object detectors on the CottoWeedDet12 data set, achieving 95.22% for YOLOv4 and 89.72% for Scaled-YOLOv4. Secondly, Cao et al. [12] (2024) provide an in-depth investigation on the DCA-ResNet plant leaf recognition model and achieved 96.79%, 96.64%, and 96.44% recognition accuracy of three public datasets.

Research efforts continue to improve and provide more effective techniques for detecting and controlling weeds. Table 1 summarizes recent research studies on weed detection providing more details on used methodologies, key features, and reported accuracies that offers a big picture of the key points regarding the integration of technology in weed detection and crop protection. The increasing global crop losses highlight the importance of incorporating classical ML and DL algorithms in detecting and eliminating weeds, while encouraging farmers to reduce pesticide usage. DL techniques have proven to be effective in weed detection, exploiting sensor data to achieve higher recognition accuracy compared to traditional methods such as K-means, SVM, and ANN. Recent advancements include CNN models such as ResNet, InceptionV3, and Xception, as well as hybrid models like AgroAVNET and GWN. Techniques like YMufT and DCA-ResNet have shown promising results in enhancing performance. Ongoing research is focused on developing technologies for real-time analysis and adaptability to new challenges, effectively supporting farmers in managing weeds and protecting crops.

3 Methodology

Convolutional Neural Networks (CNNs) have revolutionized the field of computer vision, providing robust solutions for a variety of tasks, including image classification, segmentation, and particularly, object detection. The process of locating and identifying various objects in digital images is termed object detection, being one of the most essential components of image analysis. Rather than production of class labels only as in

Table 1. Summary of Recent Advancements in Weed Detection

Authors	Datasets	Methods	Accuracy (%)
Mahajan et al. (2021) [1]	Flavia	Support vector machine with adaptive boosting	95.85
Bakhshipour et al. (2018) [2]	Not mentioned (private datasets)	Artificial Neural Networks (ANN) Support Vector Machines (SVM)	92.92 95.00
Naresh et al. (2016) [3]	Flavia	MLBP (Modified Local binary patterns)	97.55
Yang et al. (2021) [4]	Flavia	LBP-HF (local binary pattern histogram Fourier) and MTD (multiscale triangle descriptor)	99.1
Chavan et al. (2018) [5]	Plant seedlings dataset	AgroAVNET, a hybrid VGGNET and AlexNet model	98.23
Trong et al. (2021) [6]	Plant seedlings dataset	Min-class-max-bound technique (MCMB), ResNet, yielding multi-fold training (YMufT) approach, and DNN combined	97.18
Xu et al. (2021) [7]	Plant seedlings dataset	Convolutional neural network with depth-wise separability, Xception	99.63
Olsen et al. (2019) [8]	Deepweeds	InceptionV3 ResNet-50	95.1 95.7
Ferreira et al. (2019) [9]	Deepweeds	Deep Clustering for Unsupervised Learning of Visual Features (DeepCluster) and Joint Unsupervised Learning of Deep Representations and Image Clusters (JULE)	95
Hu et al. (2020) [10]	Deepweeds	Graph Weeds Net (GWN)	98.1

(continued)

Table 1. (*continued*)

Authors	Datasets	Methods	Accuracy (%)
Dang et al. (2023) [11]	CottoWeedDet12	YOLOv4, Scaled-YOLOv4	95.22, 89.72
Cao et al. (2024) [12]	Flavia plant leaves, Flavia plant leaf diseases, Plant Village leaf dataset	DCA-ResNet	96.79, 96.64, 96.44

classification, it is required to also provide the position of the objects in the image under taking the work. In recent years, many deep learning models for object detection have also been developed. Some of the prominent ones include: convolutional neural network (CNN) [8, 13, 14], InceptionV3 [8, 15], R-CNN Architecture [16], Faster R-CNN [17], You Only Look Once (YOLO) [18–20], Single Shot MultiBox Detector (SSD) [21]. The model finds weed patterns and features in the images. Two main models, InceptionV3 and YOLOv8, were selected. Figure 1 shows the structure of the system with features such as InceptionV3 and YOLOv8, for weed detection. This architecture helps combine these models to assist in weed identification, distinguishing it from the crops, which facilitates its use in controlling weeds in farmland.

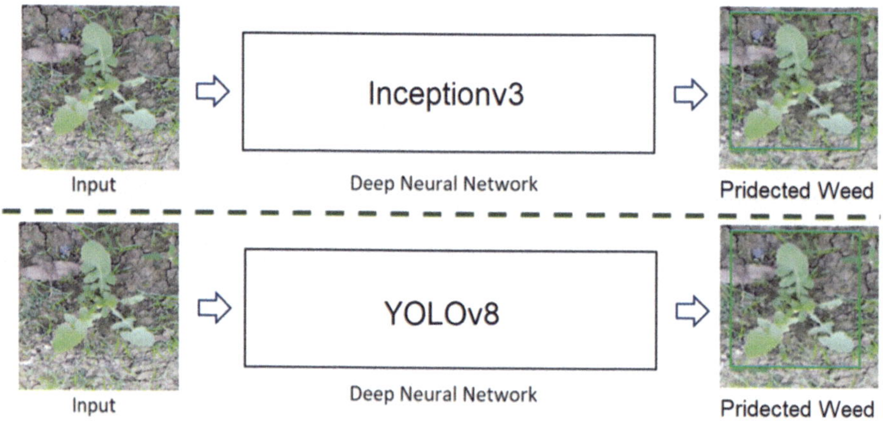

Fig. 1. Integration of InceptionV3 and YOLOv8 models for weed detection

3.1 InceptionV3

Developed by Google, InceptionV3 is a deep learning architecture that is particularly efficient for image recognition tasks. The more advanced such an architecture has its 'Inception modules' incorporated into it to allow it to extract features from images at more than one spatial feature, thereby improving its ability to recognize intricate designs within an image. InceptionV3 has found numerous applications in computer imaging, such as image classification, object recognition, and even in generating images.

The structure of InceptionV3 is given with a sequence of interconnecting systems of Inception such that the last one has Inception in it. All of these Inception modules are made up of a few alternate feature-extraction branches, each of which is able to extract image features at certain spatial frequency bands or areas.

- 1 × 1 branch: A dimensionality reduction is achieved by the use of a 1 × 1 convolution on the input feature, and this assists in reducing the computational cost, but crucial information does not get lost.
- 3 × 3 branch: A local 3 × 3 convolution is carried out in the 3 × 3 convolutional branch to capture extras that are not extracted from the large features by making use of the patch size of the images.
- Branch of the 5 × 5: Utilizes a larger kernel size than normal, with a 5 × 5 conv being utilized on the images. This would help recover the spatial features at an even larger scale.
- Average pooling branch: Ingres a morphological closure operation on the images using simple 3 × 3 average pooling. This aids in cognition by reducing the dimension of the input features and in aggregation of spatial patterns.

These branches process the input data simultaneously, and the outputs produced are processed together to create an informative and comprehensive set of features. This process of multi-scale feature extraction makes InceptionV3 be able to learn a wide range of patterns effectively, making it a good candidate for advanced image recognition.

Several key architectural innovations contribute to the effectiveness of InceptionV3. One significant feature is factorized convolutions, which reduce computational complexity by breaking down larger convolutions (such as 5 × 5) into smaller, more manageable operations (like two 3 × 3 convolutions). Additionally, the architecture employs asymmetric convolutions (for example, a 3 × 1 convolution followed by a 1 × 3 convolutions) to further decrease computational load while maintaining the model's ability to capture spatial dependencies.

InceptionV3 also includes auxiliary classifiers positioned at intermediate points within the network. These classifiers improve convergence during training by providing additional gradient signals and act as regularizes to reduce overfitting. Another important feature is the use of batch normalization layers throughout the network, which stabilize and accelerate the training process by normalizing the inputs of each layer, thereby mitigating issues related to internal covariate shift.

Furthermore, InceptionV3 incorporates label smoothing, a regularization technique that prevents the model from becoming overly confident in its predictions, thus improving generalization performance. The input to the InceptionV3 model is an image in RGB format, which includes resizing it to the required dimensions (299 × 299 × 3 pixels) and normalizing pixel values. The output would be a probability distribution over the predefined classes (weed vs. crop). This model might output whether the image contains weed or a crop. Figure 2 depicts the detailed structure of the InceptionV3 model used in this process, showcasing its various inception modules, convolutional layers, and pooling operations that contribute to its ability to accurately classify and differentiate between weeds and crops.

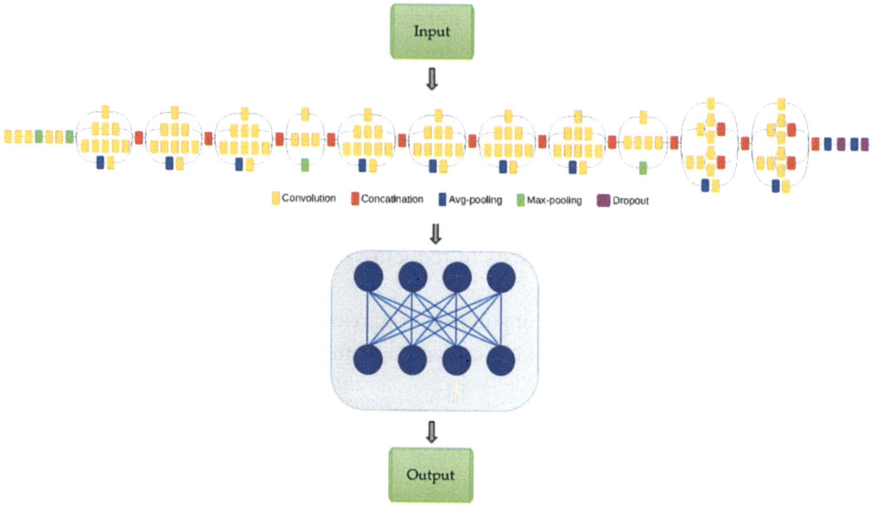

Fig. 2. Architecture of the InceptionV3 model

3.2 YOLO

The YOLO (You Only Look Once) model, created by J. Redmon et al. [18] (2016) was specifically designed to overcome the shortcomings of earlier object recognition models such as Fast R-CNN. Despite Fast R-CNN being a top-performing model at the time, it struggled with real-time applications due to its slow prediction speed of 2–3 s per image. YOLO revolutionized this process by making final predictions in just a single forward pass through the network, enabling real-time object detection. Figure 3 illustrates the YOLO architecture, which begins by resizing the input image to 448 × 448 × 3 pixels while maintaining its aspect ratio through padding. The resized image is then processed by a convolutional neural network (CNN) that includes 24 convolutional layers and 4 max-pooling layers, followed by 2 fully connected layers.

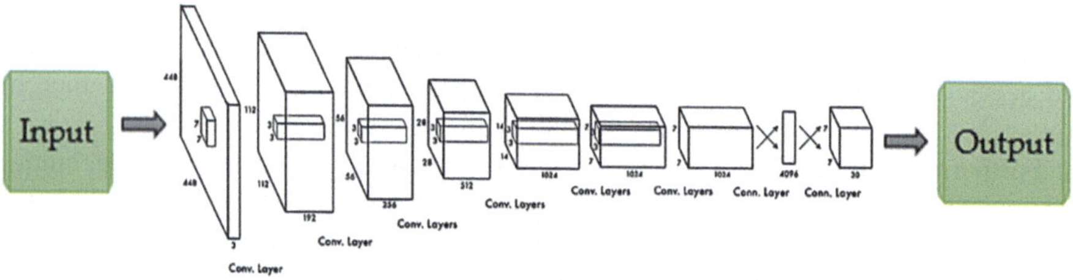

Fig. 3. Architecture of YOLO model

To manage computational complexity and reduce the number of channels, the architecture uses 1x1 convolutions in conjunction with 3 × 3 convolutions. The final layer of YOLOv1 converts the (1, 1470) vector from the last fully connected layer into a (7,

7, 30) cuboidal output. The model predominantly employs the Leaky ReLU activation function, except in the final layer, where a linear activation function is used. Batch normalization is applied to stabilize the learning process, and dropout is used to mitigate overfitting [19]. Initially trained on the ImageNet-1000 dataset, YOLO also has a variant called Fast YOLO, which reduces the number of layers from 24 to 9 and uses fewer filters while maintaining the core parameters of the original YOLO.

The YOLO model uses a sum-squared error loss function, which simplifies the optimization process but treats classification and localization tasks with equal importance. The loss function consists of five components: the first two address the mean-squared error for localization, focusing on how accurately the bounding boxes align with the ground truth. A specific component accounts for the square root of the difference in width and height, ensuring that the loss function penalizes size deviations more heavily for smaller boxes.

The remaining components relate to classification errors, including the sum-squared error between predicted confidence scores and ground truth for each bounding box within each cell, the mean-squared sum across cells that do not contain bounding boxes, and the sum-squared error of class predictions within these grid cells [18]. The YOLO architecture divides the input image into an $S \times S$ grid. If an object's bounding box center falls within a grid cell, that cell is tasked with detecting the object. Each grid cell predicts bounding boxes and their corresponding confidence scores, which indicate both the likelihood of an object being present and the accuracy of the bounding box coordinates compared to the ground truth. During testing, the model multiplies conditional class probabilities by individual box confidence predictions. The confidence score is calculated as:

$$Score = P_r(Object) * IoU_{pred}^{truth} \tag{1}$$

If no object exists in the grid cell, the confidence score is zero. If an object is present, the confidence score equals the IoU between the ground truth and predicted boxes. Each bounding box prediction includes five elements: *(x, y)* coordinates for the center of the box relative to the grid cell boundaries, *(w, h)* for the width and height relative to *(x, y)*, and the confidence score, which indicates the likelihood of an object being present within the box.

In January 2023, Ultralytics introduced YOLOv8 [20], the latest and most advanced model in the YOLO series. Although a formal paper has yet to be released and some features are still being added, initial comparisons suggest that YOLOv8 outperforms its predecessors. Integrating the trained YOLOv8 model into our application through Roboflow's API allows for seamless automation of weed detection. This integration not only streamlines the deployment process but also leverages Roboflow's advanced tools for visualizing predictions and evaluating model accuracy, ultimately enhancing the effectiveness and precision of agricultural practices.

3.3 Dataset

In this study, the "Crop and Weed Detection" dataset was used, available on Kaggle [22], which focuses on identifying various crop and weed species with fine-grained accuracy.

This dataset consists of 1,300 color images of sesame crops and different types of weeds, each labeled for use in machine learning applications. The images are 512 × 512 pixels and have been formatted for YOLO-based object detection models. The dataset includes detailed annotations such as bounding boxes and semantic masks, which are essential for accurately recognizing and differentiating between multiple weed species and crops. This is crucial for improving weed detection and control systems.

The dataset covers a wide range of weed species commonly found in agricultural environments. While the specific names of all 74 weed species are not explicitly listed in the source, the diversity of the dataset ensures comprehensive coverage, making it highly valuable for machine learning applications in agriculture. Figure 4 provides examples of images from this dataset. To train our crop and weed detection model, we divided the dataset into a training set (70%), a validation set (20%), and a test set (10%). This split allows for effective model training, evaluation, and testing, ensuring robust performance across different subsets of data.

Fig. 4. Samples of images from the dataset

3.4 Performance Evaluation Criteria

Evaluating the performance of deep learning models is a critical step that offers valuable insights into their effectiveness and accuracy. There is various evaluation metrics commonly employed in machine learning, particularly in the context of weed detection, which are discussed below [23].

Accuracy is one of the most widely used metrics to assess model's performance. It is calculated by comparing the number of correct classifications (both true positives (*TP*) and true negatives (*TN*)) to the total number of classifications made by the model (true positives (*TP*), true negatives (*TN*), false positives (*FP*) and false negatives (*FN*)). The formula for accuracy is as follows:

$$Accuracy = \frac{TP + TN}{TP + TN + FP + FN} \tag{2}$$

Precision is another important metric, especially in classification tasks. Precision measures the number of true positive (*TP*) classifications relatively to the total number of positive classifications made by the model. It becomes extremely beneficial in cases where it is important to avoid false positives. The formula for calculating precision is given as follows:

$$Precision = \frac{TP}{TP + FP} \tag{3}$$

The recall or sensitivity is the one that is used to evaluate the capability of the model to detect all true positive cases present in the dataset. Its importance cannot be overemphasized in a situation that needs qualitative or positive instances in large numbers. The recall can be computed using the following expression:

$$Recall = \frac{TP}{TP + FN} \tag{4}$$

These metrics are core in doing any model evaluation especially in domains like weed detection where a good compromise between precision, recall and other metrics is needed for feasible and functional systems development.

4 Experiments

To apply the InceptionV3 architecture in the classification of weed and crops, first the authors fetch the "inception v3" architecture from the internet and remove the top layers in order to adjust it to their requirements. Further we impose a capsule at the end of classification layers that allows shrinking the spatial dimension with respect to the feature maps obtained. This goes with fully connected with 1024 neurons layer with ReLu activation, which helps in capturing the intricacy in the dataset. To classify images into two categories (crop and weed), we add a softmax output layer. These new layers are combined with the base model layers, which are then frozen to retain the pre-trained weights. This ensures that the model benefits from the knowledge acquired during its initial training. The model is then compiled using the Adam optimizer, with categorical cross entropy as the loss function and accuracy as the evaluation metric. As seen here, this approach combines pre-trained knowledge with new knowledge that is particularly useful to our data.

The model is fit to the training data over 10 epochs during which the cumulation of steps per epoch includes the number of training samples and batch size. At the end of every epoch, the model is subjected to validation with a validation data generator, the number of steps of which is based on the previously described proportions.

These graphs incorporated with those depicting the changes in training and validation losses and accuracies show how the model was able to predict outputs over the entire training and validation process as illustrated in Fig. 5. Accuracy graph demonstrates the performance of the model over time showing a progressive increment in accuracy of the model. Loss graph on the contrary measures the efficiency of the model in terms of accuracy against time and depicts decreasing values of loss. The synthesis of these graphs clarifies the training and validation processes and the performance of the InceptionV3 model in that period. It can then be concluded that the model was able to learn and make correct predictions since the accuracy of the model and the loss was in a reasonable region.

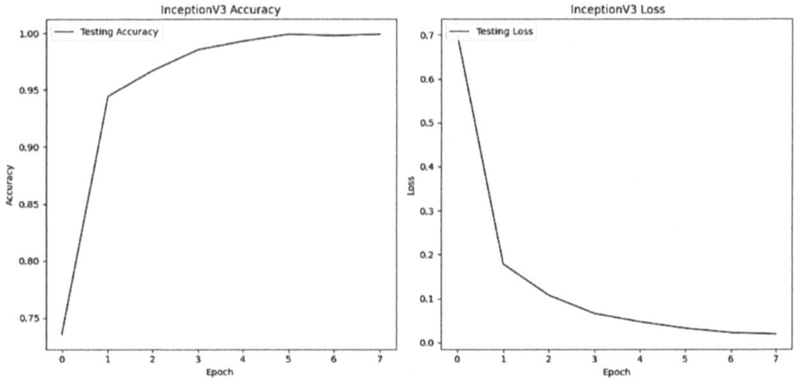

Fig. 5. Accuracy and loss graphs for InceptionV3

In the case of crop and weed detection dataset preparation with Roboflow infrastructure, it appeared that the model was trained well after 300 epochs of the YOLOv8 training. This made it possible to detect the crops and weeds accurately and efficiently under varying circumstances and conditions. The foremost goal here was to figure out the optimization of object detection with respect to the fact that the model will not only accept its findings in regards to crops and weeds but will be able to correctly identify them. Roboflow examined the model from a broader point of view, assessing box loss, class loss, object loss, accuracy, and many other metrics. These evaluations of the model filled a performance gap of models that involved evaluation for objects's detection and classification performance.

In Fig. 6, a graphical division of the box loss, class loss, and the object loss within training of the YOLOv8 model has been represented. These graphs show how the model learns and its determination to predict correct boxes, identify the objects, and locate them from the given images. Hence the model attained a baffling accuracy of 98%. This high accuracy is a result of the information derived from the model on both crops and weeds situated in agricultural images, demonstrating that YOLOv8 is appropriate for such applications as agriculture. In such applications, it is essential to have accurate recognition of agricultural objects and their classification and recognition at high precision and recall.

While analyzing the YOLOv8 and InceptionV3 models in the "Crop and Weed Detection" dataset, a number of key differentiations can be noted, as depicted in Table 2. This table contains the assessment of the models on three parameters of great importance: accuracy, precision, and recall. Each of these measurements is important in assessing the efficiency of the models in distinguishing crops from weeds and vice versa in the agricultural images.

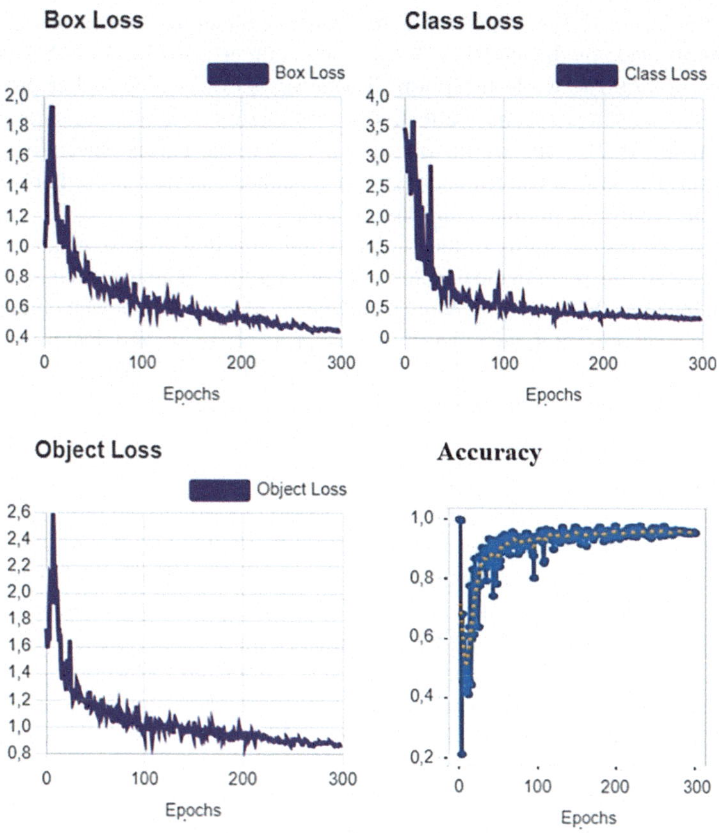

Fig. 6. Accuracy and loss graphs for YOLOv8

Table 2. Results of two selected models

Model	Accuracy (%)	Precision (%)	Recall (%)
InceptionV3	90.50	91.00	90.67
YOLOv8	98.00	97.90	95.20

For instance, accuracy is used to evaluate the ability of the model to identify both crops and weeds. The YOLOv8 achieves a remarkable accuracy of 98.00%, while the InceptionV3 stands at a lower 90.50%. The clear differences in the accuracy figures from the above analysis suggest that YOLOv8 is more dependable than a normal model of high accuracy over the entire dataset. The advantage of YOLOv8 in terms of accuracy stresses the fact that it is better at addressing the image's high level of interaction and heterogeneity, possibly owing to the structure of the architecture it is designed for object detection applications.

Precision measures the proportion of true positive identifications among all positive identifications made by the model. YOLOv8 again outperforms InceptionV3, achieving a precision of 97.90%, while InceptionV3 achieves 91.00%. The higher precision of YOLOv8 indicates that it is more effective at minimizing false positives, meaning it is less likely to incorrectly label a crop as a weed or vice versa. This is especially important in farm experience zones, where the crop might be unnecessarily removed or weeds may remain in the environment due to misclassifications.

Recall measures the ability of the model to find all actual positive instances present in the images. At this point in the development of YOLOv8, it attains a recall of 95.20% better than InceptionV3's 90.67%. More recall means more of the relevant objects, which should have been present in the image, are most probably recovered than not (reduction of the false negative rate). Especially in the agricultural use cases, high recall is preferred as it relates to how many of those pesky weeds are managed.

5 Conclusion

In this paper, we presented an overview of the methods that have been used for weed management, which rely on both traditional ML and deep learning techniques. Furthermore, we undertook a comparative analysis of two deep learning models—YOLOv8 and InceptionV3—because of their effectiveness. From the experimental results, it can be inferred that indeed YOLOv8 is quite good as compared to Inception V3 in accuracy, precision, as well as recall performance. This more likely sounds so because of the improved design and structure of the YOLOv8, which attains fast and accurate object detection.

As such, this model performs well in the separation of crops and weeds, leading to fewer false positives and false negatives. However, as far as effective image classification is concerned, InceptionV3 is considered quite a strong model as opposed to where it seems ineffective, especially in image classification. Thus, these results also imply that a crop detection model using YOLOv8 is ideal for agriculture since there is a need for a high level of accuracy and reliability in identifying crops and weeds. Thus, it has higher accuracy and precision at performance, together with a close recall for the non-growth of weed management practices, improving the effectiveness of practice.

References

1. Mahajan, S., Raina, A., Gao, X.-Z., Kant Pandit, A.: Plant recognition using morphological feature extraction and transfer learning over SVM and AdaBoost. Symmetry **13**(2) (2021). https://doi.org/10.3390/sym13020356
2. Bakhshipour, A., Jafari, A.: Evaluation of support vector machine and artificial neural networks in weed detection using shape features. Comput. Electron. Agric. **145**, 153–160 (2018). https://doi.org/10.1016/j.compag.2017.12.032
3. Naresh, Y.G., Nagendraswamy, H.S.: Classification of medicinal plants: an approach using modified LBP with symbolic representation. Neurocomputing **173**, 1789–1797 (2016). https://doi.org/10.1016/j.neucom.2015.08.090
4. Yang, C.: Plant leaf recognition by integrating shape and texture features. Pattern Recogn. **112**, 107809 (2021). https://doi.org/10.1016/j.patcog.2020.107809

5. Chavan, T.R., Nandedkar, A.V.: AgroAVNET for crops and weeds classification: a step forward in automatic farming. Comput. Electron. Agric. **154**, 361–372 (2018). https://doi.org/10.1016/j.compag.2018.09.021
6. Trong, V.H., Hyun, Y.G., Young, K.J., Bao, P.T.: Yielding multi-fold training strategy for image classification of imbalanced weeds. Appl. Sci. **11**(8) (2021). https://doi.org/10.3390/app11083331
7. Xu, Y., Zhai, Y., Zhao, B., Jiao, Y., Kong, S.: Weed recognition for depthwise separable network based on transfer learning. Intell. Autom. Soft Comput. **27**(3), 669–682 (2021). https://doi.org/10.32604/iasc.2021.015225
8. Olsen, A., Konovalov, D.A., Philippa, B., et al.: DeepWeeds: a multiclass weed species image dataset for deep learning. Sci. Rep. **9**, 2058 (2019). https://doi.org/10.1038/s41598-018-38343-3
9. dos Santos Ferreira, A., Freitas, D.M., da Silva, G.G., Pistori, H., Folhes, M.T.: Unsupervised deep learning and semi-automatic data labeling in weed discrimination. Comput. Electron. Agric. **165**, 104963 (2019). https://doi.org/10.1016/j.compag.2019.104963
10. Hu, K., Coleman, G., Zeng, S., Wang, Z., Walsh, M.: Graph weeds net: a graph-based deep learning method for weed recognition. Comput. Electron. Agric. **174**, 105520 (2020). https://doi.org/10.1016/j.compag.2020.105520
11. Dang, F., Chen, D., Lu, Y., Li, Z.: YOLOWeeds: a novel benchmark of YOLO object detectors for multi-class weed detection in cotton production systems. Comput. Electron. Agric. **205**, 107655 (2023). https://doi.org/10.1016/j.compag.2023.107655
12. Cao, J., Han, P., Zhang, W., Zhao, G., Chen, K.: Deep neural network-based plant protection strategy in rural garden landscape construction. Crop Prot. **182**, 106716 (2024). https://doi.org/10.1016/j.cropro.2024.106716
13. Bendib, I., Gattal, A., Marouane, G.: Handwritten digit recognition using deep CNN. In: Proceedings of the 1st International Conference on Intelligent Systems and Pattern Recognition, pp. 67–70. Presented at the Virtual Event, Tunisia (2020). https://doi.org/10.1145/3432867.3432896
14. Gharbi, A., Gattal, A., Bendib, I.: Basic emotion recognition in facial expressions using deep CNN. In: 2024 6th International Conference on Pattern Analysis and Intelligent Systems (PAIS), pp. 1–6 (2024). https://doi.org/10.1109/PAIS62114.2024.10541143
15. Szegedy, C., et al.: Going deeper with convolutions. In: 2015 IEEE Conference on Computer Vision and Pattern Recognition (CVPR), pp. 1–9 (2015). https://doi.org/10.1109/CVPR.2015.7298594
16. Girshick, R., Donahue, J., Darrell, T., Malik, J.: Rich feature hierarchies for accurate object detection and semantic segmentation. In: Proceedings of the IEEE Conference on Computer Vision and Pattern Recognition, pp. 580–587 (2014). https://doi.org/10.48550/arXiv.1311.2524
17. Ren, S., He, K., Girshick, R., Sun, J.: Faster R-CNN: towards real-time object detection with region proposal networks. In: Advances in Neural Information Processing Systems, vol. 28 (2015). https://doi.org/10.48550/arXiv.1506.01497
18. Redmon, J., Divvala, S., Girshick, R., Farhadi, A.: You only look once: unified, real-time object detection. In: 2016 IEEE Conference on Computer Vision and Pattern Recognition (CVPR), pp. 779–788 (2016). https://doi.org/10.1109/CVPR.2016.91
19. Redmon, J., Farhadi, A.: YOLOv3: an incremental improvement. arXiv preprint arXiv:1804.02767 (2018). https://doi.org/10.48550/arXiv.1804.02767
20. Jocher, G., Chaurasia, A., Qiu, J.: YOLO by Ultralytics (2023). https://github.com/ultralytics/ultralytics
21. Liu, W., et al.: SSD: single shot multibox detector. In: Leibe, B., Matas, J., Sebe, N., Welling, M. (eds.) ECCV 2016. LNCS, vol. 9905, pp. 21–37. Springer, Cham (2016). https://doi.org/10.1007/978-3-319-46448-0_2

22. Dabhi, R., Makwana, D.: Crop and weed detection data with bounding boxes (2020). https://www.kaggle.com/datasets/ravirajsinh45/crop-and-weed-detection-data-with-bounding-boxes
23. Kavzoglu, T.: Increasing the accuracy of neural network classification using refined training data. Environ. Model. Softw. **24**(7), 850–858 (2009). https://doi.org/10.1016/j.envsoft.2008.11.012

Deep Learning for Prediction of Leaks in Oil and Gas Pipelines

Basma Hamrouni[1(✉)], Khadra Bouanane[1], Izdihar Hadjouj[1], Ziad Legougui[1], Mohammed el Moncef moad[1], Belkacem Aoufi[3], Abdelhabib Bourouis[2], and Ahmed Korichi[1]

[1] Computer Science Department, University of Kasdi Merbah, Ouargla 30000, Algeria
besma.hamrouni1@gmail.com

[2] Research Laboratory on Computer Science's Complex Systems, University of Oum El Bouaghi, Oum El Bouaghi 04000, Algeria
bourouis.abdelhabib@univ-oeb.dz

[3] Department of Electronics and Communications, University of Kasdi Merbah, Ouargla 30000, Algeria

Abstract. Pipelines are essential infrastructure for transporting oil and gas, but leaks might possess devastating consequences, including fires, injuries, property damage and environmental pollution. Thus, ensuring pipeline integrity is essential for a safe and sustainable energy supply. This paper explores the application of Artificial Intelligence (AI), Machine Learning (ML), and Deep Learning (DL) to enhance the prediction of leaks in oil and gas pipeline systems, which is vital for ensuring environmental safety and economic stability. Through a comprehensive review and data-driven methodologies, the study demonstrates how ML algorithms, including neural networks and deep learning models, significantly outperform traditional leak detection methods in terms of accuracy. The research introduces anomaly detection models driven by machine learning and deep learning to address the issue of leakage in oil and gas pipelines. Our contribution is particularly distinguished by applying deep learning techniques, specifically Multi-Layer Perceptron (MLP) and Convolutional Neural Network (CNN), to numerical data from the oil and gas sector. In addition, various ML algorithms were employed, such as K-Nearest Neighbor, Decision Tree, Random Forest, Support Vector Machine (SVM), and Gradient Boosting, to create reliable models for predicting pipeline leaks. Among these, the SVM algorithm achieved an accuracy of 96%, notably outperforming other models. This study underscores the potential of these technologies to provide timely and precise leak detection, ultimately contributing to better management and maintenance of critical infrastructure.

Keywords: Oil and Gas · Pipelines leakage · Machine learning and deep learning

T. Marir et al. (Eds.): ICCSA 2024, ISEM 45, pp. 101–116, 2025.
https://doi.org/10.1007/978-3-031-90758-6_8

1 Introduction

Pipelines are widely regarded as the most economical and advanced technology for transporting oil and gas globally, accounting for 57.5% of global primary energy consumption [10]. They offer a safer, more environmentally friendly, and more efficient means of transporting large volumes of flammable and hazardous fluids like crude oil, refined petroleum products and natural gas, as opposed to transport by trucks. However, like all infrastructure, pipelines are susceptible to various failures, leakage being the critical concern addressed in this paper. Pipeline leaks can trigger a cascade of severe accidents, including fluid spillage, fires, and explosions, leading to injuries, fatalities, environmental damage and significant financial distress. Therefore, maintaining pipeline integrity and functionality is crucial for economic stability, environmental protection, and public safety [10].

Given the severity of these risks, detecting and preventing pipeline leaks has become a top priority in pipeline management. Several key risk factors, including pressure, temperature, third-party interference, design flaws, operational errors and particularly corrosion, are critical in assessing pipeline integrity [3]. Among these, corrosion is recognized as the most critical factor. Extensive research, expert analysis, and statistical data consistently identify corrosion as a leading cause of pipeline leaks. Numerous studies and incident analyses have shown that corrosion is the primary factor behind the majority of pipeline leakage incidents [14]. Machine learning algorithms are now widely considered as the most effective methods for building predictive models in addressing complex challenges within engineering, energy, and environmental fields [10]. ML enhances predictive accuracy, reduces dependence on traditional data analysis methods, enables autonomous information processing, and efficiently handles large-scale, high-speed, and diverse datasets [19]. In the realm of engineering materials, particularly in the oil and gas industry, ML algorithms are highly beneficial due to their ability to quickly and accurately assess mechanical properties at a reduced cost in comparison with traditional modeling techniques [4,7,10,14].

The literature on corrosion prediction in pipelines can be categorized into two main areas: Computer Vision and Deep Learning Techniques, and Numerical Data-Based Machine Learning and Deep Learning Methodologies for Corrosion Defect Prediction. The authors of [3] explored machine learning-based anomaly detection models to address leaks in oil and gas pipelines, comparing five algorithms: RF, SVM, KNN, GB, and DT. s reported by [20] focused on predicting internal corrosion rates using robust ensemble learning techniques, with the Extreme Gradient Boosting model demonstrating superior performance with an RMSE of 0.031 mm/y and a performance index of 0.61 [15] developed an SVM-based model to predict gas pipeline corrosion rates using various pipeline parameters, providing valuable insights for risk management and maintenance. However, despite the robustness of the techniques used so far, they do not fully exploit the recent advancements in deep learning, such as applying multilayer perceptron (MLP) networks and convolutional neural networks (CNNs) to numerical data.

Our contribution is distinguished by applying deep learning techniques, specifically Multi-Layer Perceptrons (MLP) and 1D Convolutional Neural Networks (CNN), to oil and gas numerical data. This approach is particularly noteworthy because deep learning is predominantly used with oil and gas image data. By extending its application to oil and gas numerical datasets, we aim to enhance prediction accuracy and uncover patterns that are not typically explored with machine learning methods.

This article builds on previous studies in Numerical Data-Based Machine Learning and Deep Learning Methodologies for Corrosion Defect Prediction by comparing algorithms such as Support Vector Machine (SVM), Random Forest (RF), Decision Tree (DT), k-nearest Neighbors (KNN), and Gradient Boosting (GB) for predicting pipeline leakage using industrial datasets. This study is the first application of 1D CNN and MLP models in the context of pipeline leak prediction with numerical data. It also highlights the crucial role of data analysis in improving decision-making for pipeline management.

2 Background

2.1 Leakage Risk in Oil and Gas Pipeline

The risk of leakage in oil and gas pipelines involves the possible release of oil, gas, or other hazardous substances. This risk is affected by several factors, including the age of the pipeline, the materials used, operational pressure, environmental conditions, and the presence of corrosion or mechanical damage. For more effective risk management, regular inspections, maintenance, and the use of advanced monitoring techniques to detect and prevent potential leaks are critical, thereby ensuring the safety and integrity of the pipeline infrastructure. According to [2], pipeline leaks in oil and gas systems are influenced by various factors, including:

Temperature Fluctuations: Can cause expansion and contraction of pipeline materials, leading to micro-fractures and leaks.

Corrosion: Both internal and external corrosion weaken pipelines, with variability making it challenging to predict and manage. Regular monitoring and corrosion-resistant materials are essential.

Pressure Variations: Sudden changes in pressure can stress pipeline walls, increasing the risk of leaks. Pressure monitoring systems help prevent catastrophic failures.

Third-Party Damage: Human activities, especially excavation, are a major cause of pipeline failures.

Incorrect Operation: Human error accounts for the majority of pipeline accidents, making it crucial to address and manage these risks effectively.

2.2 Corrosion and Leak Risk

Research consistently highlights corrosion as the most critical threat to pipeline integrity, with both internal and external corrosion posing significant risks to the safety and durability of pipelines over time as illustrated in Fig. 1. Consequently, continuous inspection is essential. This finding is strongly supported by experts in the petroleum and gas field. As a result, effective pipeline management strategies focus on regular corrosion inspections and implementing measures to enhance the safety, reliability, and longevity of pipeline infrastructure [14].

2.3 AI-Driven Corrosion Detection Techniques

The high costs and limitations associated with traditional leak detection techniques have led experts to seek alternative methods for identifying leak risks. Considering the significant impact of corrosion, our study on pipeline leaks will primarily focus on AI-driven methods for detecting and predicting corrosion. We will leverage advanced deep learning techniques, specifically 1D convolutional neural networks (CNN) and multi-layer perceptrons (MLP), to develop predictive models that accurately identify and forecast corrosion rates. These models will use comprehensive datasets encompassing parameters such as temperature, pressure, fluid composition, and other relevant factors. By targeting the level of corrosion for classification-whether high or low, we directly address the real-world conditions most critical to pipeline operation, providing a practical and aligned approach to enhancing pipeline management strategies.

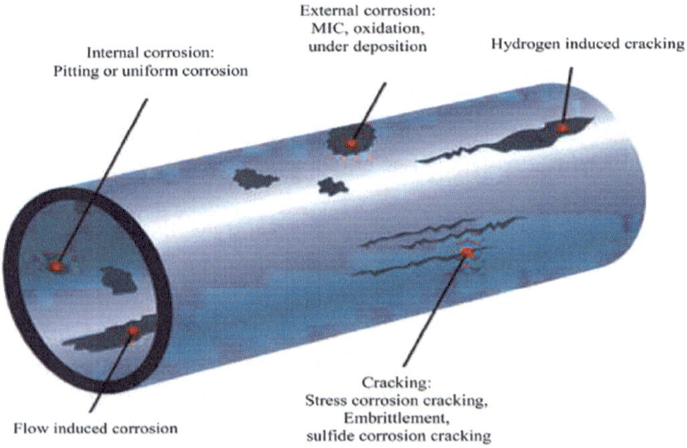

Fig. 1. Types of corrosion in pipeline [1].

3 Related Work

Considering the extensive research on detecting and predicting oil and gas pipeline corrosion using machine learning and deep learning, numerous innovative approaches and methodologies have been investigated. Below, we summarize key studies and their findings:

3.1 Machine Learning and Deep Learning Research on Corrosion Defect Prediction Using Numerical Data

The authors of [19] conducted a study on using machine learning models to address pipeline leaks in the oil and gas industry, The performance of the SVM algorithm was found to be the best, with an accuracy of 97.4%. The study by [20] focused on predicting internal corrosion rates using ensemble learning methods, their model achieving high predictive accuracy. As reported by [12] found that the decision tree model excelled in processing corrosion data with 99.9% accuracy. Similarly, authors in [15] developed an SVM-based model for predicting corrosion rates in gas pipelines, offering valuable insights for pipeline risk management and maintenance. The authors in [5] leveraged AI algorithms, including the DeWaard, Norsok, and Leak Rate models, to predict corrosion and leak rates in pipelines, enhancing accuracy with a type 2 fuzzy logic subroutine that adapts to new data. The authors in [18] assessed soil corrosivity in Nepal using an empirical model, based on factors like pH and moisture content. Authors of [9] developed a method using Infrared Light (ILI) data and hierarchical clustering to classify pipeline corrosion defects.

3.2 Images Data Set Based Deep Learning Researches for Corrosion Defect Prediction

Researchers in [17], explored deep learning techniques for detecting corrosion at the pixel level, addressing challenges with limited publicly available image datasets. Their research introduced Bayesian variants, including variational inference, Monte Carlo dropout, and an ensemble method, to provide not only predictions but also estimates of uncertainty. Authors of [6] developed a custom CNN for detecting and classifying corrosion from optical images, offering high accuracy and efficiency. In [21], authors compared different ML and DL approaches for automated corrosion detection using both laboratory-generated and real-world images, finding that deep learning techniques, especially the SSD model, outperformed traditional methods in accurately detecting and localizing corrosion in real-world scenarios.

In this section, we have exposed the related works of prediction corrosion. These studies have notably leveraged images and numerical datasets. The numerical data include critical parameters such as temperature, pressure. These variables are essential for predicting the corrosion rate of pipelines.

However, while the techniques used thus far are robust, they do not fully leverage the recent advancements in deep learning, such as Multi-Layer Perceptron (MLP) networks and 1D Convolutional Neural Networks (CNN) applied to numerical oil and gas data. By integrating these advanced methods, we can significantly enhance prediction accuracy, utilizing the full potential of datasets that encompass critical parameters. Our contribution is distinct in this regard, as it applies deep learning techniques typically reserved for image data to numerical datasets, aiming to uncover patterns and improve outcomes beyond what traditional methods achieve.

4 Methodology

This section outlines the methodology used to evaluate predictive modeling techniques for corrosion classification within oil and gas pipeline systems. The study utilized a dataset containing relevant features, temperature, pressure, and material composition. Our method focuses on the application of deep learning techniques, specifically 1D Convolutional Neural Networks (CNN) and Multi-Layer Perceptrons (MLP), to oil and gas numerical data. This approach is particularly significant because deep learning has traditionally been applied to image data in the oil and gas industry. By extending these techniques to numerical datasets, we aim to enhance prediction accuracy and uncover patterns that are often missed by conventional machine learning methods.

Morever, to conduct a comparative analysis of the results, five advanced machine learning models were applied: K-Nearest Neighbour (KNN), Decision Tree (DT), Random Forest (RF), Gradient Boosting (GB), and Support Vector Machine (SVM). These models were selected for their distinct capabilities in handling complex data patterns. The dataset was preprocessed to handle missing values and normalize the data. Each model was then trained and tested. Performance metrics including recall, precision, accuracy and F1-score were computed to assess the effectiveness of each model. The models' performance was compared to determine the best algorithm for predicting high and low corrosion scenarios, as illustrated in Fig. 2).

Our methodology consists of the following steps:

4.1 Data Collection

A publicly dataset accessible on GitHub[1] was utilized for the analyses presented in this study. This dataset, initially designed for regression analysis targeting corrosion defects, comprises 10,293 instances across eight features, including critical variables such as wellhead temperature, pressure, and gas composition. These variables are essential for evaluating the structural integrity of pipeline systems. Before its application, the dataset underwent several preprocessing techniques, including normalization and outlier removal, to ensure the accuracy and reliability of subsequent analyses.

[1] https://github.com/Abimbola-ai/Oil-and-gas-pipeline-leakage.

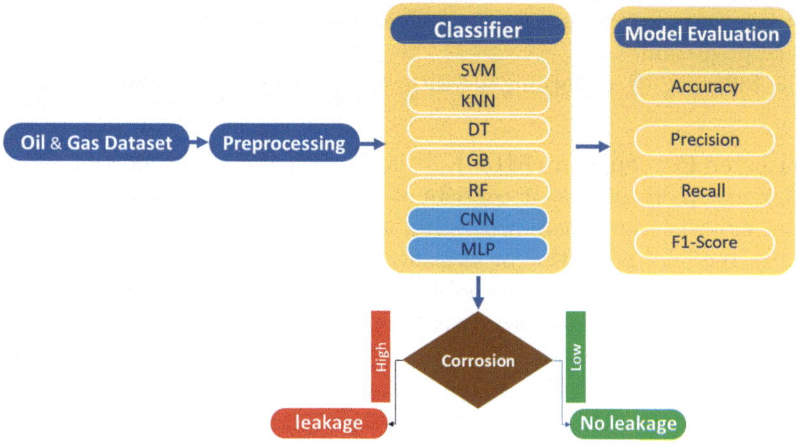

Fig. 2. Adopted research methodology.

Label Binarization: Converting categorical data into a format that can be easily used by algorithms is essential. This typically involves transforming labels into a binary format, which is particularly useful for categorical target variables in classification problems. In our methodology, we focus on the 'CR-corrosion defect' parameter, which quantifies corrosion defects within oil and gas pipeline components. Originally a continuous variable, it was converted into a binary format to simplify analysis, and enhance predictive modeling and risk assessment. This transformation involved setting a threshold of 0.211. Values above this threshold are classified as 'high', indicating a significant risk of corrosion, while values below are classified as 'low', signifying minimal risk [3]. This resulted in a balanced dataset, with the 'high' category consisting of 5, 491 samples and the 'low' category containing 4,801 samples. Feature scaling is essential to standardize the range of features, ensuring that all features contribute equally and avoiding the dominance of any one feature in the model.

4.2 Classification and Model Design

Following steps involved in preprocessing and cleaning the data, machine learning models were constructed. The dataset was split into training and testing subsets to assess the model's performance, with accuracy as the key metric. The standard methods typically employed for dataset splitting include the 7 : 3 split and 10-fold cross-validation. Using a 70/30 split, training is conducted using seventy percent of the data, whereas thirty percent is kept for testing purposes. The accuracy of data analysis was tested through experiments involving various machine learning and deep learning algorithms. We employed deep learning techniques, specifically Multi-Layer Perceptrons (MLP) and 1D Convolutional Neural Networks (CNN), to oil and gas numerical data. This approach is particularly noteworthy because deep learning is predominantly used with oil and gas image data. By extending its application to numerical datasets, we aim to enhance prediction.

The following deep-learning algorithms were used in our experiment:

Multi-layer Perceptron (MLP). It is an artificial neural network with multiple layers of perceptrons. It consists of an input layer that receives data, one or more hidden layers where processing is performed, and an output layer that generates the final prediction or classification as illustrated in Fig. 3. The number of hidden layers can be adjusted based on specific requirements. The output layer classifies and provides the predicted result based on the computations performed by the hidden layers. This setup is illustrated in Fig. 3, which shows the structure of a Multilayer Perceptron [13].

Convolutional Neural Network (CNN). A Convolutional Neural Network (CNN) [8] is a specialized feedforward neural network, for processing data that has a known grid-like topology. Characterized by its convolutional computation and deep architecture, making it a cornerstone algorithm in deep learning. CNNs excel in translation-invariant classification and can handle both supervised and unsupervised learning through parameter sharing of the convolutional kernels and the sparsity of interlayer connections. Key structural features of CNNs include the local receptive field, weight sharing, and pooling layers [11,16].

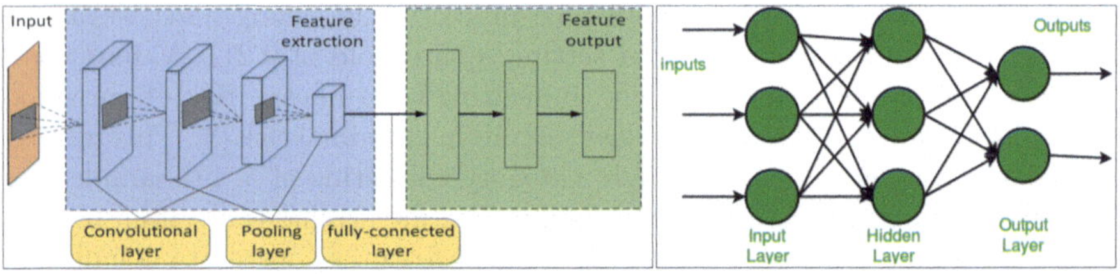

Fig. 3. Schematic diagram of MLP and 1D CNN structure [13,16].

Convolutional Neural Networks (CNNs) are adept at extracting deep features from data, with distinct types tailored for different tasks. The 3D CNN integrates temporal and spatial dimensions, making it ideal for video processing and behavior analysis. The 2D CNN is designed for two-dimensional data, excelling in image recognition by capturing spatial features and structures. Meanwhile, the 1D CNN is optimized for one-dimensional sequences as illustrated in Fig. 3, such as time series or text, focusing on extracting local features and patterns from sequential data [11,16].

In addition, our methodology incorporates five advanced machine learning models to perform a comparative analysis of the results. These include K-Nearest

Neighbor (KNN), which classifies data based on the majority class among the nearest neighbors to a given point; Decision Tree (DT), which uses a tree-like structure of decisions to make predictions based on feature values; Random Forest (RF), an ensemble method that combines the results of multiple decision trees to enhance accuracy and robustness; Gradient Boosting (GB), a technique that builds models sequentially to correct errors made by previous models; and Support Vector Machine (SVM), which identifies the hyperplane that best separates different classes by maximizing the margin between them. Each of these models was analyzed for its effectiveness in handling the dataset.

4.3 Model Evaluation

The evaluation process incorporates precision, accuracy, recall and F1-score metrics, ensuring reliable and actionable predictions for pipeline maintenance and safety [3].

Accuracy indicates the proportion of correctly predicted samples out of the total number of samples in the testing set.

$$\text{Accuracy} = \frac{TP + TN}{TP + FP + FN + TN} \tag{1}$$

Precision refers to the proportion of correctly predicted positive samples out of all the instances predicted as positive.

$$\text{Precision} = \frac{\text{TP}}{\text{TP} + \text{FP}} \tag{2}$$

Recall measures the percentage of actual positive samples that were correctly identified among all the true positive instances.

$$\text{Recall} = \frac{\text{TP}}{\text{TP} + \text{FN}} \tag{3}$$

5 Results and Discussion

In this work, we used 7 different machine-learning algorithms to model leaks in oil and gas pipelines. The objective of this analysis was to identify which model demonstrates the highest accuracy, precision, recall, and F1-score in predicting high and low corrosion scenarios. Each model was rigorously trained and tested using a well-curated dataset, and the results were systematically recorded and analyzed.

5.1 Results of Data Prepossessing

The statistical operations including removing attributes missing values, Standard Scalar (SS), Min-Max Scalar, means, and standard division have been applied to the dataset. The results of these operations are reported in Table 1.

The dataset has 10,293 instances with 8 input attributes and one output label. It includes daily metrics for gas (MMCFD), oil (BOPD), water (BWPD), wellhead temperature (°C), and pressure (psi), along with BSW, CO_2 content, gas gravity, and corrosion defects.

Wellhead Temperature (°C): Measures the heat at the wellhead.
Wellhead Pressure (psi): Fluid force at the wellhead.
MMCFD: Daily gas volume in millions of cubic feet.
BOPD: Barrels of oil produced per day.
BWPD: Water is produced daily with oil.
BSW: Percentage of solids and water in crude oil.
CO_2 mol.: Mass of a CO_2 molecule, used for gas stream evaluation.
Gas Gravity: Relative gas density compared to air.
CR: Corrosion defects, indicating potential leaks or failures.

Table 1. Summary Statistics of Oil Well Operations.

Parameter	Count	Mean	Std	Min	25%	50%	75%	Max
Wellhead Temp. (C)	10292	57.35	9.43	41.07	49.22	57.36	65.41	73.87
Wellhead Press (psi)	10292	1361.74	559.27	382.00	880.00	1364.90	1848.25	2317.23
MMCFD-gas	10292	8.85	4.97	0.23	4.57	8.88	13.09	17.54
BOPD	10292	1103.56	565.39	129.47	611.64	1106.08	1589.71	2087.43
BWPD	10292	4636.56	2685.80	40.61	2295.52	4591.99	6997.44	9314.26
BSW (%)	10292	44.87	25.71	0.13	22.89	45.08	67.21	89.26
CO_2 mol	10292	2.52	1.04	0.68	1.61	2.52	3.41	4.30
Gas Gravity	10292	0.82	0.06	0.71	0.77	0.82	0.87	0.93
CR-corrosion defect	10292	0.21	0.04	0.00	0.19	0.21	0.23	0.41

Summary and data analysis of the dataset statistics:
The dataset reveals key insights into pipeline conditions, with the Wellhead Temperature averaging 57.35 °C and showing moderate variability, indicating diverse geothermal conditions. Wellhead Pressure varies significantly, ranging from 382.00 to 2317.23 psi, which could impact extraction efficiency and safety. Gas and oil production levels exhibit substantial standard deviations, reflecting the differing productivity of wells. Water production notably surpasses oil output, emphasizing the need for effective water management. Variability in Basic Solid and Water Content and CO_2 levels highlights potential corrosion risks, while Gas Gravity remains consistent.

Data visualization plays a crucial role in summarizing and presenting large amounts of data clearly and understandably. The histogram in Fig. 4 shows the frequency distribution of this specific phenomenon within a set range, while the correlation heatmap in Fig. 5 illustrates the correlation among dataset features.

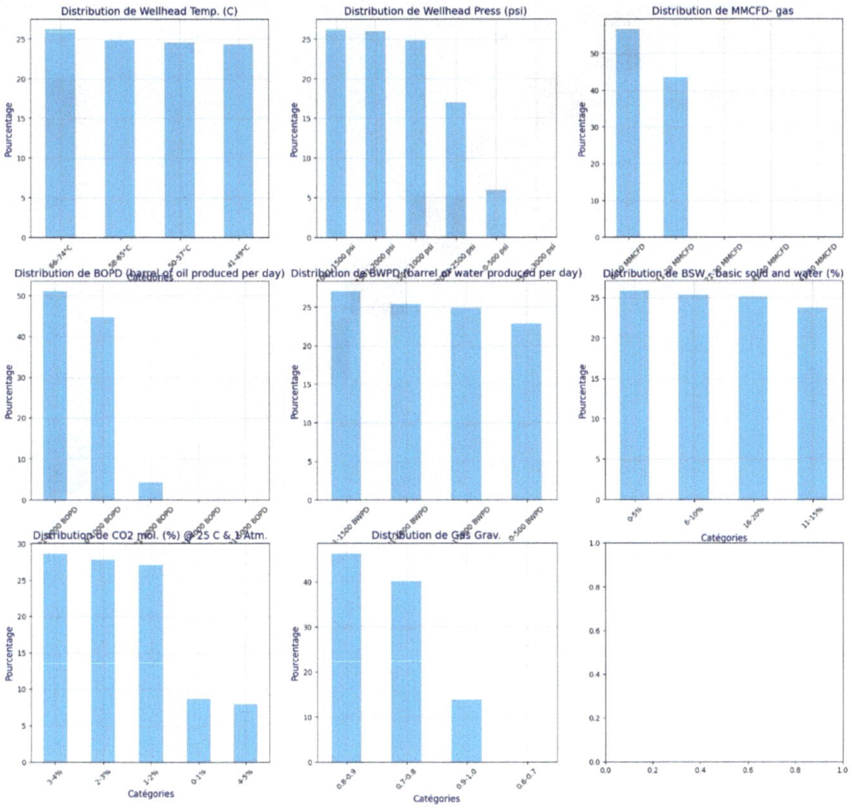

Fig. 4. Histograms.

The detailed analysis of key features, with their respective data distributions, is provided below.

The analysis categorizes and visualizes pressure conditions, highlighting a notable concentration in the $1001 - 1500$ psi range, which is critical for understanding pressure-induced risks on pipeline materials. The analysis categorizes and examines MMCFD values, revealing a significant concentration in the $6 - 15$ MMCFD range, lower in the $0 - 5$ MMCFD range, and much less in the $16 - 20$ MMCFD range. The distribution of barrels of oil produced per day (BOPD) shows that most operations fall within typical production levels, with outputs exceeding 2000 barrels per day being rare. Additionally, the temperature distribution is categorized into four key intervals, providing insights into operational conditions.

Visualizing histograms of oil and gas production highlights the variability in well productivity, helping identify patterns that can guide operational decisions. This data analysis is beneficial for optimizing resource allocation, improving extraction efficiency, and enhancing safety measures in the oil and gas industry.

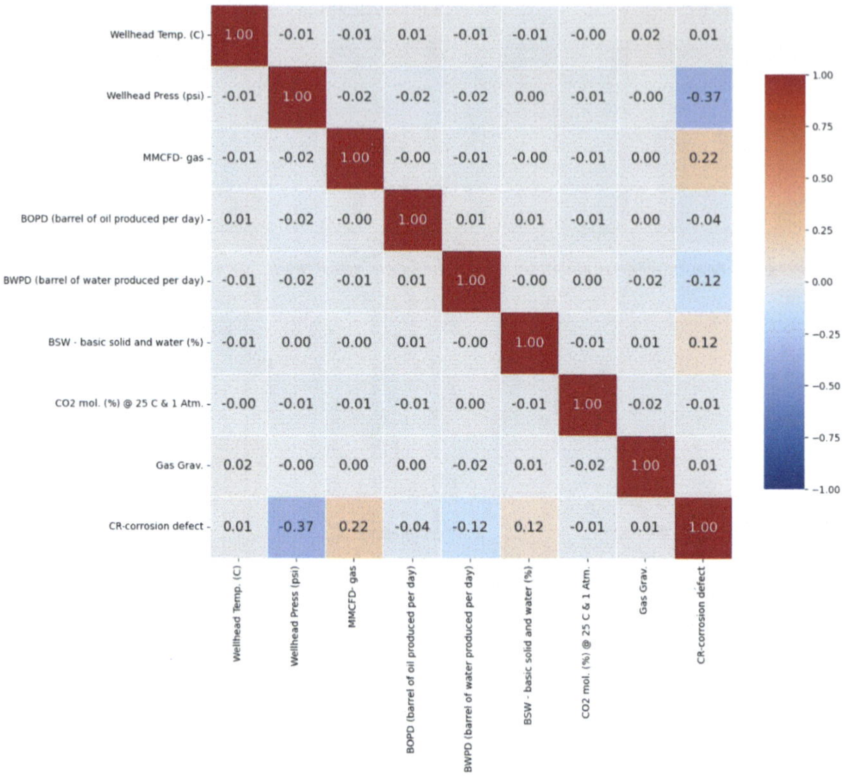

Fig. 5. Correlation Matrix.

A heat map of the correlation matrix was generated to explore the relationships between the influencing factors in the dataset and the output (y), as illustrated in Fig. 5. The analysis of pipeline integrity data reveals key correlations between corrosion defects and various operational parameters. A notable negative correlation of -0.37 with Wellhead Pressure indicates that higher pressures within the pipeline system may lead to fewer or less severe corrosion defects, possibly due to material resistance or reduced corrosive interactions. On the other hand, a positive correlation of 0.22 with MMCFD suggests that higher gas throughput rates may aggravate corrosion, likely due to increased flow rates and the impact of gas composition on the internal pipeline surfaces. Additionally, slight positive correlations with BWPD and BOPD imply a rise in corrosion incidents associated with higher outputs of these fluids.

5.2 Classifiers Performance Results

Table 2 presents the parameters of the models used in our analysis, providing a detailed overview of the settings and configurations applied to optimize their performance.

Table 2. Model Parameters for Various Classification Models.

Model	Parameters
K-Nearest Neighbors	n_neighbors = 21
Gradient Boosting	n_estimators = 200, learning_rate=0.1, max_depth = 3
Random Forest	n_estimators = 100
Support Vector Machine	kernel = poly, degree = 3, gamma = scale, coef = 0.1
MLP Classifier	hidden_layers = (6, 5), lr=init=0.01, random_state = 101
CNN	filters = 32, kernel_size = 2, layers = (128, 64), optimizer = Adam

The classification report summarized in Table 3 provides a comprehensive view of the performance metrics for various machine learning models, including KNN, GB, DT, RF, SVM, ANN, MLP, and CNN. Notably, the SVM and CNN models exhibit superior performance across all metrics, demonstrating their robustness in handling the task. Specifically, SVM achieved the highest overall accuracy and F1-scores of 0.96 for both classes, highlighting its exceptional balance between precision and recall. Similarly, CNN performed exceptionally well, with an accuracy of 0.94 and very high precision and recall values, underscoring its effectiveness in feature extraction and classification in complex datasets.

Table 3. Performance metrics of the various Models.

	Low Corrosion			High Corrosion			Accuracy
	Precision	Recall	F1-score	Precision	Recall	F1-score	
KNN	0.91	0.86	0.88	0.89	0.93	0.91	0.89
GB	0.91	0.87	0.89	0.9	0.93	0.91	0.9
DT	0.82	0.82	0.82	0.85	0.85	0.85	0.83
RF	0.91	0.90	0.91	0.92	0.92	0.92	0.91
SVM	0.96	0.97	0.96	0.97	0.97	0.97	**0.96**
MLP	0.93	0.96	0.94	0.95	0.91	0.93	0.93
CNN	0.92	0.96	0.94	0.96	0.92	0.94	**0.94**

In contrast, DT models underperformed, with DT achieving the lowest accuracy of 0.83. Traditional models like KNN, GB, DT, and RF showed competitive but varied results. Among these, RF stood out with an accuracy of 0.91, indicating its strength.

5.3 Performance Comparison with Previous Methods

In comparing our work to Aljameel's study [3], which utilized the same dataset, we conducted a thorough data analysis that revealed crucial insights for the oil and gas sector. This analysis was instrumental in uncovering new patterns and risks within the data that were previously unexplored. In addition, we explore the application of advanced deep learning algorithms to this numerical data set. We implemented Convolutional Neural Networks (1D CNN) and Multi-Layer Perceptrons (MLPs) to assess their effectiveness in predicting pipeline leaks, which was not addressed before. The 1D CNN provided promising results, though it did not outperform the Support Vector Machine (SVM), which remained the most accurate model, achieving a notable accuracy of 96%. Our findings indicated that while the CNN and MLP models contributed valuable insights, the SVM still offered superior performance. However, we observed improvements in the K-Nearest Neighbors (KNN) classifier compared to Aljameel's results, suggesting its potential for further development. Overall, our research supports the effectiveness of established methods while highlighting the potential of newer techniques for enhancing pipeline leak detection.

6 Conclusion

The analysis revealed that machine learning models, including Support Vector Machines (SVM), K-Nearest Neighbors (KNN), Convolutional Neural Networks (CNN), Gradient Boosting (GB), and Decision Trees (DT), are highly effective in predicting and managing corrosion levels in pipelines. The 1D CNN provided promising results, though it did not outperform the Support Vector Machine (SVM), which remained the most accurate model, achieving a notable accuracy of 96%. Some models excelled in terms of predictive accuracy, precision, recall, and F-score, highlighting their potential for real-world applications in pipeline integrity threat prediction.

The perspectives of this work include enhancing pipeline maintenance by integrating advanced machine learning models in real-world applications, particularly SVM and CNN, to provide early warnings for potential corrosion or leaks. Additionally, improving the model by applying other AI techniques to enhance predictive accuracy and reliability. Furthermore, we are also working on an IoT system that provides the necessary data for these analyses, enabling real-time monitoring and more informed decision-making regarding pipeline integrity.

Acknowledgments. This study was not funded by any organization or grant.

Disclosure of Interests. The authors have no competing interests to declare.

References

1. Agala, A., Khan, M., Starr, A.: Degradation mechanisms associated with metal pipes and the effective impact of LDMS and LLMS in water transport and distribution. Proc. Inst. Mech. Eng. C J. Mech. Eng. Sci. **237**(8), 1855–1876 (2023). https://doi.org/10.1177/09544062221133948
2. Al-Sabaeei, A.M., Alhussian, H., Abdulkadir, S.J., Jagadeesh, A.: Prediction of oil and gas pipeline failures through machine learning approaches: a systematic review. Energy Rep. **10**, 1313–1338 (2023). https://doi.org/10.1016/j.egyr.2023.08.009

3. Aljameel, S.S., et al.: An anomaly detection model for oil and gas pipelines using machine learning. Computation **10**(8), 138 (2022). https://doi.org/10.3390/computation10080138
4. Arbaoui, A., Ouahabi, A., Jacques, S., Hamiane, M.: Wavelet-based multiresolution analysis coupled with deep learning to efficiently monitor cracks in concrete. Frattura ed Integrità Strutturale/Fracture and Structural Integrity **15**(58), 33–47 (2021). https://doi.org/10.3221/IGF-ESIS.58.03
5. Aslam, N.: Artificial intelligence based algorithm for predicting pipeline leak and corrosion detection (Dec 2018), US Patent App. 15/840,535
6. Bastian, B.T., Jaspreeth, N., Ranjith, S.K., Jiji, C.: Visual inspection and characterization of external corrosion in pipelines using deep neural network. NDT & E International **107**, 102134 (2019). https://doi.org/10.1016/j.ndteint.2019.102134
7. Behnood, A., Golafshani, E.M.: Artificial intelligence to model the performance of concrete mixtures and elements: a review. Arch. Comput. Methods Eng., 1–24 (2021). https://doi.org/10.1007/s11831-021-09644-0
8. Bengio, Y., Goodfellow, I., Courville, A.: Deep learning, vol. 1. MIT press Cambridge, MA, USA (2017)
9. Cui, B., Wang, H.: Analysis and prediction of pipeline corrosion defects based on data analytics of in-line inspection. J. Infrastructure Preservat. Resilience **4**(1), 14 (2023). https://doi.org/10.1186/s43065-023-00081-w
10. Dudley, B., et al.: Bp statistical review of world energy 2016. British Petroleum Statistical Review of World Energy, Bplc. editor, Pureprint Group Limited, UK (2019). https://www.bp.com/content/dam/bp/business-sites/en/global/corporate/pdfs/news-and-insights/speeches/bp-statistical-review-of-world-energy-2016-spencer-dale-presentation.pdf
11. Gao, D., Liu, X., Yang, Q.: Hybrid neural network method of the lithium-ion battery remaining useful life prediction. In: Li, K., Coombs, T., He, J., Tian, Y., Niu, Q., Yang, Z. (eds.) LSMS/ICSEE -2021. CCIS, vol. 1468, pp. 159–167. Springer, Singapore (2021). https://doi.org/10.1007/978-981-16-7210-1_15
12. Ismail, M., May, Z., Asirvadam, V.S., Nayan, N.A.: Machine-learning-based classification for pipeline corrosion with monte carlo probabilistic analysis. Energies **16**(8), 3589 (2023). https://doi.org/10.3390/en16083589
13. Joy, J., Kannan, A., Ram, S., Rama, S.: Speech emotion recognition using neural network and mlp classifier. Ijesc **2020**, 25170–25172 (2020)
14. Karami, M.: Review of corrosion role in gas pipeline and some methods for preventing it. J. Pressure Vessel Technol. **134**(5), 5 (2012). https://doi.org/10.1115/1.4006124
15. Luo, Z., Hu, X., Gao, Y.: Corrosion research of wet natural gathering and transportation pipeline based on SVM. In: ICPTT 2013: Trenchless Technology, pp. 964–972 (2013). https://doi.org/10.1061/9780784413142.10
16. Mou, J., Yang, Q., Tang, Y., Liu, Y., Li, J., Yu, C.: Prediction of the remaining useful life of lithium-ion batteries based on the 1d CNN-BLSTM neural network. Batteries **10**(5), 152 (2024). https://doi.org/10.3390/batteries10050152
17. Nash, W., Zheng, L., Birbilis, N.: Deep learning corrosion detection with confidence. npj Mater. Degradation **6**(1), 26 (2022). https://doi.org/10.1038/s41529-022-00232-6
18. Poudel, A., Dahal, K.P., Kc, D., Bhattarai, J.: A classification approach for corrosion rating of soil to buried water pipelines: a case study in budhanilkantha-maharajganj roadway areas of Nepal. World J. Appli. Chem. **5**(3), 47 (2020). https://doi.org/10.11648/j.wjac.20200503.12

19. Rachman, A., Ratnayake, R.C.: Corrosion loop development of oil and gas piping system based on machine learning and group technology method. J. Qual. Maint. Eng. **26**(3), 349–368 (2019). https://doi.org/10.1108/JQME-07-2018-0058
20. Seghier, M., Höche, D., Zheludkevich, M.: Prediction of the internal corrosion rate for oil and gas pipeline: implementation of ensemble learning techniques. J. Nat. Gas Sci. Eng. **99**, 104425 (2022). https://doi.org/10.1016/j.jngse.2022.104425
21. Shirsath, K., V.A.Parjane, Meherkhamb, A., Ratnaparkhi, S., Kanade, S.: Deep learning for automated corrosion detection using supervised machine learning & CNN algorithm. Inter. Res. J. Modernization Eng. Technol. Sci. **4**, 1307–1312 (2022). https://www.irjmets.com/uploadedfiles/paper/issue_6_june_2023/41642/final/fin_irjmets1686218467.pdf

FNN-LSTM Model for Remaining Useful Life Estimation

Samira Abderrezek[1,2(✉)], Abdelhabib Bourouis[1], and Naouel Ouafek[2]

[1] ReLa(CS)2 Laboratory, University of Oum El Bouaghi, P.O. Box 358, 04000 Oum El Bouaghi, Algeria
bourouis.abdelhabib@univ-oeb.dz
[2] Ecole Normale Superieur Constantine, El Khroub, Algeria
{Abderrezak.samira,ouafek.naouel}@ensc.dz

Abstract. In this paper, we present a hybrid model combining Feedforward Neural Network (FNN) and Long Short-Term Memory (LSTM) for the prediction of the Remaining Useful Life (RUL) of turbofan engines. The proposed model leverages the strengths of FNN in capturing non-linear relationships and the temporal capabilities of LSTM to handle sequential data. Additionally, we applied a Kalman filter during the preprocessing phase to reduce noise, which significantly enhanced the model's performance. We evaluated our model on the C-MAPSS dataset, a widely used benchmark for RUL prediction in aero-engines. Our experimental results demonstrate that the FNN-LSTM model, with the Kalman filter preprocessing, achieves superior performance, providing accurate and reliable RUL predictions, thereby contributing to more effective maintenance decision-making and improved operational efficiency.

Keywords: LSTM · RUL prediction · Deep learning · FNN · C-MAPSS dataset · Denoising

1 Introduction

Accurate estimation of Remaining Useful Life (RUL) for critical machinery components is crucial for prognostics and health management in aerospace, manufacturing, and energy production [9,13]. It helps in scheduling maintenance on time, potentially preventing catastrophic failures, minimizing downtime, and optimizing schedules. RUL prediction is particularly critical for aero engines in the aerospace industry, ensuring safety and long-term aircraft operability. Reliable RUL prediction systems facilitate the transition from corrective maintenance to predictive maintenance, leading to cost savings, improved safety, and reduced equipment degradation [5].

Over the past two decades, RUL prediction has garnered significant interest from academia and industry. Traditional approaches involved physical models or statistical techniques, but with the availability of real-time sensor data

T. Marir et al. (Eds.): ICCSA 2024, ISEM 45, pp. 117–129, 2025.
https://doi.org/10.1007/978-3-031-90758-6_9

and advances in machine learning, the focus has shifted towards data-driven approaches, especially deep learning. Long Short-Term Memory (LSTM) deep learning models [12] are particularly effective in handling complex temporal dependencies and pattern modeling in degradation data.

Despite the promise of deep learning models for RUL prediction, there are challenges. These models are often computationally expensive and susceptible to noise in sensor data, which can impact predictive performance and RUL estimation quality.

To address these challenges, a noise-robust deep learning model is proposed for RUL prediction. The proposed hybrid model combines a Feedforward Neural Network (FNN) with an LSTM architecture and integrates a Kalman filter to remove sensor data noise. The Kalman filter is a statistical tool known for state estimation and is well-suited for denoising sensor data [15]. This approach enhances both robustness and accuracy in RUL predictions.

The proposed noise-robust deep learning model combines a Feedforward Neural Network (FNN) and Long Short-Term Memory (LSTM) architecture with a Kalman filter for sensor data denoising. The Kalman filter, a widely used statistical tool for state estimation, helps the model focus on critical degradation patterns by filtering out sensor noise, thereby improving both robustness and accuracy of RUL predictions. This hybrid approach balances model complexity and performance, offering competitive results with manageable computational costs. Validated on the NASA Commercial Modular Aero-Propulsion System Simulation (C-MAPSS) benchmark dataset [11], the model demonstrates improved reliability and safety, contributing to predictive maintenance by providing more accurate RUL estimates and reducing downtime.

The subsequent sections will present related works on RUL prediction, describe the methodology of using a Kalman filter model with FNN-LSTM, and report performance evaluation of the adopted model on benchmark datasets. This approach overcomes previous model limitations by incorporating a denoising step during pre-processing and has practical applicability in real-world RUL estimation tasks.

2 Related Work

Commonly, the RUL models are divided into three classes: data-driven, physical, and hybrid models [14]. Recently, deep learning-based models have shown excellent performance in improving accuracy. Although much progress has been achieved, there are several challenges in terms of model complexity, computation efficiency, noisy data robustness, and scalability for industrial applications.

Li et al. proposed a hybrid CNN-Bidirectional LSTM model that combines CNN's spatial feature extraction with LSTM's temporal modeling [6]. However, managing both CNN and LSTM components leads to high computational overhead. The hybrid architecture requires significant computational resources, limiting its use in real-time applications where computational efficiency is crucial. Furthermore, there is limited emphasis on handling sensor noise during data preprocessing, making the model less robust under noisy conditions, leading to decreased accuracy in RUL estimation.

Borst et al. extended the CNN-LSTM ensemble proposed by Li et al. to apply it to the C-MAPSS dataset [2]. They introduced two variants: Adaptive Time Window (ATW) and subnetwork learning. The ATW variant yielded better RUL prediction accuracy for early maintenance planning, but the subnetwork learning approach did not show significant improvements. Their study mostly focuses on the controlled NASA C-MAPSS dataset, with limited exploration of real-world noisy datasets. The reliance on complex architectures may limit scalability and deployment in industrial environments with limited computational resources.

Xia et al. proposed an integrated approach using LSTM layers and traditional neural networks to learn temporal information directly from raw sensor data, avoiding handcrafted feature extraction [18]. However, the LSTM's vulnerability to overfitting, especially with noisy or sparse data, can decrease the model's real-world performance.

Chui et al. developed a hybrid model that combined Recurrent Neural Networks (RNN) and LSTM optimized with a genetic algorithm for both short-term and long-term predictions [3]. This method significantly enhanced prediction accuracy compared to RNN and LSTM models independently. However, the addition of a genetic algorithm significantly increases the computational cost and complexity of the system, making implementation difficult in resource-sensitive industrial environments. Additionally, the proposed method does not account for noisy sensor data, which is a major limitation in real industrial applications.

Li et al. presented a Wiener-based RUL prediction model employing an improved Kalman filter (KF) to reduce the effect of noise on sensor data [7]. It effectively reduces random noise; thus, it offers a more stable and accurate model. However, by using a Wiener degradation process assuming a linear pattern in degradation, its application becomes limited against the real-world systems characterized by complex non-linear degradation processes. This limits scalability for general use in industrial applications because the model cannot easily handle different machinery with any type of non-linear degradation.

Despite the great achievements of deep learning models in RUL prediction, some challenges remain. Among those are the following:

Model Complexity: Many of these models are unnecessarily complex to add computationally, especially those hybrid models with architectures such as CNN-LSTM or those optimized by genetic algorithms-which makes deployment difficult in real-time or resource-constrained applications.

Noise Handling: Most methods either inadequately treat noise in sensor readings or depend on linear models, which are inappropriate for nonlinear degradation processes. Scalability and generalization: Most of these models perform well only on benchmark datasets such as NASA C-MAPSS but lack robustness in real-world industrial settings, where most of the time the data is irregular and noisy.

Those can be overcome by proposing a new method based on combining a Feedforward Neural Network (FNN) for spatial feature extraction with LSTM layers that will capture temporal dependencies. By this proposal, the model architecture is simplified in front of CNN-LSTM hybrid models and offers a computationally efficient solution without an accuracy penalty. For robustness against noisy environments, we propose incorporating the Kalman filter, which reduces noise at the preprocessing step while retaining important features of the signal. In this respect, batch normalization and dropout are also used to widen the generalization capability and reduce the possibilities of overfitting.

By focusing on a reduction of complexity, increasing robustness against noise, and improving scalability, our approach tries to bridge the gap between the high accuracies of existing models and the demands of practical industrial applications.

3 Proposed Method for RUL Estimation

This work presents a sequential neural network FNN for RUL prediction of aircraft engines of data using multivariate time series. It embeds the sequential processing capability of LSTM while integrating regularization mechanisms into the architecture to enhance the accuracy and robustness of predictions. Figure 1 illustrates the proposed method, carefully crafted to capture complex data dynamics while minimizing bias and noise inherent to the sensors.

3.1 Preprocessing Phase

A) Data Normalization: The need to bring all data into one scale to ensure that different features with their own dimensions and units do not interfere with the results of the analysis. This process of normalization eliminates the dimensional biases among the features, enabling its full comparison. For this work, we use the min-max normalization given by the formula 1:

$$x' = \frac{x - x_{\text{min}}}{x_{\text{max}} - x_{\text{min}}} \tag{1}$$

B) Denoising Using Kalman Filter: In our approach, we have used the Kalman Filter to denoise the multivariate time series data. The Kalman Filter is quite a powerful recursive algorithm for estimating the state of a dynamic system based on a series of noisy observations. For this purpose, we applied the Kalman filter which is initialized with two critical parameters: transition covariance and observation covariance. The transition covariance determines the internal process noise of the system, or, in other words, how much noise between states there exists, while the observation covariance does the same but for the noise in measured data. Here, we set the *transition_cov_scale* to 0.7 and the *observation_cov_scale* to 0.02, both optimized to reach a good balance between the system's ability to track sudden changes and suppress noise in measurements.

By using this method the noise in the data is reduced while keeping the important features of the signal, see Fig. 2.

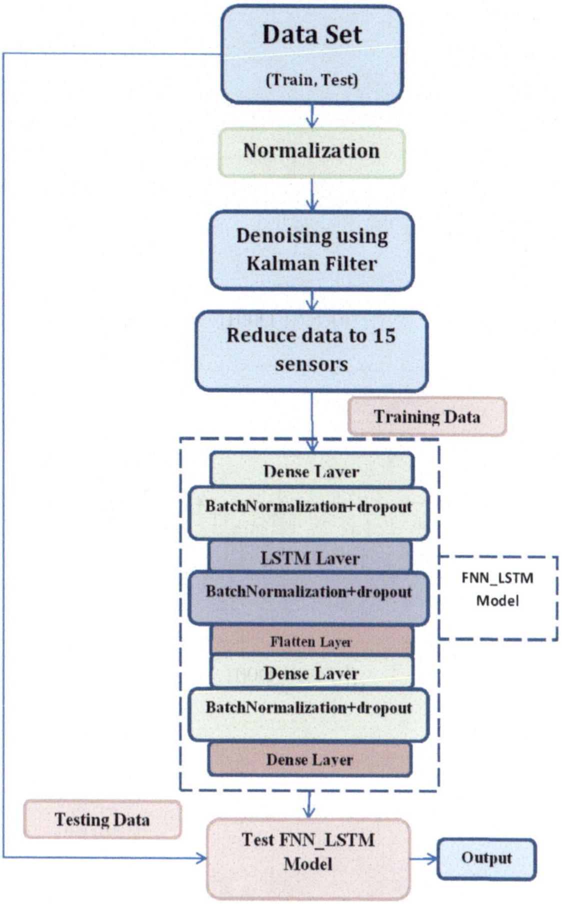

Fig. 1. Overview of the proposed method.

The filtered data was then normalized by using a scaling function.

The scaling ranged between [0,1]. To scale columns, each column should be treated separately and have uniform normalization. This step is quite necessary as preparation for the data before feeding it to the model in training; in such a way, all the values would fall within a good distribution and fit the model requirements.

We initialized the filter with the mean of the initial state, derived from the first row of the data, and set both the transition matrix and the observation matrix to identity matrices, assuming that the system dynamics follow an unchanging state model. Then, the Kalman Filter smooths the data by iteratively updating its estimates of the true system state based on the noisy observations and the covariance matrices. This will yield a denoised version of the original dataset that can then be used for further processing. The parameters

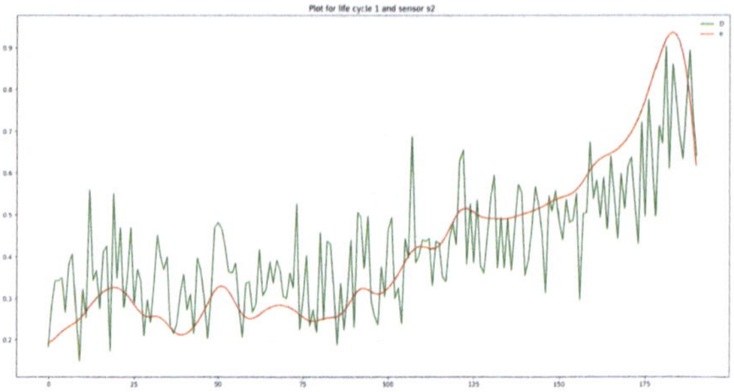

(a) S2 FD001.

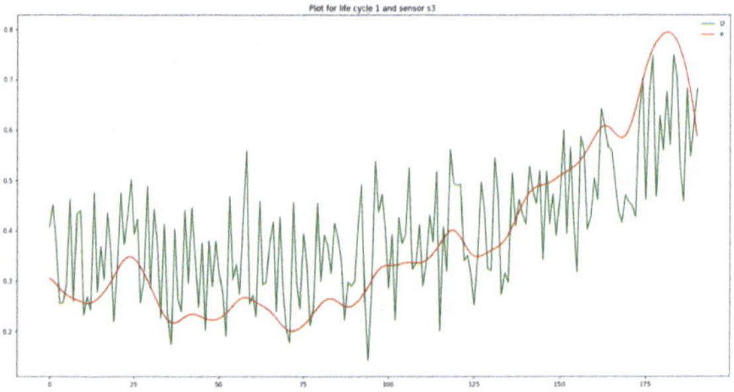

(b) S3 FD001.

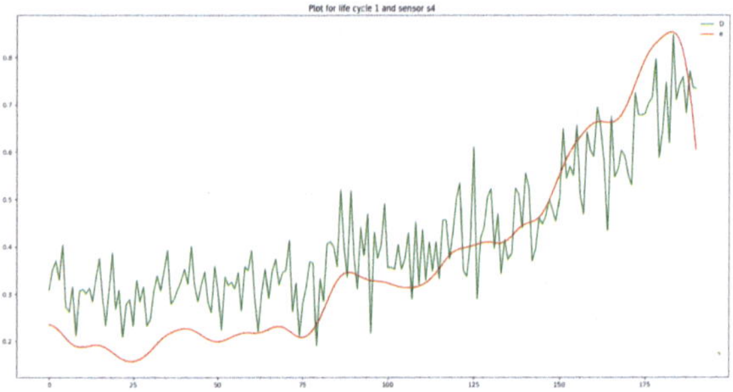

(c) S4 FD001.

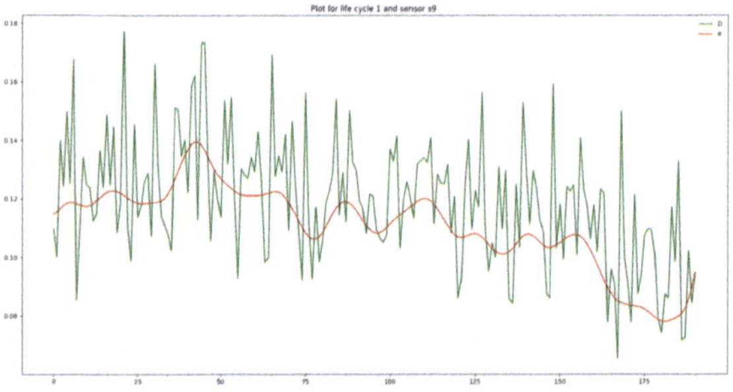

(d) S9 FD001.

Fig. 2. Denoising using Kalman Filter: Green: original data, Red: after denoising. (Color figure online)

$transition_{c}ov_{s}cale$ and $observation_{c}ov_{s}cale$ are tuned to 0.7 and 0.02, respectively, to fine-tune the model for the fine details of the data denoising process.

C) Data Pre-selection: The parameters of some turbojet sensors, which are not related to the degradation process in use, must be excluded. Careful selection of features that are highly correlated with the life cycle can improve the efficiency of network training. For example, the number of unique values in the FD001 dataset, detailed in Table 1, shows that sensors 1, 5, 10, 16, 18 and 19 remain constant during the whole cycle life. In the approach proposed these sensors had not been considered.

Table 1. Unique value (in grey) in **FD001**.

id	100
cycle	362
setting1	158
setting2	13
setting3	1
s1	1
s2	310
s3	3012
s4	4051
s5	1
s6	2
s7	513
s8	53
s9	6403
s10	1
s11	159
s12	427
s13	56
s14	6078
s15	1918
s16	1
s17	13
s18	1
s19	1
s20	120
s21	4745
RUL	362

3.2 Model Architecture

The proposed architecture combines several layers to capture the nonlinear relationships and temporal dependencies of the data in Fig. 1.

The architecture will be discussed in detail for each layer:

- **Dense Layer**: It is the first layer of this model, a Dense layer with 256 units and ReLU for activation. Using such a layer allows for the modeling of higher-dimensional, complex nonlinear relationships among input features. BatchNormalization follows for stabilizing and speeding up the training process while a Dropout layer with 0.2 reduces overfitting.
- **LSTM Layer(128 units):** The next layer would be an LSTM layer consisting of 128 units with ReLU activation that will further allow the model to grasp the temporal dependencies built in the sequence. Batch normalization is once more applied in this layer, this time with a 0.2 Dropout to regularize the model.
- **Flatten Layer:** This layer flattens the outputs of the LSTM layer for input to a following Dense layer to efficiently combine information extracted from temporal sequences.
- **Dense Layer (120 units):** Added another Dense Layer with 120 units and ReLU for refining learned representations. Applied batch normalization and dropout to 0.2 to help with stability and generalization.
- **Output Layer**: A Dense layer with a single unit and linear activation completes the network for the RUL prediction.

3.3 Motivation and Hyperparameters Values

In this architecture, Each layer is chosen to capture the nuances that exist in the input data. Furthermore, this architecture can capture nonlinear and temporal relationships through the combination of FNN and LSTM layers, which indeed holds great importance in RUL prediction. Besides, the use of batch normalization and Dropout at each step ensures better generalization and much more stability of the model reflected by the exceptional performance obtained during the evaluation.

Algorithm 1 groups the different layers.

Hyperparameters:

Sequence_length: $L \leftarrow 110$

nb_features: $F \leftarrow 19$

seed_value: $s \leftarrow 42$

dropout rate: $d \leftarrow 0.2$

Algorithm 1: FNN-LSTM Model for RUL Prediction

Algorithm 1. FNN-LSTM Model for RUL Prediction

```
Input: Sequence length L, number of features F, training data X, labels Y
Initialize model parameters: learning rate α, seed value s, dropout rate d
Step 1: Dense Layer (FNN)
Input the training data X with shape (L, F) to a fully connected layer
Apply ReLU activation: f(x) = max(0, x)
Apply Batch Normalization to stabilize the learning process
Apply Dropout with rate d for regularization
Step 2: LSTM Layer
Pass the output of the Dense layer to an LSTM layer with N units
Apply ReLU activation
Apply Batch Normalization
Apply Dropout with rate d
Step 3: Output Layer
Flatten the LSTM output
Add a Dense layer with 120 units and ReLU activation
Apply Batch Normalization
Apply Dropout with rate d
Add a final Dense layer with 1 unit for the regression task
Step 4: Model Compilation
Compile the model using Adam optimizer and loss function as Mean Squared Error (MSE)
Set metrics for evaluation: Mean Absolute Error (MAE), Root Mean Squared Error (RMSE), and R^2 score
Step 5: Model Training
Train the model on training data X, Y for T epochs
Monitor the performance on the validation set
Step 6: Model Evaluation
Evaluate the model performance on the test data
Output predictions Ŷ and calculate evaluation metrics (RMSE, MAE, R^2)
```

4 Results

Table 2 illustrates the performance of several models in terms of RMSE and R^2 score related to the remaining useful life (RUL) prediction. The result shows that the introduced architecture outperforms the other tested models in **FD001** subset of CMAPSS that is detailed as follows:

- **The LSTM model** proposed in [21] achieved the RMSE and R^2 of 16.14 and 0.798 respectively, while the model proposed in [20] reached an RMSE of 14.72. Although the LSTM is able to learn the temporal dependencies in the sequence data, from the above results it appears that its performance might be restricted, implying that it may not capture all the complicated relationships in data.

- A BiLSTM layer was shown to deliver better performance in [17] with an RMSE of 13.65 and an R^2 of 0.855, and in [1] with an RMSE of 10.15 and an R^2 score of 0.91. The improvements could be attributed to the fact that the BiLSTM model captures contextual information in both temporal directions, thus enhancing the prediction performance.
- **The CNN model** in [10] achieved less performance than the LSTM and BiLSTM methods with an RMSE of 18.45 and R^2 of 0.736. The underlying reason could be related to the fact that CNNs capture spatial patterns much better than temporal dependencies in the time series data. [16] using CNN with attention mechanism gives 10.60 of RMSE and 0.93 of R^2 score.
  By using the ATW method, the prediction accuracy is improved too in [2] by giving an RMSE of 11.09.
- **Proposed Architecture:** with the incorporation of both FNN and LSTM produces the best result with an RMSE value of **8.16761** and an R^2 score of **0.94297** with FD001 sub data and an RMSE value of **11.86356** and an R^2 score of **0.897503** with FD003 sub-data. This proves that the proposed model is highly reliable for RUL prediction. This architecture seems to combine optimally the nonlinear relationships modeled by the FNN layers with the temporal dependencies modeled by the LSTMs. The systematic use of batch normalization and dropout probably resulted in better model generalization given the significant increase in the R^2 score. Predicted RUL versus true RUL is plotted in Fig. 3. Also, Table 3 compares the proposed approach applied to FD003 and the others in the literature. It shows that it gives good performance in terms of RMSE.

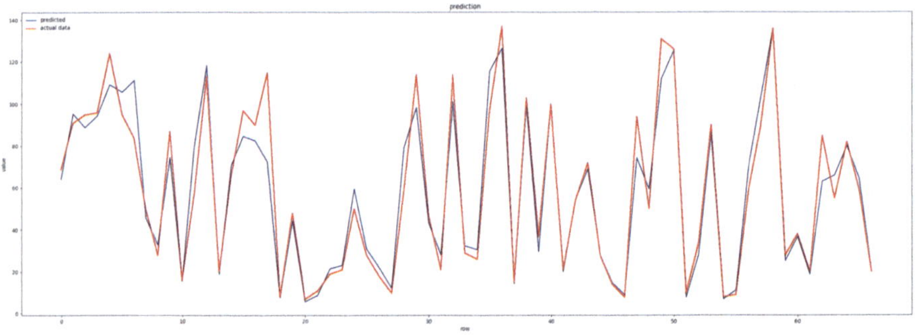

Fig. 3. Predicted RUL (Blue) vs True RUL (Red).

Table 2. Evaluation and Benchmarking Results for **FD001**.

Model	RMSE	R^2
[10]	18.45	0,736
[8]	17.89	/
[21]	16.14	0,798
[17]	13.65	0,855
[20]	14.72	/
[18]	14.57 (± 1.04)	/
[4]	13.95 (± 0.43)	/
[6]	12.58	/
[19]	11.18	/
[3]	11.19	/
[16]	10.60	0.93
[1]	10.15	0.92
[2]	11.09	/
Proposed Architecture(FNN_LSTM)	**8.16761**	**0.94297**

Table 3. Evaluation and Benchmarking Results for **FD003**.

Model	RMSE
[10]	19.8174
[17]	13.74
[8]	16.22
[3]	11.47
[16]	12.31
[4]	12.72 (± 0.73)
[2]	11.09
[19]	10.25
Proposed Architecture(FNN_LSTM)	**11.86356**

5 Conclusions

In this research, a model that merges a feedforward neural network (FNN) with an LSTM layer is developed to predict aircraft engines remaining useful life (RUL). The FNN captures intricate feature relationships, while the LSTM handles temporal dependencies. Also, batch normalization and Dropout are incorporated to enhance stability and generalization. In terms of data preprocessing, a Kalman filter is used to decrease noise while retaining important signal features. Our experimental results demonstrate that our model outperforms benchmark models such as LSTM, BiLSTM, and CNN, achieving an RMSE of 8.16 and an R^2 of 0.94. This underlines the significance of a well-designed architecture and regularization in time-series modeling, especially for predictive maintenance, where accurate RUL predictions are crucial.

In future research, it is crucial to integrate attention mechanisms or advanced models such as Transformers to improve performance. Thorough testing of the approach on various datasets is essential to assess its generalizability, and optimizing hyperparameters is crucial. Additionally, data augmentation techniques should be applied to improve generalization and reduce overfitting. Adding noise, window shift, and sensor reading synthesis can greatly improve the learning capabilities of a model with limited data.

Acknowledgments. This work is fully funded by the General Direction of Scientific Research and Technological Development (DGRSDT) of the Algerian Ministry of Higher Education and Scientific Research (MERS).

Disclosure of Interests. The authors declare that they have no competing interests.

References

1. Abderrezek, S., Bourouis, A.: Convolutional autoencoder and bidirectional long short-term memory to estimate remaining useful life for condition based maintenance. In: 2021 International Conference on Networking and Advanced Systems (ICNAS), pp. 1–6. IEEE (2021). https://doi.org/10.1109/ICNAS53565.2021.9628958
2. Borst, N., Verhagen, W.: Introducing CNN-LSTM network adaptations to improve remaining useful life prediction of complex systems. Aeronautical J. **127**(1318), 2143–2153 (2023). https://doi.org/10.1017/aer.2023.84
3. Chui, K.T., Gupta, B.B., Vasant, P.: A genetic algorithm optimized RNN-LSTM model for remaining useful life prediction of turbofan engine. Electronics **10**(3), 285 (2021). https://doi.org/10.3390/electronics10030285
4. Da Costa, P.R.D.O., Akcay, A., Zhang, Y., Kaymak, U.: Attention and long short-term memory network for remaining useful lifetime predictions of turbofan engine degradation. Inter. J. Prognostics Health Manag. **10**, 034 (2019). https://research.tue.nl/files/146883988/ijphm_19_034.pdf
5. Lei, Y., Li, N., Guo, L., Li, N., Yan, T., Lin, J.: Machinery health prognostics: a systematic review from data acquisition to rul prediction. Mech. Syst. Signal Process. **104**, 799–834 (2018). https://doi.org/10.1016/j.ymssp.2017.11.016
6. Li, C.A.: A deep learning framework with integration of cnn and bi-lstm for remaining useful life prediction (2020)
7. Li, Y., Huang, X., Ding, P., Zhao, C.: Wiener-based remaining useful life prediction of rolling bearings using improved kalman filtering and adaptive modification. Measurement **182**, 109706 (2021). https://doi.org/10.1016/j.measurement.2021.109706
8. Liao, Y., Zhang, L., Liu, C.: Uncertainty prediction of remaining useful life using long short-term memory network based on bootstrap method. In: 2018 IEEE International Conference on Prognostics and Health Management (ICPHM), pp. 1–8. IEEE (2018). https://doi.org/10.1109/ICPHM.2018.8448804
9. Okoh, C., Roy, R., Mehnen, J., Redding, L.: Overview of remaining useful life prediction techniques in through-life engineering services. Procedia Cirp **16**, 158–163 (2014). https://doi.org/10.1016/j.procir.2014.02.006

10. Sateesh Babu, G., Zhao, P., Li, X.-L.: Deep convolutional neural network based regression approach for estimation of remaining useful life. In: Navathe, S.B., Wu, W., Shekhar, S., Du, X., Wang, X.S., Xiong, H. (eds.) DASFAA 2016. LNCS, vol. 9642, pp. 214–228. Springer, Cham (2016). https://doi.org/10.1007/978-3-319-32025-0_14
11. Saxena, A., Goebel, K., Simon, D., Eklund, N.: Damage propagation modeling for aircraft engine run-to-failure simulation. In: 2008 International Conference on Prognostics and Health Management, pp. 1–9. IEEE (2008). https://doi.org/10.1109/PHM.2008.4711414
12. Shi, Z., Chehade, A.: A dual-LSTM framework combining change point detection and remaining useful life prediction. Reliability Eng. Syst. Safety **205**, 107257 (2021). https://doi.org/10.1016/j.ress.2020.107257
13. Si, X.S., Wang, W., Hu, C.H., Zhou, D.H.: Remaining useful life estimation – a review on the statistical data driven approaches. Eur. J. Oper. Res. **213**(1), 1–14 (2011). https://doi.org/10.1016/j.ejor.2010.11.018, https://www.sciencedirect.com/science/article/pii/S0377221710007903
14. Sikorska, J.Z., Hodkiewicz, M., Ma, L.: Prognostic modelling options for remaining useful life estimation by industry. Mech. Syst. Signal Process. **25**(5), 1803–1836 (2011). https://doi.org/10.1016/j.ymssp.2010.11.018
15. Son, J., Zhou, S., Sankavaram, C., Du, X., Zhang, Y.: Remaining useful life prediction based on noisy condition monitoring signals using constrained kalman filter. Reliability Eng. Syst. Safety **152**, 38–50 (2016). https://doi.org/10.1016/j.ress.2016.02.006
16. Tan, W.M., Teo, T.H.: Remaining useful life prediction using temporal convolution with attention. AI **2**(1), 48–70 (2021). https://doi.org/10.3390/ai2010005, https://www.mdpi.com/2673-2688/2/1/5
17. Wang, J., Wen, G., Yang, S., Liu, Y.: Remaining useful life estimation in prognostics using deep bidirectional lstm neural network. In: 2018 Prognostics and System Health Management Conference (PHM-Chongqing), pp. 1037–1042. IEEE (2018). https://doi.org/10.1109/PHM-Chongqing.2018.00184
18. Xia, M., Zheng, X., Imran, M., Shoaib, M.: Data-driven prognosis method using hybrid deep recurrent neural network. Appl. Soft Comput. **93**, 106351 (2020). https://doi.org/10.1016/j.asoc.2020.106351
19. Zhang, Y., Hutchinson, P., Lieven, N.A., Nunez-Yanez, J.: Remaining useful life estimation using long short-term memory neural networks and deep fusion. IEEE ACCESS **8**, 19033–19045 (2020). https://doi.org/10.1109/ACCESS.2020.2966827
20. Zhao, S., Zhang, Y., Wang, S., Zhou, B., Cheng, C.: A recurrent neural network approach for remaining useful life prediction utilizing a novel trend features construction method. Measurement **146**, 279–288 (2019). https://doi.org/10.1016/j.measurement.2019.06.004
21. Zheng, S., Ristovski, K., Farahat, A., Gupta, C.: Long short-term memory network for remaining useful life estimation. In: 2017 IEEE International Conference on Prognostics and Health Management (ICPHM), pp. 88–95. IEEE (2017). https://doi.org/10.1109/ICPHM.2017.7998311

An Intelligent Approach to Predicting of Heart Problems

Soltane Merzoug(✉) and Hadil Goufi

Laboratory of Vision and Intelligence (LAVIA), Echahid Cheikh Larbi Tebessi University, Tebessa, Algeria
soltane.merzoug@univ-tebessa.dz

Abstract. Early detection of heart diseases is a crucial factor for successful healthcare. In recent years, the medical field has witnessed the emergence of various methods to predict heart diseases before they occur, based on machine learning and deep learning. Heart diseases remain a major cause of death worldwide, particularly coronary artery disease, which is one of the most dangerous diseases. It accounts for a significant portion of patient cases and deaths related to cardiovascular diseases, as this disease is latent and does not clinically manifest. Therefore, it is necessary to diagnose and treat it as early as possible. The objective of this work is to develop a predictive model for coronary artery disease capable of detecting its early onset, often fatal.

Our research project aims to predict coronary artery disease using deep learning techniques. We utilized a convolutional neural network to achieve this objective. It is believed that this will contribute to improving early diagnostic rates and reducing complications associated with this disease.

Keywords: prediction · machine learning · deep learning · cardiovascular diseases · coronary artery disease · convolutional neural network

1 Introduction

Currently, heart diseases are one of the most significant problems threatening human life. They are the leading cause of death globally, resulting in approximately 17.9 million deaths annually. Among these deaths, it is estimated that 7.4 million are due to coronary heart disease [1], which is the leading cause of premature deaths in developed countries. This disease can persist for a long time in a latent form without clinically manifesting. This is what makes coronary heart disease a highly important global health issue, as it accounts for a significant portion of patient cases and deaths related to cardiovascular diseases. Identifying individuals at risk of developing coronary heart disease is crucial for prompt intervention and prevention efforts, which could prevent premature deaths. The diagnostic process for this disease requires an extensive battery of tests, such as blood pressure measurement, glucose level assessment, vital signs monitoring, chest pain evaluation, electrocardiograms, and maximum heart rate determination, among others. In addition to these analyses, clinical studies, patients' medical history, and their responses

T. Marir et al. (Eds.): ICCSA 2024, ISEM 45, pp. 130–145, 2025.
https://doi.org/10.1007/978-3-031-90758-6_10

to questions are also taken into account. However, conducting all these diagnoses is often prone to errors or delays in testing. Moreover, this approach incurs high costs and takes time for evaluations. In this context, significant research has been conducted to predict heart diseases, with a focus on coronary heart disease, using artificial intelligence techniques, including machine learning and deep learning. These methods rely on patient data, medical history, and a series of analyses to establish predictive models. In the scope of this project, our main objective is to develop a deep learning-based approach to predict coronary heart disease using patient data. To achieve this goal, we have implemented a specific method by constructing a model.

2 Problem and Context

Sudden death has become one of the most common causes of death at present. Without prior warning, the patient finds themselves among the deceased without even knowing in advance that they have a heart condition. This is essentially what distinguishes coronary heart disease. The World Health Organization has classified this cardiovascular disease type as the most dangerous and deadly, owing to its very high mortality rate. Today, this disease has become the leading cause of premature death in developed countries and in most countries worldwide.

This is likely due to the fact that diagnosing this type of disease by experts takes a lot of time and coincides with a lack of experts with the necessary knowledge. Ultimately, it is a disease with complex symptoms that are difficult to explain by the patient and diagnose by the doctor.

We can say that this disease is latent and does not appear clinically. There is a main cause for this disease, which is the accumulation of fats in the blood (i.e., LDL cholesterol). This process, the accumulation of fat, begins in adolescence, which over time leads to hardening of the arteries, namely the coronary arteries that supply the heart with blood. This often occurs due to the buildup of plaque in the artery walls, gradually reducing their diameter, slowing blood flow, and potentially leading to complete blockage.

Automatic methods such as deep learning can address this issue by training a model to make early predictions on individuals at risk of developing coronary heart disease, thus avoiding a significant portion of the suffering associated with the disease and the possibility of certain death due to this disease by identifying risk factors such as: advanced age, male gender, heredity, a diet high in saturated fat, hypercholesterolemia (LDL cholesterol), smoking, high blood pressure, diabetes, sedentary lifestyle, abdominal obesity (BMI), psychosocial disorders (including stress and depression), and alcohol abuse.

3 State of the Art

Table 1. Summary of Related Work.

Reference	Techniques	Evaluation Parameters	Datasets	Accuracy	Conclusion/Future Work
[2]	XGBoost, AdaBoost, Light Gradient Boosting Machine, CatBoost, Random Forest, Ridge, Decision Tree, Logistic Regression, K Neighbors, SVM-Linear Kernel, Naive Bayes, and deep neural networks (DNN3 and DNN4)	Accuracy, Precision, Recall, F-1 score, and AUC curve	UCI	ML (Extreme Gradient Boosting) 81.10% DL (DNN3) 85.41%	
[3]	Random Forest, SVM, Logistic Regression, XGBoost, Deep Learning Ensemble, and Stacking	accuracy, precision, and F1 score	UCI	Between 90% and 100%	• Exploration of different cardiac pathologies • Improvement of data quality • Continuous updating of models • Interpretation of results for healthcare professionals
[4]	ML algorithms		Local dataset	ML techniques 92% The gradient-boosting classifier 98.51%	
[5]	Logistic regression, Naïve Bayes, KNN, SVM, decision tree, random forest, MLP	Accuracy , Precision, Recall, F1-score, ROC-AUC Score	UCI	Random forest 90%	• Improvement of prediction models and exploration of new artificial intelligence techniques to enhance clinical outcomes
[6]	CNN, BN, ReLu, ADAM, ROC, AUROC	Accuracy	Local dataset	deep learning 90.43%	• Add additional demographic and clinical data
[7]	LSTM-DBN, LSTM, DBN, simple RNN, GRU, CNN, Ensemble, MLP, SVM, Logistic Regression, Random Forest, PCA, AutoEncoder	Precision, Sensitivity, F-measure, Accuracy, Fall-out, Miss rate, Specificity	Local IoT dataset	LSTM-DBN 88.42%	• There is a need for further research to address the complexity and effectiveness of the proposed method and to bridge the gap between academic and practical use of deep learning approaches and data collection

(continued)

Table 1. (*continued*)

Reference	Techniques	Evaluation Parameters	Datasets	Accuracy	Conclusion/Future Work
[8]	Logistic RegressionCV, Linear SVC, Linear Discriminant Analysis, Gaussian Mixture, Nearest Centroid, MultinomialNB,Quadratic Discriminant Analysis, GaussianNB, Random Forest Classifier, SGD Classifier, Extra Tree Classifier, Calibrated ClassifierCV, Bagging Classifier, LGBM Classifier, Ada Boost Classifier, ComplementNB, MLP Classifier, BernoulliNB, K Neighbors Classifier, Logistic Regression, Gradient Boosting Classifier	Accuracy, Precision, Recall, F1_score	Kaggle	Deep Learning, Random Forest 92.23%	• Deep Learning exhibits the highest level of accuracy among the various sets of algorithms utilized
[9]	The hybrid random forest linear model (HRFLM)		Locally collected dataset with IoT	88.7%	• The integration of IoT in healthcare services has provided patients with painless strategies to adhere to their wellness boundaries, which is a significant improvement over traditional techniques
[10]	SVM, DT, LR, KNN, NB, Relief, LASSO	Accuracy, Sensitivity, Specificity, Precision, MCC	UCI	Decision tree 85.21%	• Further generalizing the model to make it compatible with other feature selection techniques and more robust to datasets containing significant amounts of missing data • Another future strategy is to utilize Deep Learning algorithms
[11]	Random Forest, Decision Trees, K-Nearest Neighbours	Accuracy, Precision, Recall Specificity, F1 Score, ROC	Framingham heart study dataset (Kaggle)	Random Forest 96,8%	• Random Forest is a good choice for feature selection to obtain a robust prediction model • A larger dataset is needed to achieve a better training model, and emphasis should be placed on the importance of preprocessing

(*continued*)

Table 1. (*continued*)

Reference	Techniques	Evaluation Parameters	Datasets	Accuracy	Conclusion/Future Work
[12]	HY_OptGBM (LightGBM, loss function, OPTUNA)	Precision, Recall, F-score, AUROC, AUPRC and MCC, Accuracy, Sensitivity, Specificity	Framingham heart study dataset (Kaggle)	97.8%	• As datasets increase, achieving successful outcomes will require fine-tuning the presented default parameters. In addition to using a single model for predicting coronary diseases, an alternative approach is to consider constructing a prediction model by combining multiple models
[13]	LR, SVM, NB, RF, XGB, DT, NN, RBF, KNN, GBT, MLP	Accuracy, Precision (specificity), Recall (sensitivity), F-Measure	Heart Disease (UCI Repository)	Random Forest 96.28%	• Additional datasets to attempt to obtain more reliable conclusions • Optimizing the parameters of ML classifiers and deep learning methods using metaheuristic techniques
[14]	VGG-16, 1D CNN, ResNet18	Sensitivity, specificity, AUC, F1-Score, Accuracy	Local dataset	VGG-16 95%	• The results introduce a new concept that the use of deep learning algorithms with raw cardiac sound signals can facilitate the detection of obstructive coronary artery disease
[15]	BiLSTM, GRU, RSCV, LSTM, BiLSTM-GRU	Positive predicated value (PPV), Recall or Sensitivity, Specificity, Negative predicted Value(NPV),F1-Score,Accuracy	The BRFSS-2015 Dataset	BiLSTM-GRU 98.28%	• This proposed BiLSTM-GRU model will be applied to other new datasets from BRFSS surveys on heart diseases as well as other disease predictions
[16]	SVM, statistical analysis, Gray-Level Co-Occurrence Matrix (GLCM), and Gray Level Run Length Matrix (GLRLM)	Sensitivity, specificity, F1-Score, Accuracy, Geometric Mean, Precision	Local dataset	SVM 93%	• This study provides a reference for detecting CAD from iris images. In future studies, it can be explored the relationship between various heart diseases, and performance improvement can be achieved by attempting different feature extractions that detect various diseases using convolutional neural networks

3.1 Summary

The authors of the presented articles focus on the use of machine learning and deep learning techniques for the prediction of heart diseases. Over 55 techniques were used in these studies, where machine learning was the most widely utilized. So far, deep learning has shown superior accuracy in contrast to machine learning, especially in articles that compare the usage of these two subdomains of artificial intelligence: machine learning and deep learning. Different models and algorithms were employed in these studies. Machine learning techniques that achieved higher precision rates include Extreme Gradient Boosting (XGBoost), Random Forest, Support Vector Machine (SVM), Logistic Regression, decision tree, Multilayer Perceptron (MLP), Hybrid Random Forest Linear Model (HRFLM). On the other hand, deep learning techniques that achieved higher precision rates include DNN3 (3-layer network), Convolutional Neural Network (CNN), Batch Normalization (BN), Rectified Linear Unit (ReLu), Adaptive Moment Estimation (ADAM), Receiver Operating Characteristic (ROC), Area Under the Receiver Operating Characteristic Curve (AUROC), Long Short-Term Memory - Deep Belief Network (LSTM-DBN), and Visual Geometry Group (VGG). The performance of the models varies, but most approaches obtained high precision rates ranging from 81.10% to 100%. Some articles also utilized feature selection techniques such as LASSO and PCA to enhance the performance of models. The use of deep learning techniques, such as LSTM neural networks and deep belief networks, has shown promising results in the prediction of heart diseases. Commonly used metrics to evaluate model performance include accuracy, precision, F1 score, and sensitivity. The datasets used in these studies come from various sources, including the UCI Heart Disease database and other medical centers. The articles also highlight the importance of early detection of heart diseases and emphasize the potential of machine learning in identifying risk factors and early signs. Some articles have also explored the use of the Internet of Things (IoT) to collect medical data and improve the detection of heart diseases. Countries such as India, China, and Pakistan have been mentioned as having the highest number of studies on the use of ML (Machine Learning) and DL (Deep Learning) to predict heart diseases. They also highlighted coronary artery disease, which is the deadliest among heart diseases due to its high mortality rate throughout the year. Each article presents a different approach to predict the disease and evaluates the performance of the proposed models. For example, [11] uses supervised learning algorithms such as Random Forest, Decision Trees, and K-Nearest Neighbors to predict the risk of coronary artery disease. The results show high accuracy for these algorithms. [12] focuses on the use of an enhanced LightGBM model for the prediction of coronary artery disease. The model's hyperparameters are adjusted using OPTUNA, and an improved loss function is employed. The model achieves a high AUC value, surpassing other comparative models. [13] explores the use of various machine learning classifiers to predict the presence of coronary artery disease. Gradient-boosted trees and multilayer perceptron achieve good results, but Random Forest achieves the best performance with high accuracy. [14] focuses on the use of deep learning to analyze heart sound signals and predict obstructive coronary artery disease. VGG-16 is identified as the best model with superior performance compared to other models. [15] proposes a hybrid deep learning model to predict coronary artery disease by combining multiple machine learning and deep learning techniques. The model achieves impressive accuracy in predicting coronary heart diseases. [16] presents an innovative method using iris analysis for non-invasive diagnosis of coronary artery disease. The proposed model, using the SVM classifier, achieves a high accuracy rate.

Overall, these articles demonstrate the effectiveness of machine learning and deep learning techniques in predicting heart diseases, particularly coronary artery disease. Each approach has its own advantages and performance, contributing to the advancement of predicting this serious disease, improving the understanding of risk factors, and enhancing cardiovascular care.

4 Contribution

The objective is to develop a classification model capable of predicting whether there is a possibility of future exposure to coronary artery disease, using the Framingham Heart Study dataset and the algorithms that will be used for classification are CNN (Convolutional Neural Networks).

4.1 System Architecture

The diagram in Fig. 1 represents the structure of the prediction model, through which we will attempt to provide an overview of our system that predicts whether a patient is at a 10-year risk of developing coronary artery disease in the future. This is done by submitting the data to the necessary steps that allow processing, extracting the most important features, and dividing them into training data that are used to train the model and test data that are used to test the model using the CNN classification algorithm.

Dataset

The dataset named "Framingham" comprises more than 4,240 records, featuring 16 columns and 15 attributes. The objective of this dataset is to forecast whether a patient is at a 10-year risk of developing coronary heart disease (CHD) in the future [17] (Fig. 2).

Preprocessing Phase

Overall, data preprocessing involves these essential steps to transform and organize the data, enabling effective analysis and model development (Fig. 3).

Data Analysis

Data analysis involves examining and interpreting datasets to extract useful information. The following graphs illustrate visualizations of the dataset.

Figure 4 represents the two classes within the database, where 0 represents the class of patients without heart disease, and 1 represents the class of patients with heart disease. There is a significant gap between the number of patients without heart disease and those with heart disease.

Age is one of the indicators that provides important information in determining the identification of the disease. Figure 5 presents the distribution of patients based on their age, ranging from 32 to 70 years old.

Sex is also one of the risk factors directly affected in determining the disease, as males are more likely to the disease. Figure 6 represents the count of each sex, with the

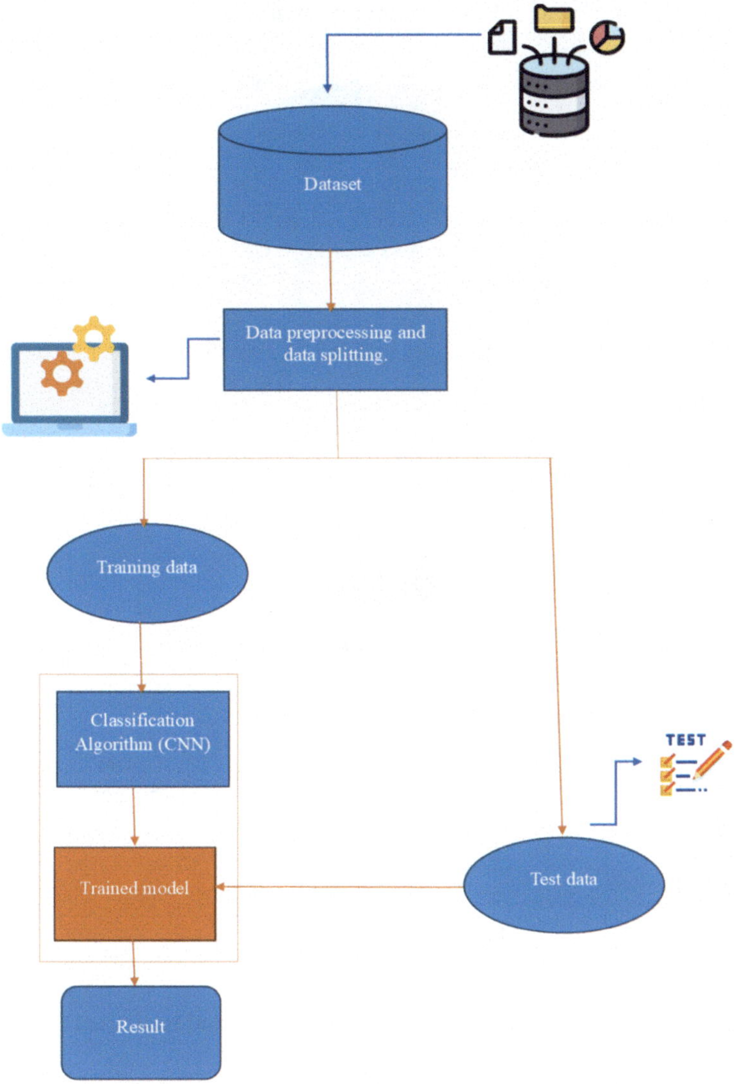

Fig. 1. Architecture of the Coronary Artery Disease Prediction System.

number of females being higher than the number of males. On the other hand, Fig. 7 represents the count of both sick and non-sick women, as well as sick and non-sick men. It appears that the number of sick men is higher than the number of sick women.

Data Cleaning

Data cleaning enhances the quality by identifying and correcting incomplete, corrupted, or inaccurate data.

Data Normalization

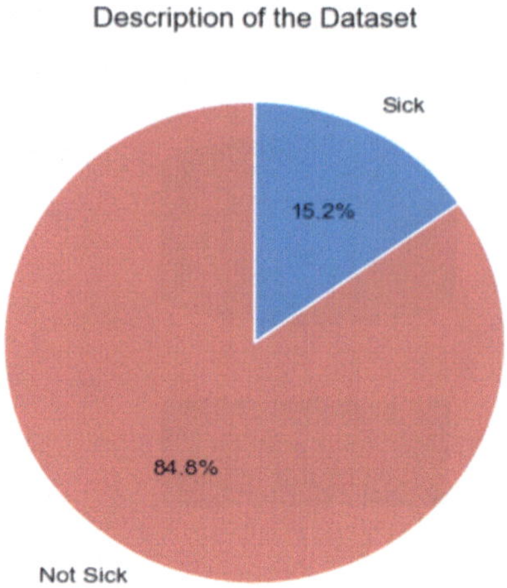

Fig. 2. Description of the Dataset.

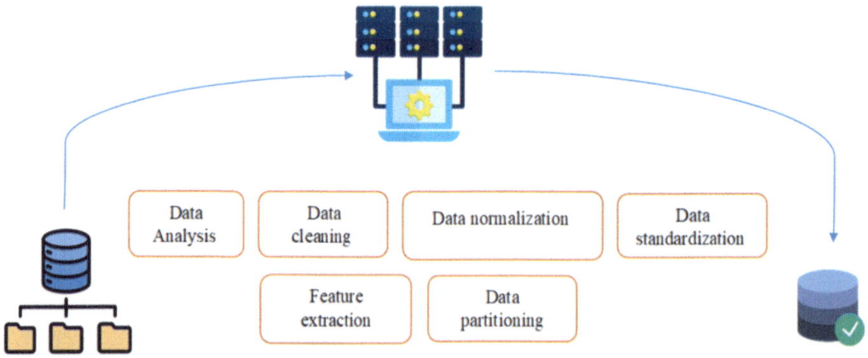

Fig. 3. Preprocessing Phase

A technique of scaling values where they are shifted and readjusted to fit within a range of [0, 1].

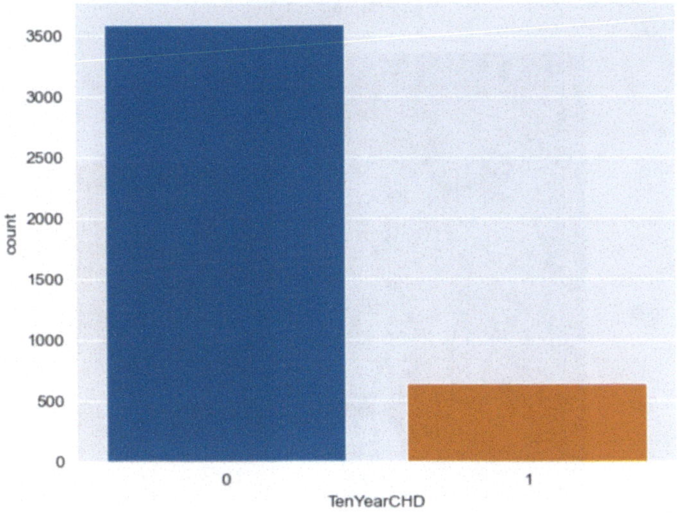

Fig. 4. Graph illustrating the two classes.

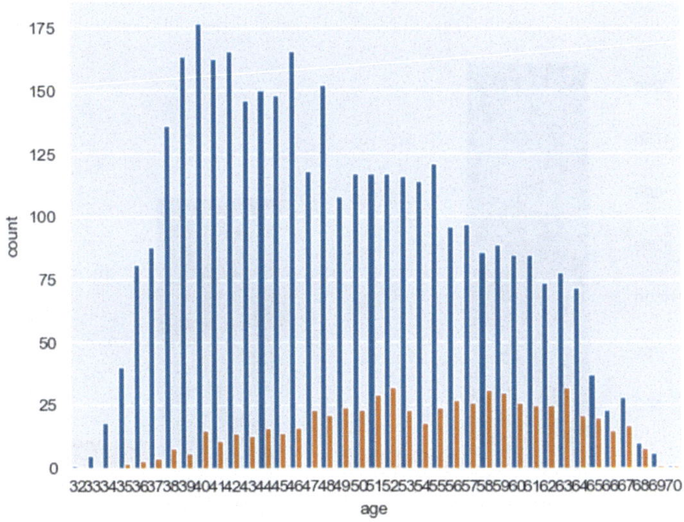

Fig. 5. Visualization of the Age/Heart Disease relationship.

Data Standardization

involves centering and scaling the values to facilitate their analysis by making them comparable and consistent.

Feature Extraction

A crucial step in data preprocessing, in this study, we will focus on identifying risk factors for coronary artery disease selected based on their direct impact on the occurrence and exacerbation of this disease.

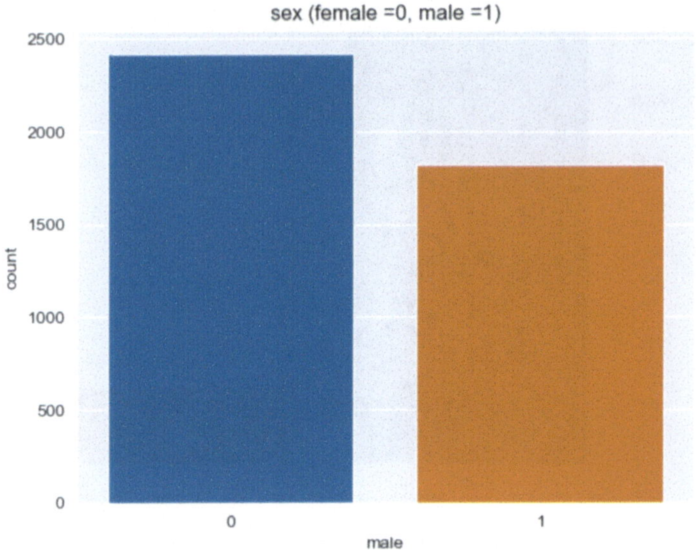

Fig. 6. Dataset Distribution by Gender.

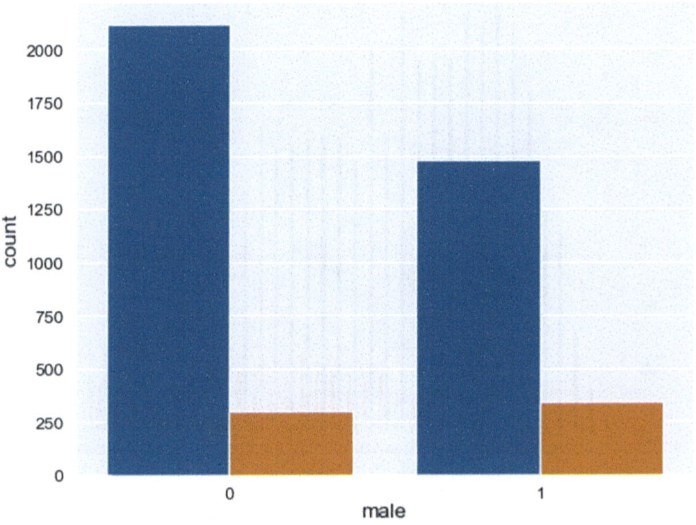

Fig. 7. Visualization of the Gender/Heart Disease relationship.

In this study, we will focus on identifying risk factors for coronary artery disease, as mentioned earlier. Table 1 represents the risk factors for coronary artery disease, which have been carefully selected based on their direct impact on the occurrence and exacerbation of this disease. The table provides the definition and detailed information for each factor separately, along with the corresponding attribute from the used dataset (Table 2).

Table 2. Risk Factors/Dataset Attributes.

Risk Factors	Dataset Attributes
Advanced age: Age is a continuous risk factor that gradually increases the incidence of coronary artery disease. This risk becomes significant from the age of 50 in men and from the age of 60 in women	Age
Male gender: Before the age of 70, men represent two-thirds of myocardial infarction cases. However, after menopause, this disparity decreases in women and completely disappears after the age of 75	Male
Hypercholesterolemia: Characterized by abnormalities in circulating lipids, it represents a major risk factor for cardiovascular diseases. It involves an elevation of Low Density Lipoprotein-cholesterol or cholesterol bound to low-density lipoproteins > 1.60 g/L (4.1 mmol/L). LDL-cholesterol directly contributes to the increase of atherosclerotic plaques and their instability leading to rupture	totChol
Abdominal obesity: Obesity is a cardiovascular risk factor, and this risk is correlated with body mass index (BMI), especially when obesity is of android type, characterized by intra-abdominal fat predominance. It is often associated with other risk factors such as hypertension (HTN) and diabetes	BMI
High blood pressure: Whether it is persistent, paroxysmal, treated, or untreated, high blood pressure is a risk factor for coronary artery disease. In fact, high blood pressure triples the risk of developing this disease	BPMeds, heartRate, sysBP, diaBP, prevalentHyp, prevalentStroke
Smoking: Smoking worsens atherosclerotic lesions by impairing endothelial function, disrupting vasomotor activity, and activating platelet aggregation. It is atherogenic and prothrombotic. It increases the risk of myocardial infarction by fivefold	currentSmoker, cigsPerDay
Diabetes: Type 1 or type 2 diabetes is associated with an increased risk of cardiovascular diseases. Cardiovascular complications can occur as early as 30 years of age in the case of type 1 diabetes, but the rising incidence of type 2 diabetes makes it a very concerning risk factor	diabetes, glucose

Now, we remove all similar attributes that play the same role, so that the set of attributes we will be working with is (Fig. 8):

```
feature_cols = ['age', 'male', 'totChol', 'BMI','prevalentHyp','heartRate','prevalentStroke','BPMeds','currentSmoker','diabetes']
```

Fig. 8. Selected attributes.

Data Partitioning

During this step, the dataset is divided into two distinct sets: a training set representing 80% of the data and a test set representing 20% of the data (Fig. 9). The training set is used to train the model, while the test set is used to evaluate the model's performance.

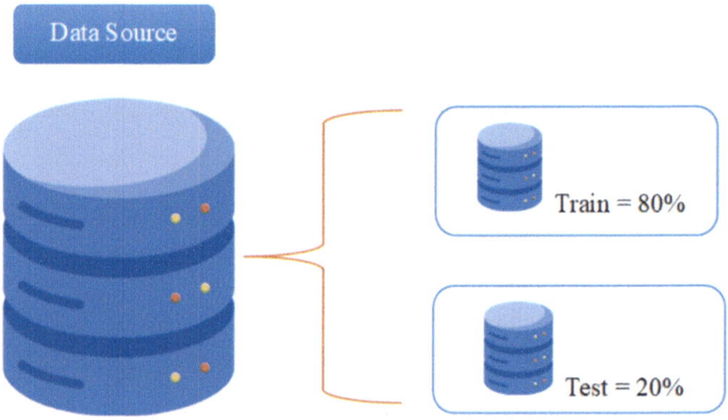

Fig. 9. Data partitioning.

Classification with Convolutional Neural Network (CNN)

In the context of coronary artery disease detection, the Convolutional Neural Network (CNN) architecture is used to learn patterns and relationships between heart attributes and coronary artery disease. This architecture, illustrated in Fig. 10, consists of input layers, hidden layers, and output layers. Each layer is composed of neurons that perform computations and transformations on the input data to produce an output. As CNN architecture is commonly used for classification tasks, it is a suitable choice for coronary artery disease detection.

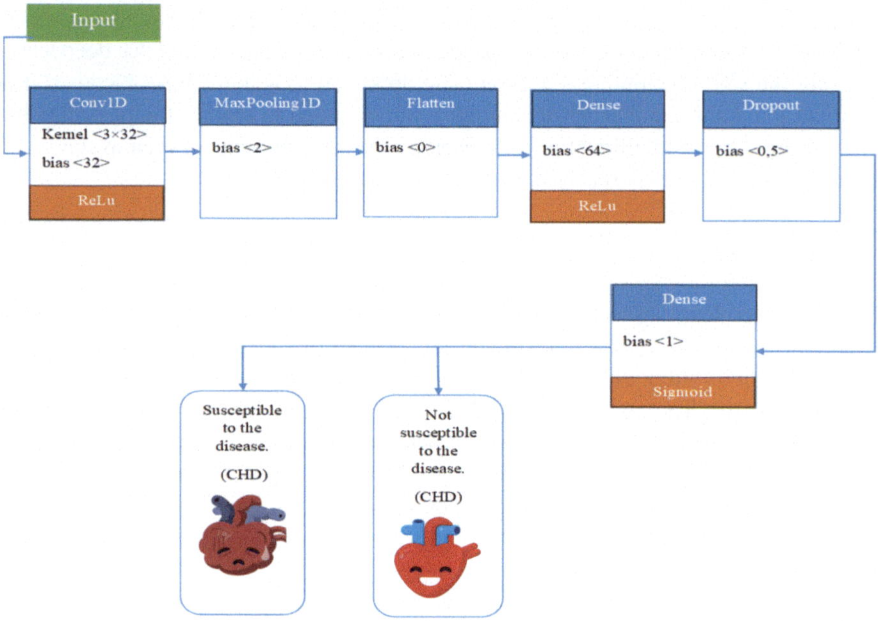

Fig. 10. Proposed CNN Architecture.

5 Result

The model results are presented in Table 3:

Table 3. Results obtained with the CNN model.

Train				Test			
Acc (%)	Prec (%)	Rec (%)	F1 Score (%)	Acc (%)	Prec (%)	Rec (%)	F1 Score (%)
85.73	85.80	85.73	80.40	86.08	84.38	86.08	80.81

The results in the table indicate the performance of the CNN model on the training and test sets, suggesting that the model has good performance with high scores for accuracy, recall, and F1 score on both the training and test sets.

5.1 Discussion

Advancements in the field of e-health, telemedicine, and big data processing are significantly transforming healthcare delivery. A crucial aspect of this transformation is making massive amounts of data accessible through computerized systems to assist physicians and healthcare professionals in their diagnoses, making them faster and more accurate.

Our goal is to develop a decision support tool for predicting cardiovascular diseases, particularly coronary artery disease. During our research, we have found that deep learning using the CNN technique is a promising approach to achieve this objective. It is important to emphasize that this technique is not intended to replace the physician but rather to serve as an intelligent tool to leverage the vast amounts of data available on patients who have already received an accurate diagnosis.

6 Conclusion

The use of deep learning techniques to predict diseases represents a promising field in modern medicine. These techniques enable the analysis of patient data, learning patterns associated with different diseases, and predicting the likelihood of future occurrence of a specific disease. The applications of deep learning in the medical field are a major advancement in healthcare, facilitating early disease detection, improved personalized diagnosis, and treatment. Deep learning algorithms can help detect early symptoms of diseases and predict the risks of developing certain conditions. This allows physicians and healthcare professionals to intervene more quickly in diagnosing and treating patients, leading to improved therapeutic outcomes.

In this project, we examined recent research in the field of cardiovascular disease prediction, with a focus on coronary artery disease. From the analysis of the reviewed studies, we found that some research proposed an approach based on comparing algorithm performances to choose the most effective one. However, this method proved to be less effective than hybrid systems using multiple algorithms. Additionally, most of these studies were focused on machine learning, although research often suggests the use of deep learning algorithms to enhance results. That's why we opted for this approach using the CNN algorithm, which allowed us to achieve a prediction accuracy of 86.08%.

Accuracy in such cases is of crucial importance as it directly impacts the lives of the individuals concerned. It is essential to understand that the numbers not only represent data but also human lives. Thus, any margin of error in result accuracy remains a concern. By continuing this work in the future, we have the potential to make significant contributions to advancing knowledge and improving healthcare in the field of cardiovascular diseases, particularly coronary artery disease, which is both dangerous and life-threatening. In the future, we hope to further increase the accuracy of coronary artery disease prediction to minimize error rates as much as possible. We aspire to ensure a higher level of confidence, thereby ensuring maximum safety for the lives of the individuals concerned.

References

1. World Health Organization: WHO (2021, 11 juin). Maladies cardiovasculaires. https://www.who.int/fr/news-room/fact-sheets/detail/cardiovascular-diseases-(cvds)#:~:text=Principaux%20faits&text=On%20estime%20%C3%A0%2017%2C7,de%20la%20mortalit%C3%A9%20mondiale%20totale
2. Rahman, S., Hasan, M.M., Sarkar, A.K.: Machine learning and deep neural network techniques for heart disease prediction. IEEE (2022). https://doi.org/10.1109/iccit57492.2022.10055902

3. Prediction of heart disease using machine learning: a systematic literature review. J. Syst. Manag. Sci. **13**(6) (2023). https://doi.org/10.33168/jsms.2023.0603
4. Al-Jamimi, H.A.: Intelligent methods for early prediction of heart disease. IEEE (2023). https://doi.org/10.1109/codit58514.2023.10284226
5. Hossain, Md.I., et al.: Heart disease prediction using distinct artificial intelligence techniques: performance analysis and comparison. Iran J. Comput. Sci. **6**, 397–417 (2023)
6. Alkhodari, M., et al.: Deep learning predicts heart failure with preserved, mid-range, and reduced left ventricular ejection fraction from patient clinical profiles. Front. Cardiovasc. Med. **8**, 755968 (2021). https://doi.org/10.3389/fcvm.2021.755968
7. Dami, S., Yahaghizadeh, M.: Predicting cardiovascular events with deep learning approach in the context of the internet of things. Neural Comput. Appl. **33**(13), 7979–7996 (2021). https://doi.org/10.1007/s00521-020-05542-x
8. Barhoom, A.M.A., Almasri, A., Abu-Nasser, B.S., Abu-Naser, S.S.: Prediction of heart disease using a collection of machine and deep learning algorithms. Int. J. Eng. Inf. Syst. **6**(4), 1–13 (2022)
9. Salunke, T., Jagade, P., Pawar, S., Rathod, P., Ghawate, N.: Heart disease detection using hybrid machine learning and IoT (software based). J. Artif. Intell. Mach. Learn. Neural Netw. **33**, 10–13 (2023). https://doi.org/10.55529/jaimlnn.33.10.13
10. Kanwal, A., Ahmad, K.T., Abid, M.K., Aslam, N.: Detection of heart disease using supervised machine learning. VFAST Trans. Softw. Eng. **10**(3), 58–70 (2022)
11. Krishnani, D., Kumari, A., Dewangan, A., Singh, A., Naik, N.S.: Prediction of coronary heart disease using supervised machine learning algorithms. In: IEEE Region 10 Conference (TENCON 2019), pp. 367–372 (2019)
12. Yang, H., Chen, Z., Yang, H., Tian, M.: Predicting coronary heart disease using an improved LightGBM model: performance analysis and comparison. IEEE Access **11**, 23366–23380 (2023). https://doi.org/10.1109/access.2023.3253885
13. Hassan, C.A.U., et al.: Effectively predicting the presence of coronary heart disease using machine learning classifiers. Sensors **22**(19), 7227 (2022). https://doi.org/10.3390/s22197227
14. Ainiwaer, A., et al.: Deep learning of heart-sound signals for efficient prediction of obstructive coronary artery disease. Heliyon **10**(1), e23354 (2024). https://doi.org/10.1016/j.heliyon.2023.e23354
15. Sharma, N., Malviya, L., Jadhav, A., Lalwani, P.: A hybrid deep neural net learning model for predicting Coronary Heart Disease using Randomized Search Cross-Validation Optimization. Decis. Anal. J. **9**, 100331 (2023). https://doi.org/10.1016/j.dajour.2023.100331
16. Özbilgin, F., Kurnaz, Ç., Aydın, E.: Prediction of coronary artery disease using machine learning techniques with iris analysis. Diagnostics **13**(6), 1081 (2023). https://doi.org/10.3390/diagnostics13061081
17. Framingham heart study dataset (2022, 19 avril). Kaggle. https://www.kaggle.com/datasets/aasheesh200/framingham-heart-study-dataset/data

Enhancing Breast Tumor Classification Using Ensemble Methods and Voting Systems

Romaissaa Koolaf, Moahemd Ramdani, and Laid Kahloul(✉)

LINFI Laboratory, Mohamed Khider Biskra Univeristy, Biskra, Algeria
{mohamed.ramdani,l.kahloul}@univ-biskra.dz

Abstract. The accurate categorization of breast tumors as either malignant or benign is crucial for effective diagnosis and treatment planning. This paper introduces a comprehensive evaluation addressing eight different machine learning models applied to the dataset used for breast cancer diagnosis in Wisconsin, which includes 569 instances featuring 30 attributes derived from digitized fine needle aspirate (FNA) biopsy samples. The algorithms assessed include Naive Bayes (NB), Logistic Regression (LR), Support Vector Machine (SVM), K-Nearest Neighbors (KNN), Decision Tree (DT), Random Forest (RF), AdaBoost, and XGBoost. Each model's performance was assessed by means of several metrics such as accuracy, precision, recall, F1 score, specificity, confusion matrix, Matthews correlation coefficient, logarithmic loss, and area under the curve (AUC). Additionally, we explored ensemble techniques, including simple and weighted methods, alongside Voting Classifier Systems like Stacking and Boosting to enhance predictive accuracy. Our comparative analysis highlights the most effective strategies for early breast cancer detection and offers insights into developing AI-driven diagnostic tools. The study supports the integration of these models into web-based platforms, including the TIASM (TIASM: **T**échnique de l'**IA** pour le **S**outien de la **M**édecine) platform, emphasizing the critical role of model selection in medical decision support systems. These findings underscore the capabilities of machine learning to improve breast cancer prediction and patient outcomes.

Keywords: Breast Cancer Prediction · Breast cancer classification · breast cancer tumor · benign · malignant · Machine Learning · Wisconsin Breast Cancer Dataset

1 Introduction

According to [30], among women around the world, breast cancer ranks as the most frequently diagnosed form of cancer and poses a significant global health challenge. According to the International Union Against Cancer (UICC), it ranks as the cancer most frequently identified in women across 157 countries, constituting a quarter of all cancer cases. Additionally, it is the second most prevalent

T. Marir et al. (Eds.): ICCSA 2024, ISEM 45, pp. 146–160, 2025.
https://doi.org/10.1007/978-3-031-90758-6_11

form of cancer overall and represents the leading cause of cancer mortality in the female population. In 2022 alone, globally, breast cancer accounted for approximately 2.3 million new diagnoses annually, highlighting its pervasive impact. Alarmingly, despite advances in healthcare, an estimated 669,418 individuals succumbed to this disease in the same year. Authors in [1] report that in Algeria, the breast cancer incidence rate is 57.8 per 100,000, according to Globocan 2020. Furthermore, Globocan 2020 reports a 5-year breast cancer prevalence rate of 179.2 per 100,000 in Algeria. Tumors are classified as benign or malignant, and the work reported in [29] shows that a benign tumor is not harmful to the human body and seldom leads to fatality. It grows in a specific area and has restricted growth. In contrast, a malignant tumor is more perilous and can result in death. It grows rapidly due to abnormal cell growth. According to [12], diverse types of breast cancer result from the spread of compromised cells and tissues across the body. Authors of [26] showed that breast cancer has had a significant impact on women's lives, so that mortality rates could be minimized through raising awareness, early-stage detection, and diagnosis.

Machine Learning (ML) and Deep Learning (DL) achieved remarkable successes in healthcare application (diagnosis as well as prediction [5,8,11,24,27, 32]). In the literature, several works have applied machine learning models for the diagnosis and prediction of breast cancer. In [28], the authors highlight that the traditional cancer detection method relies on regression processes, whereas recent machine learning approaches are primarily concerned with optimizing model architecture. These models are crafted to forecast unseen data, yielding promising results in both training and testing phases. Recent studies [10] have elucidated that the machine learning process relies on three fundamental stages: data preprocessing, feature selection or extraction, and the classification phase. In [31], contemporary research has underscored that feature extraction is instrumental in the machine learning process, facilitating cancer diagnosis and outcome prediction by distinguishing between benign and malignant tumors. Data mining techniques and machine learning algorithms are pivotal for the diagnosis and prediction of diverse breast cancer types [23], including classification, regression, and clustering, which provide valuable insights into breast cancer patients. These algorithms [6,17] operate based on a training dataset, enabling us to assess the probability of predicting numerous types of breast cancer. In [2,20], numerous studies have underscored the critical role of predicting breast cancer using various techniques. Researchers have utilized various data mining classification approaches to scrutinize the WDBC dataset for breast cancer diagnosis [9,22], resulting in significant findings. In recent literature [18], various data mining techniques and ensemble methods have been extensively employed for classification tasks. In the next section, we will present more related works and discuss them.

This paper explores the capabilities of machine learning models in unraveling the intricacies of breast cancer, proposing a novel ensemble classification method. We assess the performance of different classification algorithms, including Naive Bayes (NB), K-Nearest Neighbors (KNN), Support Vector Machines

(SVM), Decision Trees (DT), Logistic Regression (LR), Random Forest (RF), AdaBoost, and XGBoost. Besides the previous algorithms, innovative mechanisms are exploited to enhance results and improve the correctness of the diagnosis, namely: ensemble methods, boosting approaches, stacking techniques, and voting. The advocated approach performance is assessed using the openly accessible Dataset of Wisconsin Breast Cancer.

The remainder of this paper is laid out in this manner: Sect. 2 examines current literature on breast cancer tumor detection and underscores the significance of proficient performance measures in medical diagnostics. Section 3 provides a detailed explanation of the proposed system, subsequently leading to the experimentation and discussion of results in Sect. 4. Finally, Section 5 brings the paper to a close and recommends areas for future research directions for breast cancer tumor classification.

2 Related Work

According to [25], numerous studies have applied machine learning (ML) techniques within the healthcare domain to detect breast cancer (BC). Due to the algorithms' ability to deliver satisfactory results, researchers have increasingly utilized them to tackle complex medical challenges. Employing ML-based approaches to predict breast cancer remains an ongoing area of research. Various machine learning algorithms, including NB, KNN, SVM, DT, and LR, have been developed and refined for breast cancer diagnosis. The study outlined in [21] demonstrates that Quang et al. utilized both supervised and unsupervised classification models for breast tumor classification. They introduced a method combining scaling with PCA to achieve more effective feature determination. Their findings indicated that an ensemble voting approach yielded the most accurate predictions. After performing feature selection, they trained and tested several classification models on the dataset. Notably, four model-ensemble voting classifiers, logistic regression (LR), support vector machine (SVM), and AdaBoost exhibited superior performance. These models achieved accuracies of approximately 90%, as evidenced by their results in terms of precision and recall, the area under the receiver operating characteristic curve (AUC-ROC), F1 score, and computational efficiency. The authors Mandal et al. of [16] applied logistic regression (LR), Naïve Bayes (NB), and decision trees (DT), examining the runtime intricacies of each classifier. Their findings revealed that LR achieved the highest accuracy among the classifiers. On the other hand, Borges et al. in [7] assessed the effectiveness of Bayesian networks and DT, determining that Bayesian networks outperformed DT, achieving an accuracy of 97.80%. In the domain of breast cancer classification, various machine learning approaches have demonstrated significant promise. Islam et al. in their study reported in [14] applied multiple supervised learning techniques to the Wisconsin dataset, resulting in artificial neural networks (ANN) attaining the peak accuracy of 98.57%. Similarly, Amrane et al. in [3] utilized KNN and NB, resulting in cross-verification accuracies of 97.51% and 96.17%,respectively. Research in [13] indicates that feature scaling is crucial in machine learning algorithms, as it standardizes the magnitude of data characteristics. This normalization process allows most ML algorithms to calculate Euclidean distances between data points effectively. Nanglia et al. explored an ensemble model in [19], combining KNN, SVM, and DT, attaining an accuracy percentage of 78%. Mentioned in [4], Ara et al. conducted feature selection and applied machine learning algorithms for breast cancer data classification into benign and malignant categories, achieving a notable accuracy of 96.5% using RF and SVM. Additionally, the authors Jabbar et al. of [15] constructed an ensemble model incorporating Bayesian networks and Radial Basis Function, which collectively yielded a classification accuracy of 97.42%.

In fact, the novelty of the current work is to enhance the results obtained using several machine learning algorithms, largely exploited in tumor classification, by other mechanisms, namely: ensemble methods, boosting approaches, and stacking techniques. This combination of algorithms and approaches improves the correctness of the diagnosis and provides more trust in the response of our system.

3 Developing a Tumor Classification System

3.1 Algorithm Suitability Analysis

Motivation for Using the Eight Algorithms: Using all eight algorithms-Naive Bayes, Logistic Regression, SVM, K-Nearest Neighbour, Decision Tree, Random Forest, AdaBoost, and XGBoost-provides a comprehensive analysis aimed at capturing diverse insights and enhancing model robustness. The WDBC dataset, with its 569 instances and 30 measurable features extracted from digitized fine needle aspirate (FNA) biopsy images, presents a complex classification task that can benefit from diverse modeling perspectives: (i) Diverse Insights: Each algorithm brings unique strengths to the classification task. For instance, Naive Bayes offers a probabilistic approach that is useful when features are conditionally independent, while SVM is efficient in dealing with high-dimensional spaces, making it suitable for the dataset's feature-rich nature, (ii) Pattern Identification: The algorithms capture various patterns and interactions within the data. Decision Trees and Random Forests excel in handling non-linear relationships and determining feature importance, while K-Nearest Neighbour captures local data structures, (iii) Offsetting Weaknesses: By employing all algorithms, this approach offsets individual model weaknesses. For example, the ensemble of models helps reduce the overfitting tendency of Decision Trees and leverages the regularization strengths of XGBoost, (iv) Ensemble Learning Potential: This strategy enables ensemble learning techniques like stacking and boosting, which combine the strengths of different models to improve prediction accuracy. AdaBoost and XGBoost, in particular, are known for their boosting capabilities, enhancing model performance through iterative refinement. While this comprehensive approach requires significant computational power and time for experimentation, it is well-suited for those interested in understanding how different algorithms complement each other, leading to a robust and accurate predictive model designed to distinguish benign from malignant breast cancer tumors.

Focused Analysis with Top Algorithms: Focusing on the top three or four algorithms (i.e., Random Forest, XGBoost, SVM, and Logistic Regression) simplifies model management and reduces computational costs while maintaining strong performance. This targeted approach is particularly efficient in environments with limited resources, focusing on high-performing algorithms that deliver reliable classification results without unnecessary complexity: (i) Efficient Resource Use: By concentrating on Random Forest, XGBoost, and SVM, the analysis capitalizes on algorithms known for their effectiveness in handling complex datasets like WDBC. Random Forest's ability to reduce overfitting through averaging and XGBoost's prowess in handling high-dimensional data ensure robust performance, (ii) Interpretable Models: Logistic Regression and Decision Trees provide interpretable models, allowing for a straightforward understanding of feature importance and decision-making processes. This is crucial for gaining insights into which features most significantly impact classification, aiding in clinical interpretation and decision support, (iii) Proven Performance: Both Random Forest and XGBoost have consistently demonstrated strong performance across many datasets, including those related to breast cancer classification. Their ability to handle non-linear relationships and interactions in the data makes them reliable choices, (iv) Adaptability to Data Characteristics: These algorithms are well-suited to the nature of the WDBC dataset, where the features are standardized and the class imbalance is managed through appropriate techniques. SVM, with its capacity to handle imbalanced data and find optimal hyperplanes, complements the dataset's classification requirements. This focused analysis strikes a balance between performance and simplicity, targeting algorithms that are most likely to provide accurate and actionable insights into breast cancer classification.

Analysis of Features and Their Impact on Tumor Classification: Understanding the features that differentiate benign from malignant tumors is crucial for developing effective classification algorithms. The features in the breast cancer dataset primarily relate to the dimensions, morphology, and surface characteristics of cell nuclei. The dataset includes 30 features essential for tumor classification, grouped into mean, worst, and standard error measurements for: (i) Size Features (Radius, Perimeter, Area): Larger sizes may indicate malignancy,(ii) Shape Features (Smoothness, Compactness, Concavity, Concave Points, Symmetry): Irregular shapes often correlate with malignancy, and (iii) Texture Feature: Variation in pixel intensity could signal irregular cell structures.

In fact, the suitability of each feature for specific algorithms is summarized as follows: (1) Radius Features: Important for size estimation, suitable for Logistic Regression (LR), Random Forest (RF), and XGBoost due to their capabilities in modeling linear relationships and interactions. (2) Texture Features: Capture detail and variation, well-suited for Support Vector Machine (SVM) and Decision Tree (DT) due to their ability to model complex patterns. (3) Perimeter Features: Correlate with radius and area, best for RF and XGBoost, which can handle correlation effectively through ensemble methods. (4) Area Features: Indicate tumor size, significant for DT, RF, and XGBoost because of their feature importance analysis and robust modeling capabilities. (5) Smoothness Features: Capture subtle variations, suitable for LR and AdaBoost, which benefit from linear assumptions and a focus on difficult cases. (6) Compactness Features: Indicate density, well-suited for SVM and RF due to their ability to capture non-linear interactions and assess feature importance. (7) Concavity Features: Reflect tumor shape, best for DT and XGBoost, which excel in modeling non-linear relationships and interactions. (8) Concave Points: Critical for identifying malignant cases, highly relevant for DT and RF due to their strengths in feature importance and interaction modeling. (9) Symmetry Features: Capture variations in symmetry, suitable for Naive Bayes (NB) and LR, which handle less correlated and linear features well. (10) Fractal Dimension Features: Capture complexity, useful for SVM and AdaBoost, which are effective at modeling high-dimensional and hard-to-classify instances.

3.2 The Global Architecture of the System

This paper proposes a new approach for classifying breast tumors using multiple algorithms. The WBCD (Wisconsin Breast Cancer Dataset), a dataset for breast cancer diagnosis[1], is utilized for evaluation purposes. In Fig. 1, the dataset is fed into various classification models to assess and categorize tumors as either malignant or benign. This approach aims to enhance the accuracy and reliability of breast tumor classification, thereby improving clinical diagnoses. Eight classification methods are used and compared, namely: Logistic Regression (LR), K-Nearest Neighbour (KNN), Decision Tree (DT), Naive Bayes (NB), Support Vector Machine (SVM), Random Forest (RF), AdaBoost, and XGBoost.

Once the classifiers are prepared, Fig. 2 illustrates how the system can be used. The user needs to input the set of features for the tumor to be diagnosed, then they can select any classifier and view its decision regarding the tumor. In addition to these individual models, the user can access the final decision, which incorporates other approaches, including ensemble methods (weighted and simple ensembles), boosting techniques, the stacking method, and voting. The final decision obtained in stage 3 is based on voting among the intermediary decisions provided in stage 2.

3.3 Evaluation Metrics

The evaluation metrics are based on four fundamental values: (i) True Positives (TP): The number of positive instances correctly classified, (ii) True Negatives (TN): The number of negative instances correctly classified, (iii) False Positives (FP): The number of negative instances incorrectly classified as positive, and (iv) False Negatives (FN): The number of positive instances incorrectly classified as negative.

- Accuracy: This ratio indicates how well the model predicts positive cases out of all positive predictions, reflecting its effectiveness in correctly classifying

[1] https://www.kaggle.com/datasets/uciml/breast-cancer-wisconsin-data.

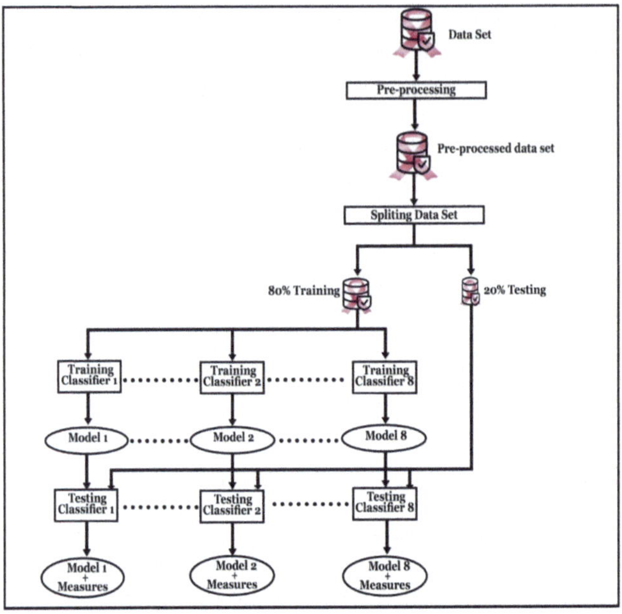

Fig. 1. Building the classifier

both classes. However, accuracy can be misleading in imbalanced datasets, as it does not differentiate between the costs of false positives and false negatives.

$$Accuracy = \frac{TP + TN}{TP + TN + FP + FN} \tag{1}$$

- Recall: This metric calculates the proportion of positive cases that the model correctly identifies among all actual positive instances, reflecting its effectiveness in detecting malignant tumors. It is essential for reducing false negatives and enhancing early detection.

$$Recall = \frac{TP}{TP + FN} \tag{2}$$

- Precision: This metric represents how many of the model's positive predictions are correct. It indicates how many of the predicted malignant tumours were actually malignant. It is crucial for reducing false positives and ensuring predicted malignant cases are accurately identified.

$$Precision = \frac{TP}{TP + FP} \tag{3}$$

- F-measure (F1-score): Serves as a metric that merges Precision and Recall into one measure. It balances the trade-off Among precision and recall, It is particularly effective in scenarios with imbalanced class is imbalanced.

$$F1\_score = 2 \times \frac{Precision \times Recall}{Precision + Recall} \tag{4}$$

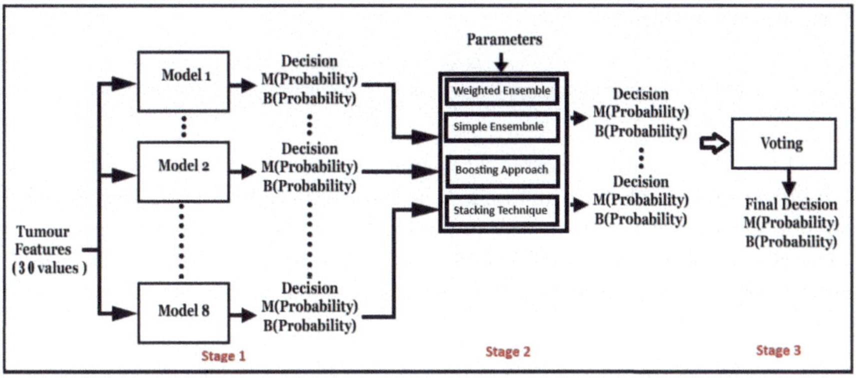

Fig. 2. Using the system to make a classification of a tumor

- Specificity (True Negative Rate): is defined as the proportion of correctly identified negative instances among all actual negative cases. It measures the ability of the model in pinpointing benign tumours correctly, thus avoiding unnecessary anxiety and procedures.

$$Specificity = \frac{TN}{TN + FP} \tag{5}$$

- The Area Under the Curve (AUC): is a metric used to assessing the performance of a binary classifier. It represents the degree or measure of separability, showing how well the model can effectively tell apart the classes. The AUC is computed from the Receiver Operating Characteristic (ROC) curve, which illustrates the True Positive Rate (TPR, commonly referred to as Recall) in relation to the False Positive Rate (FPR) across different threshold settings." $(FPR = (FP + TN)/FP)$. A perfect model has an AUC of 1, indicating perfect separability.
- In our case, we are classifying tumours into benign and malignant categories, we have separate classifiers for each class. We have to decide about the three best classifiers based on their individual F1-scores for their respective classes. In our case, *combined F1-score* is used instead of F1-score. The combined F1 score is the average of the F1 scores for the benign and malignant classes.

$$Combined\_F1\_score = \frac{F1\_score_M + F1\_score_B}{2} \tag{6}$$

- Matthews Correlation Coefficient (MCC): serves as an indicator of binary classification quality, considering both true and false positives as well as negatives. Its values range from -1 to 1, with 1 signifying a perfect prediction, 0 indicating a prediction no better than random chance, and -1 reflecting complete disagreement between the prediction and actual outcomes. This metric offers a comprehensive evaluation of model effectiveness, particularly valuable for datasets with imbalanced classes.

$$MCC = \frac{(TP \times TN) - (FP \times FN)}{\sqrt{(TP + FP)(TP + FN)(TN + FP)(TN + FN)}} \tag{7}$$

- Log Loss (Logarithmic Loss): This metric evaluates measures how well a classification model performs based on the predicted probabilities, which lie within the interval of 0 to 1. A lower log loss indicates a better model with better probability estimates. While informative for probability calibration, it is less relevant for classification outcomes in medical diagnostics.

$$LogLoss = \frac{-1}{N}\sum_{i=1}^{N}[y_i log(p_i) + (1 - y_i)log(1 - p_i)] \tag{8}$$

Such that: N = Total number of samples, y_i = Actual class label for sample i, p_i= Predicted probability of the positive class for sample i.

3.4 Tumor Classification Stages

The user introduces the features (30 values) of a given instance (a tumor) then, as shown in Fig. 2, the system proceeds through three stages: (i) the eight algorithms are executed and their results are obtained for the input case, (ii) four specific methods are applied to enhance the correctness of the diagnosis (ensemble methods and voting methods), (iii) finally a hard voting is done if the stage two provides contradictory results. In this section, the last two stages are discussed.

Ensemble Methods: Two approaches are exploited as follows:

- Weighted Ensemble: Each model's influence on the final prediction is determined by its respective performance metric, such as F1-Score or AUC. The overall prediction is calculated as a weighted sum of the predictions from individual models. The weight of a model i is computed as the sum of the metrics: $Wieght_i = F1Score_i + AUC_i + Precision_i + Recall_i + Specificity_i + MCC_i$ These weights are normalized: $Normalized\_Wieght_i = \frac{Weight_i}{\sum_{j=1}^{n} Wieght_j}$. Thus, for a given input, each model predicts a probability of being malignant. The final probability is computed as: $P(Malignant) = \sum_{i=1}^{n} Normalized\_Wieght_i \times P_i(Malignant)$. Indeed, the **Log Loss** is ignored due to its limited relevance for classification in medical diagnostics, and the **Accuracy** is also ignored as it can be misleading in imbalanced datasets, overstating performance by favoring the majority class.
- Simple Ensemble: It computes the average of the probabilities predicted by each model, treating all models equally. The probabilities from all models are averaged: $P(Bengin) = \frac{1}{n}\sum_{i=1}^{n} P_i(Bengin)$, idem for malignant probability.

Voting Classifier Systems: Two mechanisms are proposed as follows.

- Stacking (Stacked Generalization): It trains multiple base classifiers and a meta-classifier to learn how to best combine the base predictions.
- Boosting: it is an iterative approach that trains classifiers sequentially. Each new classifier focuses on correcting the misclassifications of the prior ones. most commonly used algorithms in this class of methods are: AdaBoost and Gradient Boost. For AdaBoost, weights are assigned to each training instance, and updated based on the error of the classifier. The ultimate prediction is a weighted majority vote. However, in the case of gradient boosting, it models the residuals of previous predictions and minimizes a loss function using gradient descent.

Hard Voting: A majority-based voting, commonly referred to as plurality voting, is a prevalent method applied in ensemble classification. For each test instance (denoted as x), the classification results from each of the classifiers are collected. These classification results typically consist of predicted class labels assigned by each classifier for the test instance x. The ultimate prediction for the test instance x is determined by selecting the class label that garners the majority of votes (i.e., the majority) among the predictions of all the classifiers.

In Sect. 4, we will present a case study (classification of an instance) and show how all the previous stages are applied to make a final diagnosis.

4 Results and Discussion

4.1 The Web Platform Interfaces

The user has access to the website and the platform. The first web interface is shown in Figure 3. From this first interface, the user has access to several pieces of information about the team (About Us), breast cancer, science, and can also subscribe to the site. Most importantly, the user can access the AI tools to make predictions or analyses. In this case, the user will be transferred to the platform where they can make a query and perform tumor classification, as shown in Fig. 4.

Fig. 3. The web interface of the system

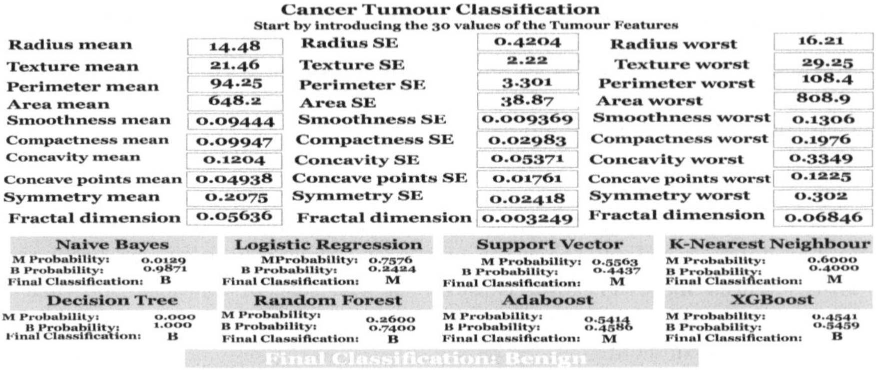

Fig. 4. The platform of AI classifiers for Breast Cancer Tumor: Predictions

4.2 Stages of the Diagnosis

1. Individual Classifiers: From the obtained results in Fig. 4, Fig. 5 shows the performance of each individual model. Metrics encompass precision, recall, F1 score, accuracy, specificity, confusion matrix, AUC, MCC, and log loss.

Model	Acc	Prec	Rec	F1	Spec	AUC	MCC	Log Loss	Confusion Matrix
Naive Bayes	0.9211	0.9231	0.8571	0.8889	0.9583	0.9891	0.8292	0.4999	[69,3; 6,36]
Logistic Regression	0.9649	0.9750	0.9286	0.9512	0.9861	0.9960	0.9245	0.0773	[71,1; 3,39]
Support Vector Machine	0.9737	1.0000	0.9286	0.9630	1.0000	0.9947	0.9442	0.0815	[72,0; 3,39]
K-Nearest Neighbour	0.9561	0.9744	0.9048	0.9383	0.9861	0.9823	0.9058	0.4091	[71,1; 4,38]
Decision Tree	0.9298	0.9250	0.8810	0.9024	0.9583	0.9196	0.8483	2.5294	[69,3; 5,37]
Random Forest	0.9649	1.0000	0.9048	0.9500	1.0000	0.9934	0.9258	0.1194	[72,0; 4,38]
AdaBoost	0.9649	0.9750	0.9286	0.9512	0.9861	0.9901	0.9245	0.4363	[71,1; 3,39]
XGBoost	0.9561	1.0000	0.8810	0.9367	1.0000	0.9934	0.9076	0.0964	[72,0; 5,37]

Fig. 5. Performance Metrics of Individual Classifiers

Based on the effectiveness, it is evident that all models display strong predictive capabilities with specific areas of excellence. The Support Vector Machine (SVM) and Random Forest (RF) models notably achieve perfect specificity and Area Under the Curve (AUC), illustrating their superior effectiveness in distinguishing between benign and malignant classes without false positives. SVM, with a precision of 1.000, also suggests that every instance predicted as malignant was correct. Logistic Regression and AdaBoost demonstrate a balanced profile with high precision and recall, reflecting consistent performance across various metrics. Notably, the Naive Bayes model, despite its simplicity, shows commendable accuracy and MCC, proving effective for preliminary analysis. The Decision Tree model, although slightly lower in F1 score and MCC, offers straightforward interpretability that could be beneficial in clinical settings where decision rationale is crucial. The log loss values across the models are relatively low, with Naive Bayes showing the highest, which could imply a slightly higher rate of misclassification under certain threshold settings. The confusion matrices further substantiate the models' effectiveness, with a greater count of true positives and true negatives indicating robust predictive performance across all models.

2. Ensemble Methods, Stacking, and Boosting: Figure 6 summarizes the performance of ensemble methods. The Weighted Ensemble method combines individual model predictions based on their performance metrics. The Simple Ensemble method averages the probabilities across models.

Method	Malignant Probability	Benign Probability	Final Classification
Weighted Ensemble	0.4031	0.5969	Benign
Simple Ensemble	0.3978	0.6022	Benign
Stacking	0.4031	0.5969	Benign
Boosting	0.4031	0.5969	Benign

Fig. 6. Results of ensemble methods, stacking and boosting

3. Hard Voting: In this scenario, all four methods diagnose the tumor as benign.

4.3 Comparison with Related Work

Table in Fig. 5 presents a comparison between our system and eight previous related works. Our proposed system demonstrates qualitative superiority by utilizing a greater number of algorithms and implementing advanced voting mechanisms. Additionally, it shows nearly quantitative advantages, as the achieved accuracy surpasses that of several existing studies.

5 Conclusion

The study presented in this paper evaluates various machine learning algorithms and ensemble techniques for breast cancer classification using the Wisconsin Diagnostic Breast Cancer (WDBC) dataset. The findings indicate that ensemble methods, particularly the weighted ensemble technique, provide superior classification performance compared to individual models. This underscores the potential of advanced machine learning techniques to enhance the accuracy of cancer diagnosis and support clinical decision-making processes. In addition to exploring machine learning algorithms for breast tumor classification, this paper introduces a user-friendly web platform. It allows users to conduct their own diagnosis by inputting the values of 30 required features. Future research directions include: (i) Expanding Research on Feature Extraction: Investigating advanced feature extraction methods to improve the quality of input data, (ii) Implementing Deep Learning Algorithms: Exploring deep learning techniques that analyze tumor images directly, which could potentially enhance classification accuracy, (iii) Creating a Local Dataset: Collaborating with medical professionals to develop a localized dataset, thereby improving the model's applicability to diverse populations, (iv) Integrating with the TIASM Platform: Deploying the web platform within the TIASM framework as a new application for AI techniques in healthcare, potentially broadening its impact and accessibility. By addressing these future directions, this research aims to contribute to more accurate and accessible cancer diagnostics while supporting ongoing advancements in medical machine learning.

References

1. Globocan 2020: Algeria, international agency for research on cancer, https://gco.iarc.fr/today/data/factsheets/populations/12-algeria-fact-sheets.pdf, https://www.uicc.org/what-we-do/thematic-areas/breast-cancer Accessed 9 May 2024
2. Abdelhafiz, D., Yang, C., Ammar, R., Nabavi, S.: Deep convolutional neural networks for mammography: advances, challenges and applications. BMC Bioinform. **20**, 1–20 (2019)
3. Amrane, M., Oukid, S., Gagaoua, I., Ensari, T.: Breast cancer classification using machine learning. In: 2018 Electric Electronics, Computer Science, Biomedical Engineerings' Meeting (EBBT), pp. 1–4. IEEE (2018)
4. Ara, S., Das, A., Dey, A.: Malignant and benign breast cancer classification using machine learning algorithms. In: 2021 International Conference on Artificial Intelligence (ICAI), pp. 97–101. IEEE (2021)
5. Baissi, F., Abdelbaki, E.A., Kahloul, L., Mohammedi, A., Ammari, A.: Paludism diagnosis using deep learning. In: Proceedings of the International Conference on Emerging Intelligent Systems for Sustainable Development (ICEIS 2024), pp. 246–260. Atlantis Press (2024). https://doi.org/10.2991/978-94-6463-496-9_19
6. Bishop, C.M., Nasrabadi, N.M.: Pattern recognition and machine learning, vol. 4. Springer (2006)
7. Borges, L.R.: Analysis of the wisconsin breast cancer dataset and machine learning for breast cancer detection. Group **1**(369), 15–19 (1989)
8. Dakhia, Z., Belouaar, H., Belaiche, L., Kahloul, L.: Precision unleashed: Enhancing vital sign predictions using long-short-term memory networks. In: 2023 International Conference on Innovation and Intelligence for Informatics, Computing, and Technologies (3ICT), pp. 397–401. IEEE (2023). https://doi.org/10.1109/3ICT60104.2023.10391537

9. Dora, L., Agrawal, S., Panda, R., Abraham, A.: Optimal breast cancer classification using gauss-newton representation based algorithm. Expert Syst. Appl. **85**, 134–145 (2017)
10. Eltalhi, S., Kutrani, H.: Breast cancer diagnosis and prediction using machine learning and data mining techniques: a review. IOSR J. Dental Med. Sci. **18**(4), 85–94 (2019)
11. Fellah, K.M., Tigane, S., Kahloul, L.: Diabetic retinopathy detection using deep learning. In: International Symposium on Modelling and Implementation of Complex Systems, pp. 234–246. Springer (2022). https://doi.org/10.1007/978-3-031-18516-8_17
12. Hou, R., et al.: Prediction of upstaged ductal carcinoma in situ using forced labeling and domain adaptation. IEEE Trans. Biomed. Eng. **67**(6), 1565–1572 (2019)
13. Huang, Q., Chen, Y., Liu, L., Tao, D., Li, X.: On combining biclustering mining and adaboost for breast tumor classification. IEEE Trans. Knowl. Data Eng. **32**(4), 728–738 (2019)
14. Islam, M.M., Haque, M.R., Iqbal, H., Hasan, M.M., Hasan, M., Kabir, M.N.: Breast cancer prediction: a comparative study using machine learning techniques. SN Comput. Sci. **1**, 1–14 (2020)
15. Jabbar, M.A.: Breast cancer data classification using ensemble machine learning. Eng. Appli. Sci. Res. **48**(1) (2021)
16. Kumar Mandal, S.: Performance analysis of data mining algorithms for breast cancer cell detection using naïve bayes, logistic regression and decision tree. Int. J. Eng. Comput. Sci. **6**, 2319–7242 (2017)
17. Li, L., Wu, Y., Ou, Y., Li, Q., Zhou, Y., Chen, D.: Research on machine learning algorithms and feature extraction for time series. In: 2017 IEEE 28th Annual International Symposium on Personal, Indoor, and Mobile Radio Communications (PIMRC), pp. 1–5. IEEE (2017)
18. Naji, M.A., El Filali, S., Aarika, K., Benlahmar, E.H., Abdelouhahid, R.A., Debauche, O.: Machine learning algorithms for breast cancer prediction and diagnosis. Proc. Comput. Sci. **191**, 487–492 (2021)
19. Nanglia, S., Ahmad, M., Khan, F.A., Jhanjhi, N.: An enhanced predictive heterogeneous ensemble model for breast cancer prediction. Biomed. Signal Process. Control **72**, 103279 (2022)
20. Nemade, V., Pathak, S., Dubey, A.K.: A systematic literature review of breast cancer diagnosis using machine intelligence techniques. Arch. Comput. Methods Eng. **29**(6), 4401–4430 (2022)
21. Nguyen, Q.H., et al.: Breast cancer prediction using feature selection and ensemble voting. In: 2019 International Conference on System Science and Engineering (ICSSE), pp. 250–254. IEEE (2019)
22. Obaid, O.I., Mohammed, M.A., Ghani, M.K.A., Mostafa, A., Taha, F., et al.: Evaluating the performance of machine learning techniques in the classification of wisconsin breast cancer. Inter. J. Eng. Technol. **7**(4.36), 160–166 (2018)
23. Olson, D.L., Delen, D.: Advanced data mining techniques. Springer Science & Business Media (2008)
24. Othmane, A., Youkana, I., Kahloul, L., Bourekkache, S.: Ai-based prediction for glucose levels: a comparative study of machine learning and deep learning approaches. In: Proceedings of the International Conference on Emerging Intelligent Systems for Sustainable Development (ICEIS 2024), pp. 231–245. Atlantis Press (2024). https://doi.org/10.2991/978-94-6463-496-9_18

25. Pang, T., Wong, J., Ng, W.L., Chan, C.S.: Deep learning radiomics in breast cancer with different modalities: overview and future. Expert Syst. Appl. **158**, 113501 (2020)
26. Patil, S., Kirange, D., Nemade, V.: Predictive modelling of brain tumor detection using deep learning. J. Critical Rev. **7**(04), 1805–1813 (2020)
27. Salem, T., Belaiche, L., Kahoul, M., Kahloul, L., Belouaar, H.: One-stage deep learning approach for the detection of fractures in cervical spine. In: 2023 Third International Conference on Theoretical and Applicative Aspects of Computer Science (ICTAACS), pp. 1–5. IEEE (2023). https://doi.org/10.1109/ICTAACS60400.2023.10449644
28. Salod, Z., Singh, Y.: Comparison of the performance of machine learning algorithms in breast cancer screening and detection: a protocol. J. Public Health Res. **8**(3), jphr–2019 (2019)
29. Turkki, R., et al.: Breast cancer outcome prediction with tumour tissue images and machine learning. Breast Cancer Res. Treat. **177**(1), 41–52 (2019). https://doi.org/10.1007/s10549-019-05281-1
30. UICC: Breast cancer, uicc - what we do. https://www.uicc.org/what-we-do/thematic-areas/breast-cancer Accessed Nov 05 2016
31. Witten, I.H., Frank, E.: Data mining: practical machine learning tools and techniques with java implementations. ACM SIGMOD Rec. **31**(1), 76–77 (2002)
32. Zerari, A., Djedidi, O., Kahloul, L., Carlo, R., Remadna, I.: Paediatric bone age assessment from hand x-ray using deep learning approach. In: International Conference on Computing Systems and Applications, pp. 373–383. Springer (2022). https://doi.org/10.1007/978-3-031-12097-8_32

Skin-VGG16: A Multimodal-Based Approach for Skin Cancer Classification

Nawel Zemmal[1,2](✉), Nacer Eddine Benzebouchi[2], Nabiha Azizi[2], and Rim Chaib[2]

[1] Computer Science Department, Souk Ahras University, Souk Ahras, Algeria
n.zemmal@univ-soukahras.dz

[2] LabGED Laboratory, Computer Science Department, Badji Mokhtar Annaba University, Annaba, Algeria

Abstract. Melanoma is the most aggressive type of skin cancer, significantly affecting human health and life expectancy. Automated melanoma classification has become an important challenge; a promising technique to address this issue is the use of a Deep Convolutional Neural Network (CNN). The success of CNN techniques highly depends on having a sufficient amount of skin lesion data to build a well-performing predictive model. Due to the sensitivity of medical data, the performance is often impacted by the overfitting problem. Effective solutions to this issue are firstly, data augmentation, which permits to generate a large amount of artificial data. Secondly, using the pre-trained VGG16 model, researchers can benefit from transfer learning, where the network's knowledge from a large dataset can be transferred to the skin cancer classification task. On the other hand, to improve the diagnosis precision, it is important to merge multimodal information collected from various image representation. The idea is that different imaging modalities contain extensive information that are different and complementary to each other. Motivated by these issues, this paper presents a multimodal approach based on the cooperation of four imaging modalities and using transfer learning scenario. Each pre-trained VGG16 was trained by an imaging modality, the cross-modality fusion at the decision-making level was ensured by Sum and Product functions. The experiments were carried out on 170 dermoscopic images. After a data augmentation process, results show that the VGG16 trained with multimodal images performances well compared to VGG trained with single-modal images (between 0.98% and 0.99%). These results provide new evidence for the profit of multimodal imaging in medical applications.

Keywords: Skin cancer classification · Decision support systems · VGG16 · Convolutional Neural Network (CNN) · Melanoma

1 Introduction

Melanoma is among the most common malignancies of skin cancer in white-skinned populations. Melanoma accounts for between 2 and 3% of all cancer types; it is now the seventh most commonly diagnosed cancer in the male popula-tion and the sixth in the female population worldwide [1]. Alarmingly, in 2025, it is estimated that the number

T. Marir et al. (Eds.): ICCSA 2024, ISEM 45, pp. 161–171, 2025.
https://doi.org/10.1007/978-3-031-90758-6_12

of new melanoma cases diagnosed will increase by 5.8% and the number of melanoma deaths is expected to increase by 4.8% [2]. Early diagnosis can be established for a treatment before it spreads to other organs. To identify, an expert visually examines whether the suspicious lesion is melanoma or not. However, this traditional approach is time-consuming, and the results may be subject to the pathologist's subjectivity [3]. Facing these problems, the development of automatic diagnosis systems are of great interest. These systems have the potential to increase the diagnosis quality and speed, decrease the specialist errors and detect the skin cancer at early stage [4, 5].

Recently, automated systems based on neural network and deep learning have been used to improve diagnostic ability and have achieved great performances [5, 6]. Convolutional neural network (CNN) [7] is a deep learning model that has been widely applied for melanoma diagnosis from dermoscopic images [8, 9]. Despite the success of CNNs, they are still limited in diagnosing melanoma mainly because they need a huge amount of data to build accurate models. A very common problem in medical image analysis is that the collection of high-quality ground-truth data is sensitive and very human-dependent, which makes CNN quite challenging to apply in this field. To address the problem of insufficient medical training sets, VGG16 and data augmentation techniques for skin cancer classification was opted. We aim to overcome the challenge of limited labeled data by enhancing the model's ability to generalize and learn robust features from a smaller dataset. VGG16, a deep convolutional neural network based approach offers a powerful feature extraction capability, enabling the model to capture different patterns in dermatoscopic images. Additionally, data augmentation techniques such as rotation, flipping, and scaling can artificially increase the diversity of the training dataset, thus enhancing model's predictive performance. This approach not only addresses the scarcity of labeled data in skin cancer classification but also contributes to the development of more accurate and reliable classification models in medical image analysis [10, 11].

Furthermore, current published methods using VGG16 usually learn data representations just from a single modality, which can limit the learning ability of the deep model. Although, in some pathology, it is not possible to obtain a reliable diagnostic only by using a single image source. The premise is that various imaging modalities contain abundant information, which are different and complementary to each other. Recently, deep learning on multi-modal images, also mentioned as multi-view images has gained researcher's traction.

The focus of this paper is to investigate the importance of both data augmentation and multimodal dermoscopic imaging representation in a transfer-learning scenario for the task of automated skin lesion classification. Based on these challenges, the proposed Skin-VGG16 model includes the cooperation of different image modality; these modalities are obtained from different images representation techniques. Moreover, we explore the use of texture based image representation modality, which is more reflective of a skin cancer screening setting. The predictions are derived from a multimodal convolutional neural network where each convolutional neural network considers an imaging-modality. A cross-modality fusion at the decision-making level was ensured using different fusions functions.

The remainder of this paper is structured as follows: Sect. 2 explains the details of the proposed approach and the material used. Section 3 describes the employed experimental metrics, the experimental results and the discussion of the proposed model. Finally, the conclusion is detailed in Sect. 4.

2 Methods

The objective of our work is to design and implement a multimodal classification system for dermatological images using the pre-trained VGG16 model. This system is based on studying the cooperation of multiple modalities of skin images to effectively detect melanoma. We also aim to analyze the importance of texture as a significant source of medical data representation and demonstrate the role of classifier fusion from multiple information sources.

The general process of our system is summarized in Fig. 1.

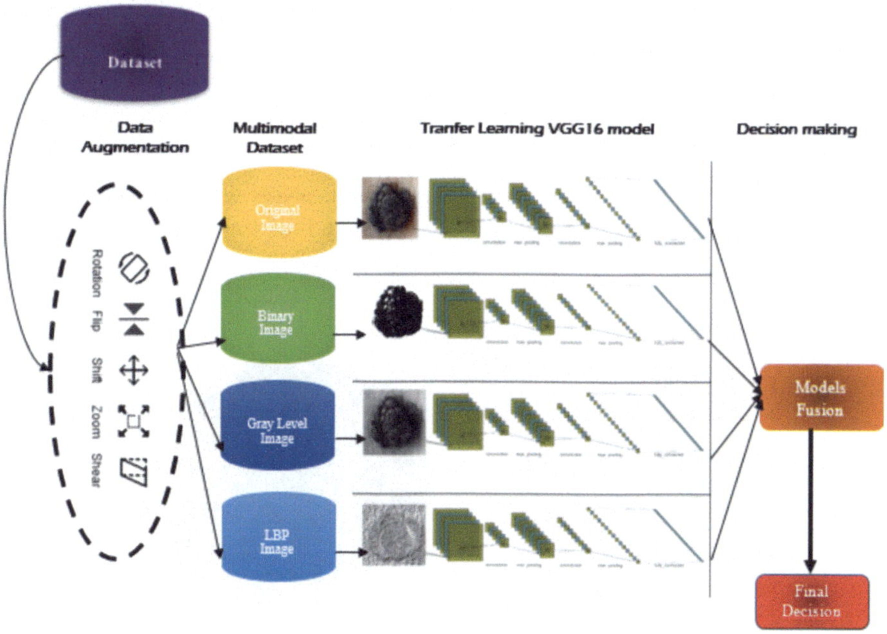

Fig. 1. The proposed system.

2.1 Data Augmentation

The dataset of 170 images is too small for designing a robust medical decision system. Deep learning typically relies on extensive datasets to ensure effective learning and mitigate the risk of overfitting to the training data. Consequently, data augmentation methods are often used during training to expand the original dataset. In our work, we employ the following data augmentation techniques: horizontal flip, vertical mirror, zoom, 180° rotation, and shear. After augmentation, the resulting dataset comprises 622 melanoma images and 895 nevus images, totaling 1,517 images.

2.2 Multimodal Data Generation

The growing diversity of image acquisition techniques has, in recent years, motivated extensive research aimed at developing increasingly effective data fusion models. In the medical field, each type of image is assigned to an expert who provides a partial diagnosis based on their specialty. Specialists then exchange their findings, leading to the final diagnosis. Handling different image varieties is challenging for a single expert. Information fusion involves combining imperfect and heterogeneous information to obtain a more comprehensive and higher-quality global insight, facilitating better decision-making.

In our work, we focus on the parallel combination of classifiers, which is our primary objective. This high-level decision fusion operates in the decision space by associating elaborated information, considered as decision proposals. The goal is to provide a global decision based on the input decisions.

In this work, we opted for three imaging modalities: binary imaging, grayscale imaging, and imaging based on the LBP method, along with the original images from the dataset.

Binary Images. For the binary representation of our database, we applied the standard OTSU algorithm, which is recognized for its effective binarization performance (see Fig. 2).

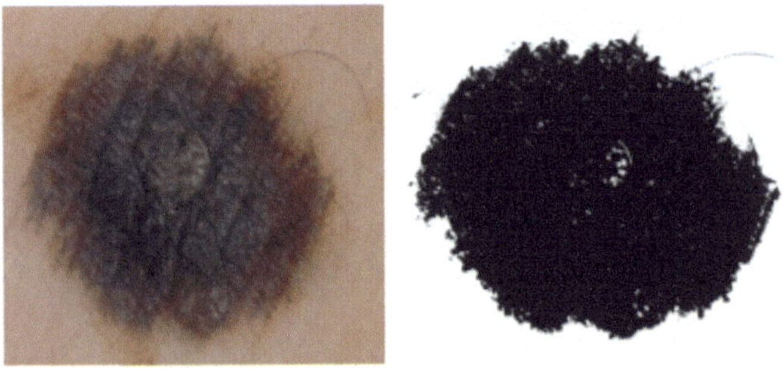

Fig. 2. Binary transformation.

Grayscale Images. Grayscale transformation is a process that converts color images into shades of gray, allowing each pixel to represent only intensity information. This transformation simplifies the image by removing color details while preserving essential features. For the grayscale transformation, we used the "EasyGraphicConverter" tool (see Fig. 3).

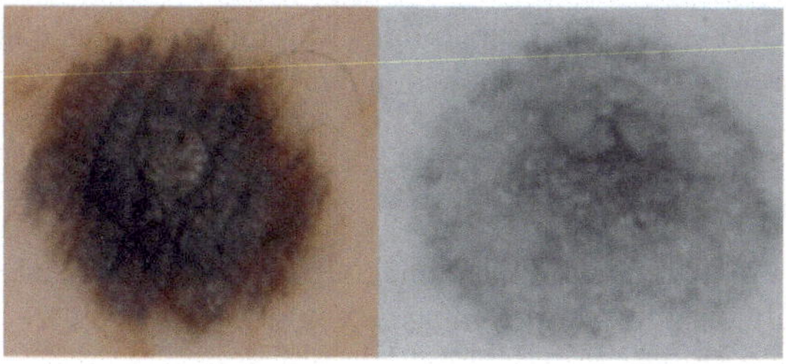

Fig. 3. Grayscale transformation.

LBP Based Images. Local Binary Patterns (LBP) are widely recognized texture descriptors in the fields of computer vision and image processing. It assigns labels to the pixels of an image by applying a threshold to the surrounding pixel neighborhood, subsequently interpreting the outcome as a binary number (see Fig. 4). The key steps of the LBP method are as follows:

1. *Grayscale Conversion:* Convert the input image to grayscale if it is a color image.
2. *Define the Neighborhood:* For each pixel in the grayscale image, define a neighborhood. In our case a 3 × 3 neighborhood around the pixel was used.
3. *Thresholding:* For each pixel p in the image, compare its intensity value with the intensity value I(p) of its neighbors. If the neighbor's intensity is greater than or equal to the center pixel's intensity, assign it a value of 1; otherwise, assign it a value of 0.

Mathematically, for a given pixel I(p) with intensity $\{p_0, p_1, p_2, \ldots, p_7\}$ and its neighbors in a 3 × 3 neighborhood:

$$LBP(p) = \sum_{i=0}^{7} s(I(p_i) - I(p)).2^i \tag{1}$$

where s(x) is the thresholding function defined as:

$$s(x) = \begin{cases} 1 \; if \;\; x \geq 0 \\ 0 \; if \;\; x < 0 \end{cases} \tag{2}$$

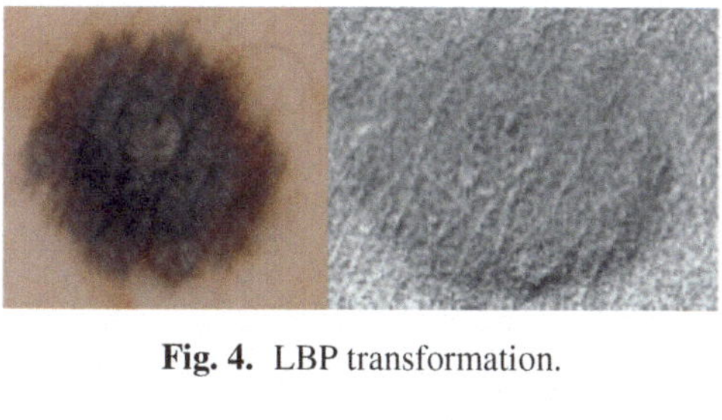

Fig. 4. LBP transformation.

Binary images can capture essential shape and boundary information of skin lesions, while gray-level images provide detailed texture and intensity features. LBP images, known for their texture description capabilities, can further enhance the model's ability to capture local patterns and variations within the skin lesion.

2.3 Transfer Learning VGG16 Model

The VGG-16 refers to a deep learning architecture based on the Visual Geometry Group Network, commonly used for various image recognition tasks.

- **Input Layer:**
  Let X represent the input image.
- **Convolutional Layers:**

VGG16 consists of several convolutional blocks, each block contain multiple convolutional layers followed by activation functions (typically ReLU) and optional batch normalization.

- *Convolutional operation*

$$Z_{i,j,k} = \sum_{m=0}^{F-1} \sum_{n=0}^{F-1} \sum_{c=0}^{C-1} X_{i+m,j+n,c} \cdot W_{m,n,c,k} + b_k \quad (3)$$

where,
$Z_{(i,j,k)}$ is the output feature map
X is the input feature map,
W is the filter weights,
F is the filter size,
C represents the number of channels,
b is the bias term.
Max Pooling Layers:

Used for down-sampling feature maps

- **Max pooling operation:**

$$P_{i,j,k} = \max_{m,n}\left(A_{i+m,j+n,k}\right) \quad (4)$$

where,
P_(i,j,k) pooled output
A: Input to the pooling layer
m,n: Pooling window size

- **Fully Connected (Dense) Layer:**
  After the convolutional and max-pooling layers, flatten the output feature maps to feed them into fully connected layers. Introduce one or more fully connected layers to learn high-level representations of the image features.

$$Z = W.X + b \quad (5)$$

where,
Z: Output of the fully connected layer
W: Weights matrix
X: Input vector
b: Bias term

- **Output Layer:**
  Introduce an output layer with a single neuron and a sigmoid activation function for binary classification or softmax activation for multi-class classification:

  - *Sigmoid activation:*

$$\sigma(z) = \frac{1}{1 + e^{-z}} \tag{6}$$

    where, $\sigma(z)$ is the probability of tumor presence.
  - *Softmax activation:*

$$\sigma(z)_j = \frac{e^{z_j}}{1 + e^{-z}} \tag{7}$$

    where,
    $\sigma(z)_j$: Probability of class j
    z_j: Input to the softmax function for class j
    K: Number of classes

3 Results and Discussion

This section describes the used dataset and the obtained results of the proposed approach.

3.1 Used Dataset

The dataset comprises 170 images (70 melanoma images and 100 nevus images) from the digital image archives of the Department of Dermatology at the University Medical Center Groningen (UMCG). All images are part of the much larger digital dermatology archives containing over 50,000 images of various types of lesions. For each image, the diagnosis was verified through medical correspondence from the Department of Dermatology. Hair is also considered a distracting element and was manually removed using Dullrazor software [12]. The corresponding database contains patient records classified into two classes: Melanoma and Nevus.

3.2 Parameters Setting

In order to summarize the parameters of our convolutional neural network, these parameters have enabled us to achieve promising results. Table 1 presents each hyper-parameter along with the corresponding value.

3.3 Experimental Settings

Several performance metrics are used to assess the effectiveness of a learning model. In the context of medical data, important metrics include accuracy, misclassification rate, F-measure, specificity, sensitivity and G-mean. The evaluation process typically involves dividing the dataset into training and testing subsets, followed by the implementation of 10-fold cross-validation. This k-fold cross-validation technique enhances the reliability of the results and helps mitigate the risk of model overfitting.

Table 1. Parameters of the used vgg16 model.

Layer Name	Description	Output Shape
Conv1_1	3 × 3 convolution, 64 filters, ReLU	(224, 224, 64)
Conv1_2	3 × 3 convolution, 64 filters, ReLU	(224, 224, 64)
Conv2_1	3 × 3 convolution, 128 filters, ReLU	(112, 112, 128)
Conv2_2	3 × 3 convolution, 128 filters, ReLU	(112, 112, 128)
Conv3_1	3 × 3 convolution, 256 filters, ReLU	(56, 56, 256)
Conv3_2	3 × 3 convolution, 256 filters, ReLU	(56, 56, 256)
Conv3_3	3 × 3 convolution, 256 filters, ReLU	(56, 56, 256)
Conv4_1	3 × 3 convolution, 512 filters, ReLU	(28, 28, 512)
Conv4_2	3 × 3 convolution, 512 filters, ReLU	(28, 28, 512)
Conv4_3	3 × 3 convolution, 512 filters, ReLU	(28, 28, 512)
Conv5_1	3 × 3 convolution, 512 filters, ReLU	(14, 14, 512)
Conv5_2	3 × 3 convolution, 512 filters, ReLU	(14, 14, 512)
Conv5_3	3 × 3 convolution, 512 filters, ReLU	(14, 14, 512)

3.4 Results and Analysis

This section outlines and evaluate the performance of the proposed method. A detailed discussion of the parameters essential to the development of the Skin-VGG16 framework is also provided.

We performed the fusion of the outputs of the four VGG16 models, which are measurement-type, using the "Sum" and "Product" combination functions. The diagnosis will be assigned to the class with the highest overall probability. Table 2 shows the obtained results of the proposed multimodal approach against single modal approaches.

Table 2. Obtained results of the proposed multimodal approach against single modal approaches.

	ACC	*MC*	*F-Score*	*Sen*	*Spe*	*G-Mean*
Original-VGG16	0.968	0.311	0.961	0.974	0.964	0.969
Binary-VGG16	0.949	0.507	0.937	0.957	0.943	0.950
Grayscale-VGG16	0.956	0.442	0.945	0.966	0.941	0.957
LBP-VGG16	**0.978**	0.211	**0.981**	0.975	**0.983**	0.979
Sum_fusion function	**0.989**	**0.105**	**0.987**	**0.992**	**0.987**	**0.990**
Product_fusion function	**0.992**	**0.008**	**0.993**	**0.991**	**0.995**	**0.994**

Based on the results shown in Table 2, we observe that local binary patterns yield highly satisfactory outcomes compared to other image modalities. This confirms that choosing texture descriptors as an additional tool is an effective approach to enhance recognition performance. Additionally, we observe that classification model performance improves across various metrics after data augmentation. Increasing the dataset allows the VGG16 classifier to enrich the information for classification and robustly classify tumors, even in different orientations, due to its invariance property. These augmented data will be retained for the fusion phase.

In terms of fusion techniques, Table 2 shows that both fusion techniques enhance model performance significantly, but the Product fusion function consistently yields better results compared to the Sum fusion function. The Product fusion achieves a higher accuracy of 99.2% and superior specificity at 99.5%, indicating a more robust classification capability. While the Sum fusion function also performs well with an accuracy of 98.9%, it falls short in specificity at 98.7%.

The confusion matrix for the proposed VGG16 model in multimodal skin cancer classification is presented in Fig. 5, which provides a detailed breakdown of the model's performance. By analyzing the confusion matrix, we can derive important performance metrics. These metrics help in understanding the strengths and weaknesses of the model in distinguishing between malignant and benign skin lesions.

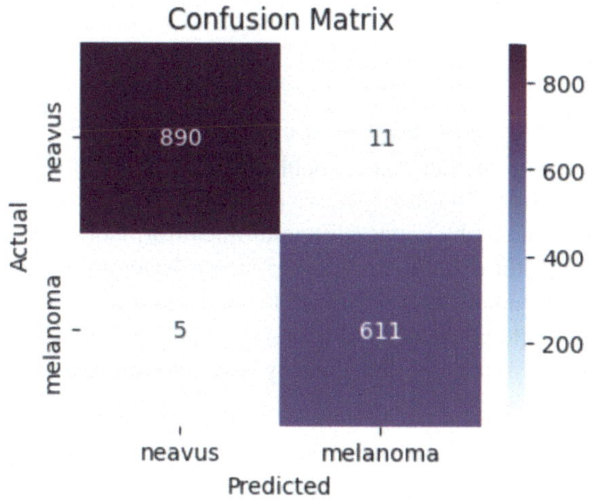

Fig. 5. Confusion Matrix of the multimodal Skin-VGG16 approach.

Combining different imaging modalities with VGG16 can enhance classification performance. This multimodal approach allows the model to capture a more comprehensive set of features.

Areas under the ROC curves (AUC) are calculated to evaluate classifier performance across different training/test partitions. A larger AUC indicates better classifier performance. Figure 6 shows the obtained AUC for the proposed approach.

When applied a pre-trained VGG16 models to multimodal skin cancer classification, demonstrate high accuracy, effective feature extraction, and adaptability to various imaging types. This approach leverages the power of transfer learning to overcome limitations in dataset size and variability in skin lesion appearances, making it a promising tool for improving skin cancer diagnosis.

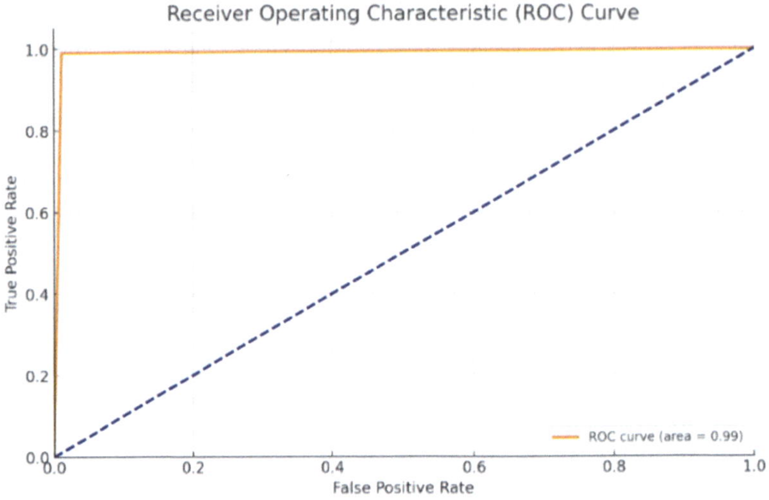

Fig. 6. ROC-Curve of the proposed approach.

4 Conclusion

Melanoma is a specific type of skin cancer characterized by rapid tumor development and metastasis. Cutaneous melanoma is a public health concern, and its early detection is crucial for patient recovery. In this paper, we proposed a multimodal approach based on a pre-trained VGG16 model by cooperating multiple information sources. We analyzed texture aspects as an additional modality containing relevant information (LBP).

Finally, to better evaluate our system, we fused the outputs of four VGG16 models. This fusion provides a global decision based on the input decisions. The main advantage of this fusion is to exploit the complementarity between different classifiers to enhance recognition.

References

1. World Health Organization: Cancer Today. International Agency for Research on Cancer (2021). https://gco.iarc.fr/today/data/factsheets/cancers/16-Melanoma-skin-fact-sheet.pdf. Accessed 16 July 2024
2. American Cancer Society: Melanoma Skin Cancer Statistics (2021). https://www.cancer.org/research/cancer-facts-statistics.html. Accessed 16 July 2024
3. Rushabh, R., Mahajan, T., Chhatlani, J., Bansode, A., Bhatia, G.: Early diagnosis of melanoma by augmenting feature extraction of epidermis using faster region-based convolutional neural networks. In: 2022 IEEE Region 10 Humanitarian Technology Conference (R10-HTC), pp. 142–147 (2022). https://doi.org/10.1109/R10-HTC54060.2022.9929864
4. Kandhro, I.A., et al.: Performance evaluation of E-VGG19 model: enhancing real-time skin cancer detection and classification. Heliyon **10**(5), e31488 (2024). https://doi.org/10.1016/j.heliyon.2024.e31488
5. Zemmal, N., Azizi, N., Ziani, A., Benzebouchi, N.E., Aldwairi, M.: An enhanced feature selection approach based on mutual information for breast cancer diagnosis. In: 2019 6th

International Conference on Image and Signal Processing and Their Applications (ISPA), pp. 1–6 (2019). https://doi.org/10.1109/ISPA48434.2019.8966803
6. Supriyanto, C., Salam, A., Zeniarja, J., Wijaya, A.: Two-stage input-space image augmentation and interpretable technique for accurate and explainable skin cancer diagnosis. Computation **11**(12), 246 (2023). https://doi.org/10.3390/computation11120246
7. Benzebouchi, N.E., Bourzama, M.O., Zemmal, N., Azizi, N.: Deep classifier fusion-based multi-classification of chest x-rays for COVID-19 detection. In: 2023 International Conference on Decision Aid Sciences and Applications (DASA), pp. 495–499 (2023). https://doi.org/10.1109/DASA.2023.1234567
8. Haggenmüller, S., et al.: Skin cancer classification via convolutional neural networks: systematic review of studies involving human experts. Eur. J. Cancer **156**, 202–216 (2021). https://doi.org/10.1016/j.ejca.2021.06.049
9. Gillstedt, M., et al.: Evaluation of melanoma thickness with clinical close-up and dermoscopic images using a convolutional neural network. Acta Dermato-Venereologica **102** (2022). https://doi.org/10.2340/actadv.v102.2681
10. Zemmal, N., Azizi, N., Sellami, M., Cheriguene, S., Ziani, A.: A new hybrid system combining active learning and particle swarm optimisation for medical data classification. Int. J. Bio-Inspired Comput. **18**(1), 59–68 (2021)
11. Zhao, J., Lui, H., Kalia, S., Lee, T.K., Zeng, H.: Improving skin cancer detection by Raman spectroscopy using convolutional neural networks and data augmentation. Front. Oncol. **14** (2024). https://doi.org/10.3389/fonc.2024.1320220
12. Giotis, I., Molders, N., Land, S., Biehl, M., Jonkman, M.F., Petkov, N.: MED-NODE: a computer-assisted melanoma diagnosis system using non-dermoscopic images. Expert Syst. Appl. **42**(19), 6578–6585 (2015). https://doi.org/10.1016/j.eswa.2015.04.034

Recent Advances in Hair Removal in Dermoscopy Images Using Deep Learning

Dalal Bardou[1(✉)], Karima Saidi[2], Hichem Rahab[3], Laishui Lv[4], and Ting Zhang[5]

[1] LMIA, Department of Computer Science, Abbes Laghrour University of Khenchela, 40004 Khenchela, Algeria
dalal.bardou@univ-khenchela.dz

[2] Mathematics and Computer Science Faculty, Batna2 University, 05000 Batna, Algeria
karima.saidi@univ-batna2.dz

[3] ICOSI Lab, Department of Computer Science, Abbes Laghrour University of Khenchela, 40004 Khenchela, Algeria
rahab.hichem@univ-khenchela.dz

[4] College of Information Engineering, China Jiliang University, Hangzhou 310018, China
21a0304181@cjlu.edu.cn

[5] College of Computer and Electronic Information, Nanjing Normal University, Nanjing 210098, China
zting135@nnu.edu.cn

Abstract. The presence of hair in dermoscopy images poses a challenge to melanoma detection tasks, as it can cause occlusion of crucial diagnostic information. In this paper, we explore the deep-learning approaches intended to address this issue. Specifically, we review notable techniques used for hair removal such as convolutional neural networks, generative models, recurrent neural networks, and autoencoders. We give an overview of these techniques and explore their application, either combined or separately, in the context of hair removal in dermoscopy images. Moreover, we highlight the challenges and discuss future directions.

Keywords: Dermoscopy images · Hair removal · Deep learning · Generative models · Neural networks

1 Introduction

Dermoscopy is a non-invasive imaging technique that helps clinicians to visualize skin lesions more clearly [1]. The emergence of high-quality digital devices in recent years, including cameras, has radically changed the way we use dermoscopy images. These developed devices have enabled dermatologists to collect precise, high-clarity images of skin lesions. These images have empowered the research community with extensive sets of skin images, aiding in the automatic detection and diagnosis of various skin conditions, such as melanoma [2, 3].

Automatic diagnosis of melanoma aims to provide clinicians with tools to efficiently detect and diagnose melanoma at an early stage. The presence of hair in dermoscopy images presents a challenging step that must be addressed before proceeding with automatic diagnosis. It is a preprocessing step for dermoscopy images [4–6].

T. Marir et al. (Eds.): ICCSA 2024, ISEM 45, pp. 172–181, 2025.
https://doi.org/10.1007/978-3-031-90758-6_13

Deep learning has emerged as a promising solution to the challenge of hair occlusion in dermoscopy images, exploiting the capability of neural networks to automatically learn and extract relevant hair features. Convolutional neural networks (CNNs) and recurrent neural networks (RNNs), two classes of deep learning models, have shown great success in tackling hair removal [6]. They have demonstrated a great ability to learn hierarchical representations of different image features and identify skin regions and hair regions. Generative models, another class of deep learning models that includes many types, such as generative adversarial networks (GANs) and autoencoders, have also been used to generate hair-free dermoscopy images [3, 7–12].

This paper summarizes the deep learning methods used for hair removal in dermoscopy images. It also discusses the application of various well-known deep learning models, including CNNs, generative models, RNNs, and autoencoders, while covering their strengths, limitations, future directions, and challenges.

2 Deep Learning for Hair Removal

Deep learning presents a powerful set of techniques that take advantage of various types of collected datasets. These techniques use neural network architectures composed of interconnected layers of neurons to process and transform input data. These neural networks are learnable in the sense that they can automatically recognize patterns and make predictions from data through the process of training [13]. Their ability to learn without being explicitly programmed makes them well-suited for many tasks such as image processing, natural language processing, and speech recognition, among others [13]. In this section, we will introduce the different deep learning techniques employed specifically for addressing the challenge of hair removal in dermoscopy images. Table 1 summarizes all the deep learning methods proposed by researchers in recent years. While most of these methods handled the problem of hair removal as one task, others developed modules for hair segmentation and others for gap inpainting, which we will detail in the coming subsections.

Table 1. The deep learning methods proposed by researchers for hair removal in dermoscopy images.

Authors	Task	Technique	Year
Attia et al. [6]	Hair segmentation	CNN-RNN	2019
Talavera et al. [7]	Hair removal	convolutional autoencoder	2020
Kim et al. [8]	Hair removal	Adversarial generative learning	2021
Li et al. [9]	Hair removal	CNN-GAN	2021
Bardou et al. [3]	Hair removal	Variational autoencoder	2022
El-Shafai et al. [10]	Hair removal	GAN-CNN	2023
Konstantinos et al. [11]	Hair segmentation	convolutional autoencoder	2023
Bardou et al. [12]	Hair removal	Combined GAN-VAE	2024

2.1 Convolutional with Recurrent Neural Networks

Convolutional Neural Networks (CNNs) are a class of deep learning models specifically designed for processing structured grid-like data, such as images. CNNs employ convolutional layers, which perform the mathematical operation of convolution. This operation allows to extract features at different spatial locations [13, 14].

On the other hand, recurrent neural networks (RNNs) are another class of deep learning employed in sequence data processing tasks [15]. RNNs use connections to form directed cycles to describe temporal behavior and represent sequential dependencies by preserving the memory of past inputs through hidden states [16].

Attia et al. [6] combined CNNs and RNNs in a hybrid approach for hair segmentation (see Fig. 1). The authors stated that conventional methods for hair segmentation can't handle varying hair characteristics such as color, texture, and thickness, which often lead to over-segmentation or under-segmentation. To address these challenges, they integrated CNNs and RNNs. Specifically, they employed a hybrid network comprising convolutional layers for feature extraction and recurrent layers to model spatial dependencies between disjointed patches of hair. This approach enabled the efficient detection of hair without the need for manual threshold tuning.

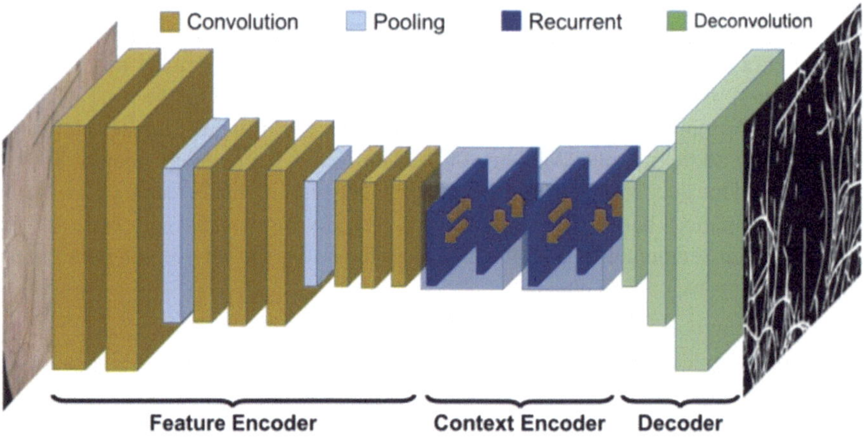

Fig. 1. Illustration of the architecture proposed by Attia et al. for hair segmentation [6].

2.2 Generative Models

Generative models are a class of deep learning models that have recently emerged and have shown great success in many fields such as natural language processing, image generation, image inpainting, and style transfer, among others [17]. Generative models encompass different deep learning architectures that aim to generate new data samples similar to the training data they were trained on. Some of the most commonly used generative models include Generative Adversarial Networks (GANs) and Autoencoders

(AEs). GANs and AEs have been particularly useful for tasks like hair removal, and we will detail how they were applied in the coming subsections.

Autoencoders

Autoencoders are a fundamental neural network architecture primarily used for unsupervised learning. They consist of an encoder and a decoder, working together to reconstruct input data by learning a condensed representation in a lower-dimensional space [3, 17] (see Fig. 2). The encoder compresses the input into this latent space, while the decoder reconstructs the original input. By minimizing reconstruction error, autoencoders capture key input features. They are used in tasks like dimensionality reduction, data denoising, and feature learning, serving as a foundation for advanced deep learning techniques. There exist many variants of autoencoders including convolutional autoencoders [18] and variational autoencoders [19–21], etc.

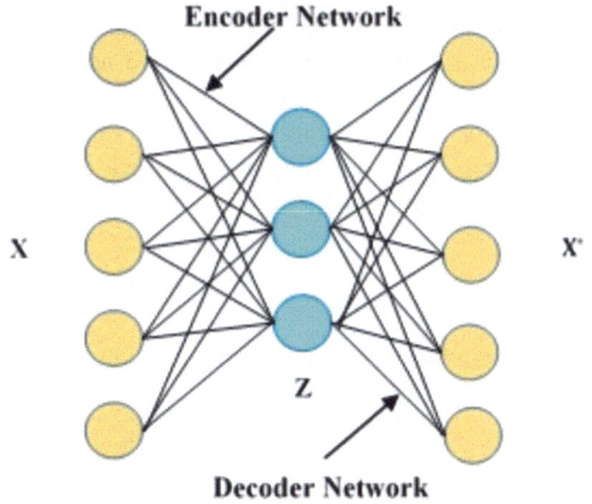

Fig. 2. The architecture of an autoencoder network, where x is the input, z is the latent representation, and x* is the output.

Convolutional autoencoders are autoencoders that use convolutional layers in their architecture and handle high-dimensional input data with spatial relationships. This design allows them to capture local patterns and spatial dependencies more effectively compared to basic autoencoders. Talavera-Martinez et al. [7] introduced a convolutional autoencoder architecture designed specifically for hair removal. This architecture employs convolutional neural networks for both encoding and decoding processes. They also adopted a more efficient objective function that integrates five loss functions, as opposed to a singular one. This chosen objective function leverages both per-pixel loss and perceptual loss for enhanced performance.

On the other hand, convolutional autoencoders were used by Konstantinos et al. [11] for hair segmentation. The authors utilized a well-known architecture named DeepLabv3+, developed by Chen et al. [22]. DeepLabv3+ is a semantic segmentation neural network architecture and is designed to perform pixel-level classification in images, particularly for tasks like object detection and segmentation. Following the hair segmentation step, the authors employed the coherent transport [23] technique to fill the hair gaps.

Variational autoencoders are a type of generative model based on neural net-works, particularly autoencoders. Like traditional autoencoders, VAEs consist of an encoder and a decoder, but they are designed to learn a latent space representation of the input data that captures its underlying structure in a probabilistic manner. So, the latent space follows a probability distribution, usually a Gaussian distribution with a mean and a variance. Bardou et al. [3] employed variational autoencoders for hair removal where hair is considered noise during the learning process. The developed model ignores hair during training and produces hair-free images. Similar to [7], the authors used a hybrid loss function with many terms.

Generative Adversarial Networks

Generative Adversarial Networks (GANs) represent a cutting-edge class of deep learning models that are unique in their architecture, comprising two neural net-works, a generator, and a discriminator [23] (see Fig. 3). The two networks are engaged in a game-like scenario. The generator aims to produce realistic data samples from random noise, while the discriminator aims to distinguish between real and generated samples. Through adversarial training, GANs continuously improve the generator's ability to produce increasingly realistic outputs [17].

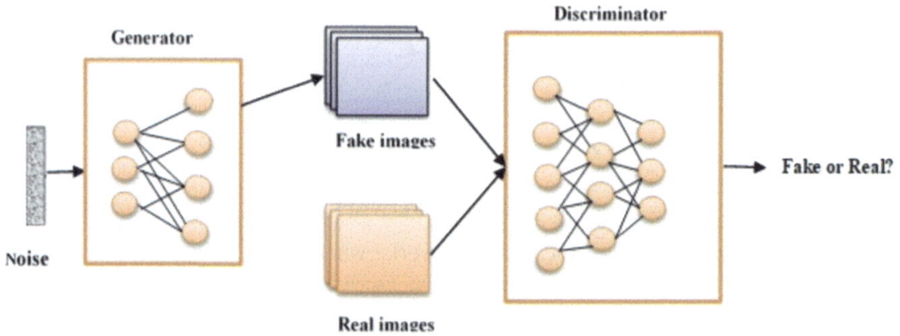

Fig. 3. The architecture of GANs.

Kim et al. [8] introduced an unsupervised feature elimination algorithm utilizing adversarial generative learning. They applied this algorithm to dermoscopy images for hair removal. Unlike traditional GANs, which typically consist of a generator and a discriminator, the authors devised a deep learning model combining a reconstructor instead of a generator to reconstruct hair-free images, along with a discriminator. Additionally, the reconstructor architecture incorporates convolutional autoencoders (Fig. 4).

Hybrid Approaches

Hybrid deep learning approaches for hair removal involve combining various deep learning techniques to effectively remove hair from images. These approaches typically integrate multiple models or methods to enhance the accuracy and performance of the hair

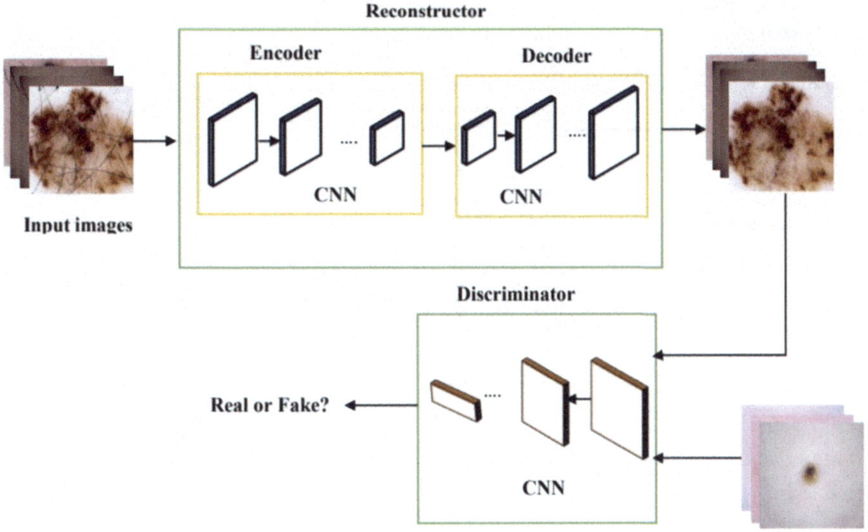

Fig. 4. The adversarial generative learning-based architecture proposed for hair removal by Kim et al. [8]. The detailed architecture of both the reconstructor and discriminator can be found in [8].

removal process. El-Shafai et al. [10] proposed a hybrid approach that uses GANs with CNNs for hair removal. As shown in Fig. 5, the authors used a generator with a modified U-net network to generate hair-free images. The U-Net network can produce precise outcomes even when training data is limited.

Alternatively, Li et al. [9] utilized deep learning techniques to independently tackle hair segmentation and gap inpainting tasks (see Fig. 6). To accomplish hair segmentation, they employed the U-net framework for labeling image pixels as either hair or skin pixels, generating hair masks through semantic segmentation. In the realm of gap inpainting, transfer learning was employed, leveraging generative adversarial networks. The authors utilized a pre-trained free-form system, based on Gated convolution and SN-PatchGAN [25], to fill in the gaps.

In [12], Bardou et al. combined VAE and GAN with disentangled representation to remove hair and generate hair masks. Their proposed model consists of a generator and a discriminator. The generator comprises one encoder that disentangles two outputs, and two decoders. One decoder is a VAE, which is used to construct hair-free images, while the other generates hair masks.

3 Discussion

While deep learning has been in existence for many years, it is noteworthy that its application specifically for hair removal began relatively recently, around 2019, with only a limited number of studies, totaling seven. Compared to the extensive research conducted in melanoma detection, the exploration of deep learning for hair removal has been relatively limited. Given the significance of accurate melanoma diagnosis and the potential impact of hair removal in enhancing diagnostic accuracy, the research

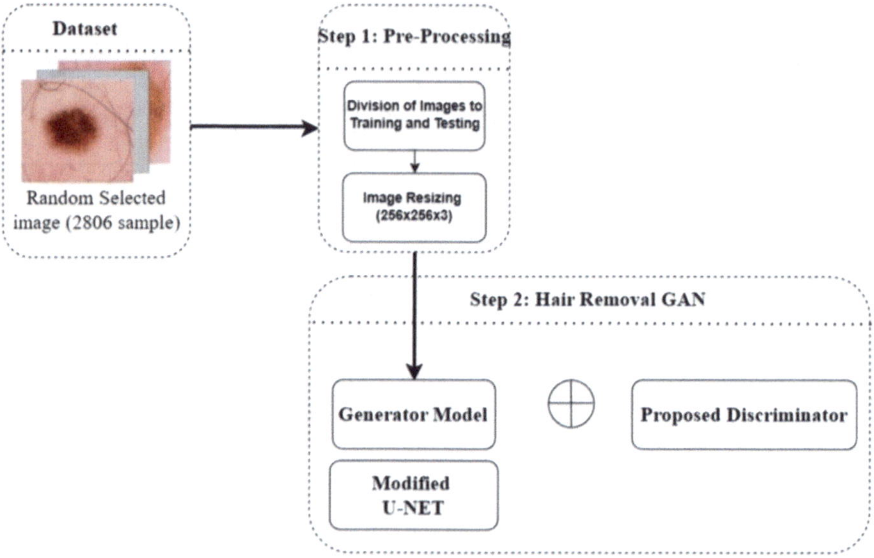

Fig. 5. The hybrid approach combining GAN and CNN proposed by El-Shafai et al. [10].

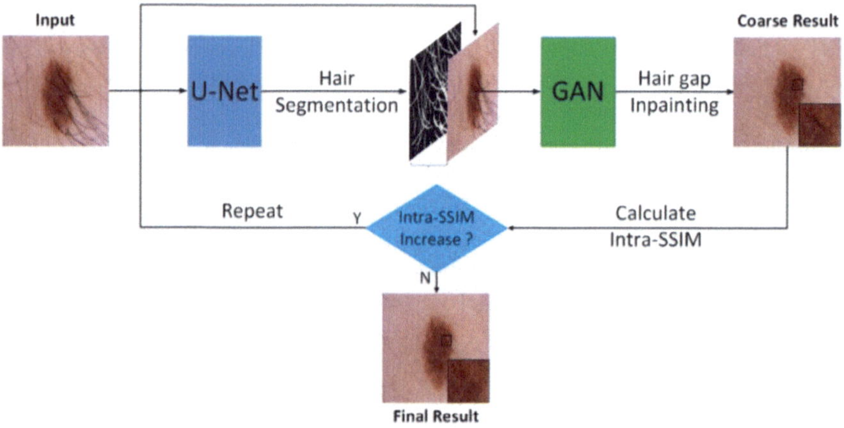

Fig. 6. The hybrid approach combining CNN and GAN proposed by Li et al. [9].

community needs to devote more attention and effort to advancing techniques for hair removal using deep learning.

The existing methods for hair removal using deep learning can be classified into two main categories: single-task and double-task approaches. In single-task approaches, the focus is on either hair segmentation or hair gap filling. Hair segmentation involves identifying and delineating the regions of hair within dermoscopy images, while hair gap filling aims to reconstruct the areas where hair has been removed, typically using techniques such as inpainting.

On the other hand, double-task approaches address both hair segmentation and hair gap filling simultaneously. These methods employ a unified model or multiple modules within a single framework to tackle both tasks in an integrated manner. Furthermore, within each approach, there can be variations in the design of the model. Some methods propose a single model to handle both tasks, while others may employ separate modules specialized for hair segmentation and gap filling.

Given the unavailability of both the code and dataset for the proposed models, direct comparisons were not possible. Nevertheless, efforts were made to compare them based on the results obtained. The variational autoencoder architecture proposed by Bardou et al. [3] efficiently removed hair; however, it resulted in generated images of reduced quality, characterized by blurriness. On the other, the convolutional autoencoder architecture proposed by Talavera-Martinez et al. [7] provided better results when compared visually. Bardou et al. [12] enhanced their approach by combining VAE with GAN to enhance the results achieved by VAE. The architecture proposed by Li et al. [9] can be considered the best since, during training, only hair pixels are used in the process after hair segmentation, and skin pixels do not interfere in the process. The solution developed by Kim et al. [8] also provided good results through the use of adversarial generative learning. El-Shafai et al. [10] proposed a similar architecture that utilizes a CNN and GAN but in a different manner. For hair segmentation, CNN was combined with RNN, and U-Net, another CNN architecture, was also employed for the same task. It is noteworthy that various combinations of deep learning architectures can be explored, demonstrating their flexibility.

4 Challenges and Future Directions

One of the most significant challenges researchers face is providing meaningful comparisons due to the unavailability of codes of data used in the proposed studies. It's crucial to ensure that data is accessible and standardized to enable effective comparisons. By making data more readily available and setting standards for its use, researchers can facilitate more efficient and reliable comparisons between different methods and approaches. Another significant challenge lies in establishing trust in artificial intelligence solutions to the extent that they can be confidently integrated into clinical practice. It is not easy to directly use AI-based tools in a sensitive field like medicine, where a small error can cost lives. Thus, it is imperative to ensure the accuracy, reliability, and safety of these tools with patients before deploying them.

For future directions, the exploration of combinations of different deep learning models, including newly emerged ones such as diffusion models, can present a promising opportunity to develop more accurate models and create hybrid models that harness the strengths of various architectures. Diffusion models, for example, could complement existing deep-learning approaches by providing more robust representations.

5 Conclusion

In this paper, we have presented the challenge of hair occlusion in dermoscopy images and its impact on melanoma detection. We have discussed the various deep-learning techniques proposed to tackle this challenge. These techniques accurately remove hair from dermoscopy images, thereby improving the reliability of melanoma diagnosis. Moreover, we have highlighted potential future directions and challenges in this field.

References

1. Karen, S., Fleming, C.: An introduction to dermoscopy. Dermatol. Nurs. **12**(4), 20–26 (2013)
2. Goyal, M., et al.: Artificial intelligence-based image classification methods for diagnosis of skin cancer: challenges and opportunities. Comput. Biol. Med. **127**, 104065 (2020)
3. Bardou, D., et al.: Hair removal in dermoscopy images using variational autoencoders. Skin Res. Technol. **28**(3), 445–454 (2022)
4. Abbas, Q., Celebi, M.E., García, I.F.: Hair removal methods: a comparative study for dermoscopy images. Biomed. Sig. Process. Control **6**(4), 395–404 (2011)
5. Koehoorn, J., et al.: Automated digital hair removal by threshold decomposition and morphological analysis. In: Mathematical Morphology and Its Applications to Signal and Image Processing: 12th International Symposium, ISMM 2015, Reykjavik, Iceland, 27–29 May 2015. Proceedings 12. Springer, Cham (2015)
6. Attia, M., et al.: Digital hair segmentation using hybrid convolutional and recurrent neural networks architecture. Comput. Methods Programs Biomed. **177**, 17–30 (2019)
7. Talavera-Martinez, L., Bibiloni, P., Gonzalez-Hidalgo, M.: Hair segmentation and removal in dermoscopic images using deep learning. IEEE Access **9**, 2694–2704 (2020)
8. Kim, D., Hong, B.-W.: Unsupervised feature elimination via generative adversarial networks: application to hair removal in melanoma classification. IEEE Access **9**, 42610–42620 (2021)
9. Li, W., et al.: Digital hair removal by deep learning for skin lesion segmentation. Pattern Recogn. **117**, 107994 (2021)
10. El-Shafai, W., El-Fattah, I.A., Taha, T.E.: Deep learning-based hair removal for improved diagnostics of skin diseases. Multimedia Tools Appl. 1–25 (2023)
11. Delibasis, K., et al.: Automated hair removal in dermoscopy images using shallow and deep learning neural architectures. Comput. Methods Programs Biomed. Update **4**, 100109 (2023)
12. Bardou, D., et al.: GAD-VAE: generative adversarial disentanglement with variational autoencoders for hair removal in dermoscopy images. Netw. Model. Anal. Health Inform. Bioinform. **13**(1), 1–18 (2024)
13. Rumelhart, D.E., Hinton, G.E., Williams, R.J.: Learning internal representations by error propagation, parallel distributed processing, explorations in the microstructure of cognition. In: Rumelhart, D.E., McClelland, J. (eds.) vol. 1, Biometrika **71**, 599–607 (1986)
14. Goodfellow, I., Bengio, Y., Courville, A.: Deep Learning. MIT Press, Cambridge (2016)
15. Sherstinsky, A.: Fundamentals of recurrent neural network (RNN) and long short-term memory (LSTM) network. Physica D **404**, 132306 (2020)
16. Boden, M.: A guide to recurrent neural networks and backpropagation. The Dallas project (2002)
17. Harshvardhan, G.M., et al.: A comprehensive survey and analysis of generative models in machine learning. Comput. Sci. Rev. **38**, 100285 (2020)
18. Masci, J., et al.: Stacked convolutional auto-encoders for hierarchical feature extraction. In: International Conference on Artificial Neural Networks. Springer, Heidelberg (2011). Med 2020; **144**(4), 500–522
19. Kingma, D.P., Welling, M.: An introduction to variational autoencoders. arXiv preprint arXiv: 1906.02691 (2019)
20. Kingma, D.P., Welling, M.: Auto-encoding variational Bayes. arXiv pre-print arXiv:1312. 6114 (2013)

21. Sharif, S.A., Hammad, A., Eshraghi, P.: Generation of whole building renovation scenarios using variational autoencoders. Energy Build. **230**, 110520 (2021)
22. Chen, L.-C., et al.: Encoder-decoder with atrous separable convolution for semantic image segmentation. In: Proceedings of the European Conference on Computer Vision (ECCV) (2018)
23. Bornemann, F., März, T.: Fast image inpainting based on coherence transport. J. Math. Imaging Vision **28**, 259–278 (2007)
24. Langr, J., Bok, V.: GANs in Action: Deep Learning with Generative Adversarial Networks. Manning Publications (2019)
25. Yu, J., et al.: Free-form image inpainting with gated convolution. In: Proceedings of the IEEE/CVF International Conference on Computer Vision (2019)

Swarm Intelligence Algorithms in Traffic Signal Control: A Critical Review of Particle Swarm, Bee Colony, and Ant Colony Approaches

Romaissa Bendaikha(✉) and Djalal Hedjazi

LaSTIC laboratory, Department of Computer Science, University of Batna 2, Batna 05078, Algeria
romaissa.bendaikha@univ-batna2.dz, romaissabendaikha@yahoo.fr

Abstract. This paper presents a critical review of Swarm Intelligence (SI) algorithms–Particle Swarm Optimization (PSO), Bee Colony Optimization (BCO), and Ant Colony Optimization (ACO)–and their applications in traffic signal control (TSC). SI techniques, inspired by natural processes, have emerged as effective tools for optimizing traffic signal timings to improve urban traffic flow, reduce delays, and enhance overall efficiency. This review highlights the distinct strengths and limitations of each algorithm, with PSO excelling in simpler scenarios, BCO offering dynamic adaptability, and ACO providing robust path-finding capabilities. The paper also explores comparative analyses and hybrid models that combine these algorithms to leverage their complementary strengths. Despite significant advancements, challenges such as scalability, data dependency, and practical implementation remain. Future research directions include bridging the gap between theoretical and real-world applications, developing hybrid and adaptive models, and incorporating advanced data-driven techniques. The insights provided aim to guide further exploration and application of SI in creating more efficient and responsive traffic management systems.

Keywords: Swarm Intelligence · Traffic Signal Control · Optimization · Ant colony · Bee colony · Particle Swarm

1 Introduction

The growing population in contemporary urban environments has resulted in more vehicles on the road and increased demand for transportation of people and goods. This has led to significant traffic congestion, especially during peak hours, causing longer travel times, higher noise levels, and deteriorating air quality. In addition, traffic congestion, which has become an significant issue in most cities [1,2], negatively impacts public health, environmental sustainability, and economic productivity. Health effects include excessive fatigue and cardiovascular, respiratory, and nervous system issues. Environmentally, congestion

T. Marir et al. (Eds.): ICCSA 2024, ISEM 45, pp. 182–196, 2025.
https://doi.org/10.1007/978-3-031-90758-6_14

increases noise, air pollution, and fuel consumption. Economically, it impedes efficient delivery of goods and services, leading to higher consumer prices [3]. Various strategies have been developed to alleviate traffic congestion through Traffic Signal Control (TSC) systems at intersections. TSC systems fall into two categories: Fixed-Time control and Adaptive TSC (ATSC). Fixed-Time control involves pre-set parameters such as cycle length and phase splits, limiting adaptability. In contrast, ATSC systems adjust these parameters dynamically based on real-time traffic data. Moreover, important progresses in AI and IoT have led to the development of numerous Intelligent Transportation Systems (ITS) aimed at enhancing traffic quality [4]. Swarm Intelligence (SI) approaches [5] are widely used and effective across various contexts. SI involves collective behaviors from group interactions, using nature-inspired algorithms in evolutionary computation. These methods, such as Particle Swarm Optimization (PSO), Ant Colony Optimization (ACO), and Bee Colony Optimization (BCO), model collaborative behavior to find optimal solutions.

Building on this foundation, this study provides a comprehensive overview of recent advancements in SI-based TSC optimization techniques. The review provides an in-depth analysis and critical assessment of significant researches and findings in the field. The central purpose is to present a detailed analysis of the subject, serving as a foundation for researchers interested in exploring new areas in TSC optimization. This paper addresses several main research questions: What are the highly influential scholarly publications in the study of SI-based TSC optimization? Which specific SI techniques have been most effective in enhancing TSC performance? What are the practical implications and real-world impacts of optimizing TSC through SI methods? How can hybrid SI approaches be further developed to improve the robustness and adaptability of TSC systems? We not only explore these questions through a detailed review of existing literature but also incorporate a comprehensive citation analysis to identify key trends and influential research within the field. By combining these approaches, we aim to provide a holistic understanding of SI-based TSC optimization and its applications, while identifying literature gaps and suggesting future research directions. The organization of this paper is as follows: Sect. 2 presents the study framework, detailing the methodologies and approaches utilized in this research. Section 3 provides the findings and analysis, where the results of the study are discussed in depth. In Sect. 4, we engage in a comprehensive discussion, interpreting the results. Section 5 addresses the limitations of the study and identifies potential opportunities for future research. Finally, Sect. 6 concludes the paper, summarizing the key findings.

2 Study Framework and Analysis

This study's framework comprises several steps, as illustrated in Fig. 1: (i) selecting and clustering scientific studies, (ii) conducting literature analysis, (iii) performing a comprehensive review, and (iv) engaging in discussion. It presents a comprehensive analysis of recent literature on SI-based approaches for optimizing TSC. The analysis involves a meticulous examination of journal papers and conference proceedings published between 2019 and 2024. The approach serves a dual purpose: providing a thorough understanding of current research in this field and reflecting recent technological advancements. Publications were sourced from multiple databases, including Scopus, Elsevier, IEEEXplore, and Google Scholar. Using Visualization Of Similarities viewer (VOSviewer), an application for building and displaying scholarly connection networks, an organization of the selected papers was carried out. VOSviewer utilizes text mining to generate networks of keywords from scholar publications. The software's co-occurrence analysis was employed to cluster studies into different groups based on authors' specified keywords. Furthermore, VOSviewer facilitated deeper bibliometric analysis by considering citations for each document, identifying cornerstone papers, and analyzing co-authorship patterns. The critical review evaluated topics covered in the selected papers, analyzing main problems addressed, approaches developed, research cases examined, and effectiveness of suggested methods. Detailed results of these analyses are presented in the following section.

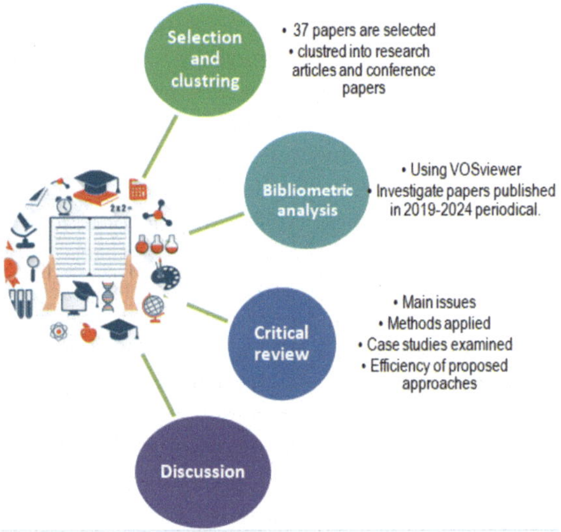

Fig. 1. Study's Framework for Comprehensive Analysis

3 Findings and Analysis

3.1 Criteria and Selection of Papers

This section details the selection and analysis of the most relevant papers on SI-based TSC optimization. We prioritized papers with advanced methodologies and practical applications, emphasizing comprehensive reviews and empirical research. Fig. 2 shows the literature's evolution since the 1960 s, reflecting the growing demand for efficient traffic solutions due to urbanization and increased traffic. The papers were sourced from Scopus, Elsevier, IEEE Xplore, and Google Scholar, covering 2019 to 2024 to ensure recent developments are included. This period underscores the rapid technological progress and evolving research trends, highlighting the relevance of the selected studies.

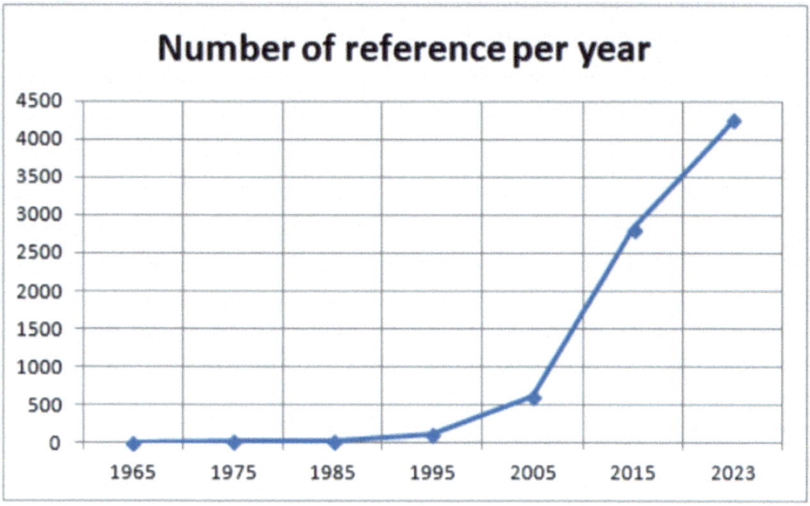

Fig. 2. Chronological Analysis of Published Works.

3.2 Categorization and Clustering

Following the previous step, the citation network of the selected papers was analyzed using VOSviewer. The software's algorithm executes keyword extraction on the terms provided by the authors and those systematically assigned by the indexing system through a process baptized "co-occurrence". For our initial examination, we focused exclusively on the authors' terms and established a minimum occurrence threshold of 3 for each term. Out of the 151 terms, only 6 met the threshold: SI, TSC, optimization, PSO, BCO and ACO. we generate a keyword network, calculating total link strength to show how often terms co-occur in publications. The co-occurrence analysis results are summarized in Table 1.
These results reveal several insights into the thematic focus of the literature. The analysis identifies 6 primary keywords: "SI", "TSC", "Optimization", "PSO", "BCO", and "ACO". "SI" appears 6 times with a total link strength of 6, indicating its significant role in the literature. This suggests that "SI" is a widely discussed topic in the context of "TSC".

"TSC" is mentioned 4 times and has a total link strength of 5, reflecting its core relevance to the studies analyzed. The keyword"Optimization" also appears

Table 1. Co-Occurrence Analysis of Keywords Using VOSviewer Software.

Keyword	Occurrences	Total Link Strength
Traffic Signal Control	6	6
Swarm Intelligence	4	5
Optimization	5	3
Particle Swarm Optimization	5	1
Bee Colony Optimization	3	0
Ant Colony Optimization	3	3

5 times with a total link strength of 3, underscoring its importance in the discussions around improving traffic management systems.

Conversely, "PSO" appears 5 times with a total link strength of 1, while "BCO" and "ACO" appear 3 times each, with link strengths of 0 and 3, respectively. This suggests that while these algorithms are mentioned, they are rarely connected with other keywords. This may indicate fragmented discussions or limited integration with related topics. Overall, although "SI" and "TSC" are prominent, specific algorithms are less integrated into broader optimization discussions. Future research should address these gaps to enhance understanding of algorithm interactions in TSC optimization.

Focusing on papers that discuss SI approaches and TSC optimization, we identified three primary clusters: PSO, BCO, and ACO. We then performed a citation analysis, considering only studies with at least one citation, resulting in a selection of 27 papers. Table 2 presents the citation counts and network links for the top 10 papers in this subset. As expected, all cited sources are articles.

Table 2. Citation analysis (based on VOSviewer software).

Rank	Document	Citations	Links	Type
1	Gao et al., 2019 [6]	45	1	Article
2	Shaikh et al., 2022 [7]	72	1	Article
3	Liao et al., 2024 [8]	2	0	Article
4	Salawudeen et al., 2022 [9]	4	0	Article
5	Suga et al., 2023 [10]	1	0	Article
6	Sibtain et al., 2021 [11]	33	0	Article
7	George et al., 2021 [12]	42	0	Article
8	Farahani et al., 2019 [13]	19	0	Article
9	jovanović et al., 2023 [14]	10	0	Article
10	Korkmaz et al., 2020 [15]	11	0	Article

Fig. 3 presents the citation network. Despite the literature growth shown in Fig. 2, most of the highly cited papers are reviews published between 2020 and 2023.

This highlights a gap and underscores the need for updated research to evaluate the current state of the art in this field.

We also conducted a co-authorship analysis of authors with at least two documents and one citation. This analysis explores author collaborations based on co-authored works. Out of 144 authors, 9 met the criteria, with 6 being interconnected. Fig. 4 illustrates these connections, highlighting collaborative clusters and key research groups. This visualization reveals influential researchers and trends in collaboration.

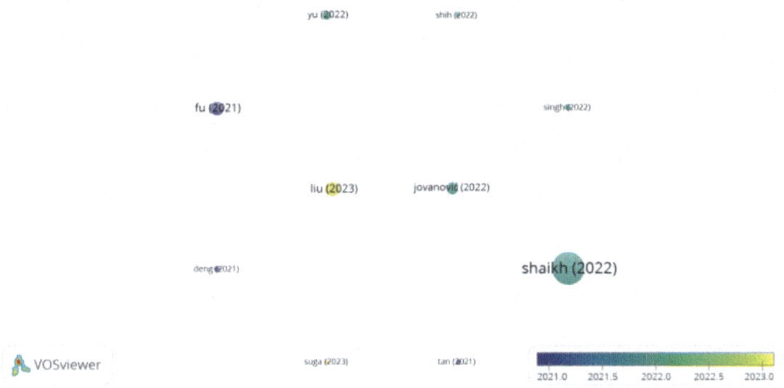

Fig. 3. Overview of Citation Relationships (*Based on VOSviewer Data*).

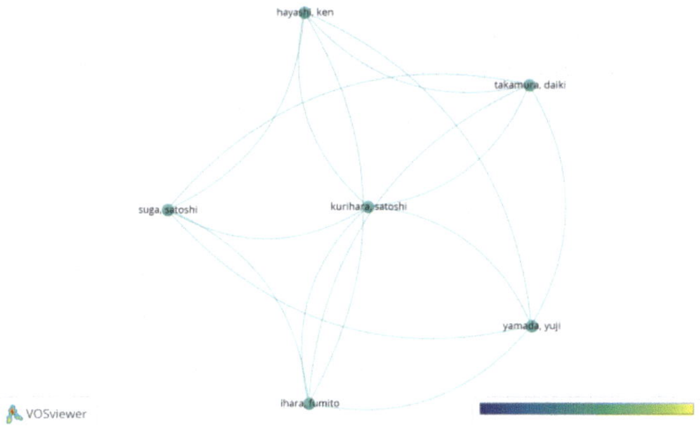

Fig. 4. Authorship Collaboration Network (*Based on VOSviewer Data*).

In addition, Table 3 details interconnected authors who have at least two publications and one citation. It includes key metrics such as publication count, co-authorship frequency, and citation numbers, providing insights into collaborative dynamics and influential research groups, as illustrated in Fig. 4. As a result, both pairs of authors, *Jovanović & Aleksandar* and *Teodorović & Dušan*, stand out with 3 publications each and 20 citations, suggesting that their work is both collaborative and influential. Their higher citation counts highlight their significant impact and recognition within the research community. *Gao & Kaizhou*, with 2 publications and 117 citations, emerges as the most cited author in this analysis. This high citation count signifies a substantial influence and suggests that *Gao & Kaizhou's* work is highly regarded and frequently referenced by other researchers.

Table 3. Collaboration Analysis of Authors *(Based on VOSviewer Data)*

Rank	Authors	Documents	Citations
1	hayashi, ken	2	1
2	ihara, fumito	2	1
3	Kurihara, satoshi	2	1
4	suga, satoshi	2	1
5	takamura, daiki	2	1
6	yamada, yuji	2	1
7	Jovanović, aleksandar	3	20
8	Teodorović, dusan	3	20
9	Gao, kaizhou	2	117

Below, we critically evaluate the selected papers, grouped into three clusters. Meta-heuristic algorithms remain prevalent in TSC. The literature highlights key topics such as objective functions (e.g., average delay, traffic congestion) and decision variables (e.g., green time, cycle time).

3.3 Critical Review

This section critically reviews three prominent SI algorithms: PSO, BCO, and ACO. Each algorithm leverages different mechanisms inspired by natural processes to address the complexities of TSC. The following subsections delve into the specific applications, successes, and limitations of these algorithms in optimizing TSC, highlighting their potential and identifying areas for further research and improvement.

Particle Swarm Optimization. Designed by Eberhart and Kennedy (1995), PSO solves optimization problems by iteratively enhancing candidate solutions based on a quality metric [16]. It employs a swarm of particles navigating the search space according to mathematical rules governing their positions and velocities. Each particle's trajectory is influenced by its best-known position and the swarm's overall best-known positions, collectively converging towards the optimal solution. PSO has demonstrated considerable success in TSC optimization [17]. A Simulated annealing PSO (Sa-PSO) was developed and applied to urban networks [18] to optimize green time and cycle times. Simulations showed that Sa-PSO significantly reduced average delay per vehicle and stop rates compared to fixed time plans.

A PSO algorithm applied to traffic networks in Sevilla and Málaga, Spain, outperformed the SUMO cycle programs generator and a Random Search algorithm, improving the number of completed trips and reducing mean trip times, enhancing overall traffic flow [19]. This research was expanded in [20], by comparing the proposed algorithm against standard PSO and Differential Evolution (DE), showing superior results. An improved PSO algorithm applied to a 4-way intersection optimized average vehicle travel time, reducing average running time by 24.1% compared to traditional timing schemes [21]. In Kumamoto Shi, Japan, PSO optimized vehicle wait time, travel distance, and time delay, outperforming the Multi-Element Genetic Algorithm (ME-GA) by increasing vehicle flow by about 4.13% [22]. In [23], a study introduces a traffic model for cars and scooters, optimizing TSC to reduce delays. It uses PSO to fine-tune signal parameters and compares it to GA. Results show PSO achieves high convergence speeds and accuracy.

A bi-level TSC model showed PSO achieving a 0.19% optimality gap, outperforming SA, with a 61.05% efficiency improvement [24]. PSO optimized traffic light timing in Malaga and Seville, reducing CO and NOx emissions by 23.3% and 29.3%, respectively [25]. In Izmir, Turkey, PSO reduced vehicle delay by 56% and increased passing vehicles by 9.3% [26]. PSO also proved effective in multi-objective TSC optimization, outperforming NSGA-II, IPSO, and GADST [27]. Comparisons with GA showed PSO's effectiveness in optimizing a network of 13 traffic lights [28].

Modified PSO (MPSO) decreased MSE from 6.3299 to 2.065 and enhanced network performance by 4.26% [29]. PSO's consistent success in TSC optimization warrants continued investigation [30]. A PSO-based simulation-optimization strategy reduced waiting times by 26% [31]. Chaotic PSO improved regional traffic signal coordination, reducing delay and queue length [32]. A multi-objective optimization model using PSO showed substantial reductions in delay and stops at an intersection in Siping City, China [33]. A dual-layer framework combining PSO and fuzzy neural networks reduced delays by 11.1%-29.0% [34].

In summary, while PSO variants show promise in TSC optimization, there are areas needing improvement. Addressing complexities, expanding applications, and incorporating advanced data-driven techniques are crucial for advancing intelligent transportation systems.

Bee Colony Optimization. The BCO algorithm, introduced by Karaboga in 2005, optimizes based on the foraging behavior of honeybees. It involves employed bees exploiting known food sources, onlookers selecting food sources based on shared information, and scouts exploring new sources. BCO is valued for its simplicity, efficiency in complex search spaces, and ability to handle high-dimensional optimization problems.

Various studies demonstrate BCO's effectiveness in TSC. For instance, [35] presents a BCO-based approach optimizing signal cycle, green light duration, and phase offset, leading to significant traffic efficiency improvements. Furthermore, [36] investigates an intersection in Lanzhou city, demonstrating BCO's favorable convergence and better performance compared to other algorithms. In [37], a MATLAB-based GUI validated an ATSC algorithm based on BCO, achieving 76.67% performance in minimizing AWT and 53.33% in clearing vehicular queues. Another study [38] proposes a centralized BCO solution for urban TSC, introducing a novel solution generation method and local search approaches, outperforming 4 meta-heuristics and the CPLEX solver. The authors found that BCO improved the network Performance Index (PI) by 2.4% and 2.7% compared to GA and hill-climbing methods [39]. In [40], a single-point multi-phase intersection model optimized with BCO shows enhanced traffic efficiency via MATLAB simulations.

Combining BCO with other metaheuristics has also been explored. For example, [41] integrates BCO with Tabu search to optimize signal control parameters, demonstrating feasibility and practicality in Zhangye City, China.

In summary, BCO effectively optimizes TSC and improves traffic flow in road networks. It dynamically adjusts signal timings based on real-time traffic, reducing delays and enhancing throughput. Future research could focus on scaling BCO, adapting it to evolving traffic dynamics, and integrating advanced technologies for modern urban transportation systems.

Ant Colony Optimisation. ACO, developed by Marco Dorigo in the early 1990s [42], is inspired by the foraging behavior of ant colonies, utilizing pheromone trails to solve optimization problems. Ants explore paths randomly, sharing pheromone information to guide the colony toward optimal solutions. This iterative process adjusts pheromone levels based on path optimality, with higher concentrations attracting more ants.

In TSC, ACO has proven effective [8]. For example, [43] introduced AVCAS+SIDRA, combining an ACO-based algorithm with SIDRA for optimizing vehicle routing and reducing fuel usage and emissions. Simulation results show AVCAS+SIDRA reduced average travel time by 25.5%, average travel speed by 19.5%, and fuel consumption by 17%. Moreover, [44] proposed a decentralized traffic management approach using ACO-based routing with intersection negotiation, enhancing routing efficiency and reducing trip times. In [45], the Inverted Ant Colony Optimization (IACO) algorithm has been introduced. It repels ants from pheromones to avoid congestion. Experiments show IACO reduces traffic density and improves network performance, cutting trip times by up to 84% for compliant vehicles and 71% for non-compliant ones, and reducing fuel consumption and CO_2 emissions by up to 49%.

In [46], the authors applied ACO to optimize signal settings, reducing total time delay by 16.5% and the number of stops by 60%. Putha et al. (2012)

found ACO outperformed GA in oversaturated networks, showing less variance and better results with more model evaluations [47]. Further, Haldenbilen et al. (2013) introduced ACORSES, generating a new ant colony each cycle and preserving optimal solutions to avoid suboptimal local minima [48]. ACOTRANS, compared with TRANSYT-7F, demonstrated up to 13.9% better network performance index (PI) in medium-sized networks. [49] developed an ACO-based adaptive algorithm using wireless sensor data, reducing average waiting time at intersections by up to 57%. Another study [50] optimized traffic signal timing using real-time data, reducing average vehicle delay time by 10.25% and average stops by 24.55%.

In [8], the researchers proposed DALM, a dynamic ACO algorithm, reducing total waiting time and emissions, especially under high traffic flow. Kouidri et al. (2024) introduced a hybrid ACO-DSOS method, reducing the average number of queued vehicles by 20% [51].

These critical reviews highlight the diverse applications and effectiveness of ACO in TSC, showing its potential to significantly enhance traffic management and reduce congestion in urban areas. The advancements and hybrid methods incorporating ACO present promising directions for future research and practical implementations in smart city infrastructures.

4 Disscussion

This study critically reviewed the applications of PSO, BCO, and ACO in TSC. The analysis of the literature reveals several key insights and trends that underscore the strengths and limitations of these SI algorithms in optimizing traffic flow. The reviewed studies were categorized into three main classes: *Algorithm-Specific Studies, Comparative and Hybrid Approaches*, and *Review Papers and Meta-Analyses.*

In the first category, many studies highlight PSO's simplicity and efficiency in converging towards optimal solutions. However, challenges such as premature convergence and sensitivity to parameter settings are noted. BCO's simulation of bee foraging behavior allows for dynamic adaptability in traffic scenarios, yet its computational complexity and the need for fine-tuning parameters can limit practical applications. ACO is admired for its robustness and ability to find optimal paths through pheromone-based communication. Its main drawbacks include slow convergence in large-scale networks and potential for stagnation.

The second category includes several studies that employ comparative analyses between the algorithms or propose hybrid models integrating elements from multiple algorithms. These hybrid approaches often show superior performance by leveraging the strengths of each algorithm. For example, combining ACO's robust path-finding capabilities with PSO's rapid convergence can yield more efficient traffic signal timings. Additionally, many comparative studies provide valuable insights into the relative effectiveness of each algorithm, with metrics such as AWT, fuel consumption, and overall traffic flow efficiency serving as benchmarks.

The final category comprises review papers that synthesize findings from multiple studies, offering a comprehensive overview of the current state of SI in TSC. These reviews underscore the evolving nature of these algorithms and identify emerging trends and gaps in the research. Some review papers highlight the need for standardized evaluation metrics and more extensive real-world testing to validate the efficacy of these algorithms.

Overall, this review emphasizes the potential of PSO, BCO, and ACO in traffic signal optimization while also acknowledging the challenges that need to be addressed to enhance their practical applications. The insights from comparative and hybrid studies, along with the synthesized knowledge from review papers, provide a solid foundation for future research in this domain.

5 Limitations and Research Opportunities

Despite promising results, the application of PSO, BCO, and ACO in TSC faces limitations such as scalability issues, data dependence, computational complexity, etc. These challenges highlight the need for future research to focus on real-world applications, developing hybrid models, and integrating machine learning to enhance adaptability and intelligence. Establishing standardized evaluation metrics and addressing computational complexity are also crucial. Bridging the gap between theoretical research and practical implementation, particularly through real-world experiments and long-term studies, will be essential in advancing SI for efficient urban traffic management.

6 Conclusion

This paper provides a comprehensive review of Swarm Intelligence (SI) algorithms in the context of Traffic Signal Control (TSC), focusing on Particle Swarm Optimization (PSO), Bee Colony Optimization (BCO), and Ant Colony Optimization (ACO). The findings underscore the significant potential of these algorithms in enhancing urban traffic management by optimizing signal timings, reducing delays, and improving overall traffic flow. PSO has shown notable success in less complex scenarios, while BCO and ACO have demonstrated robustness in varied conditions, though at the cost of higher computational resources. Despite their promise, these algorithms face challenges related to scalability, data dependence, and real-world applicability. Future research should bridge the gap between theoretical models and practical implementations, exploring hybrid models and incorporating machine learning techniques to create adaptive and intelligent TSC systems. By addressing these challenges, SI algorithms can play a pivotal role in developing efficient, responsive, and sustainable urban traffic management solutions.

References

1. Haddad, T.A., Hedjazi, D., Aouag, S.: An IoT-based adaptive traffic light control algorithm for isolated intersection. In: Senouci, M.R., Boudaren, M., Sebbak, F., Mataoui, M. (eds.) CSA 2020. LNNS, vol. 199, pp. 107–117. Springer, Cham (2021). https://doi.org/10.1007/978-3-030-69418-0_10
2. Haddad, T.A., Hedjazi, D., Aouag, S.: A deep reinforcement learning-based cooperative approach for multi-intersection traffic signal control. Eng. Appl. Artif. Intell. **114**, 105019 (2022). https://doi.org/10.1016/j.engappai.2022.105019
3. Qadri, S., Gökçe, M.A., Öner, E.: State-of-art review of traffic signal control methods: challenges and opportunities. Eur. Transp. Res. Rev. **12**(1), 1–23 (2020). https://doi.org/10.1186/s12544-020-00439-1
4. Haddad, T., Hedjazi, D., Aouag, S.: A new deep reinforcement learning-based adaptive traffic light control approach for isolated intersection. In: ISIA (2022). https://doi.org/10.1109/ISIA55826.2022.9993598
5. Wang, C., Zhang, S., Ma, T., Xiao, Y., Chen, M.Z., Wang, L.: Swarm intelligence: a survey of model classification and applications. Chin. J. Aeronaut. **38**(3) (2024). https://doi.org/10.1016/j.cja.2024.03.019
6. Gao, K., Zhang, Y., Su, R., Yang, F., Suganthan, P.N., Zhou, M.: Solving traffic signal scheduling problems in heterogeneous traffic network by using meta-heuristics. IEEE Trans. Intel. Trans. Syst. **20**, 3272-3282 (2019). https://doi.org/10.1109/TITS.2018.2873790
7. Shaikh, P.W., El-Abd, M., Khanafer, M., Gao, K.: A review on swarm intelligence and evolutionary algorithms for solving the traffic signal control problem. IEEE Trans. Intell. Transport. Syst. **23**, 48–63 (2022). https://doi.org/10.1109/TITS.2020.3014296
8. Liao, S., Wu, Y., Ma, K., Niu, Y.: Ant colony optimization with look-ahead mechanism for dynamic traffic signal control of IoV systems. IEEE Internet Things J. **11**, 366-377 (2024). https://doi.org/10.1109/JIOT.2023.3286799
9. Salawudeen, A.T., Umoh, I.J., Sadiq, B.O., Oyenike, O.I., Mu'azu, M.B.: An adaptive ant colony optimisation for improved lane detection in intelligent automobile vehicles. Int. J. Bio. Ins. Comput. **19**, 108–123 (2022). https://doi.org/10.1504/IJBIC.2022.121225
10. Suga, S., et al.: Traffic information interpolation method based on traffic flow emergence using swarm intelligence. Artif . Life Rob. **28**, 367–380 (2023). https://doi.org/10.1007/s10015-022-00847-7
11. Sibtain, D., Murtaza, A., Ahmed, N., Hadeed, Gulzar, M.: Multi control adaptive fractional order PID control approach for PV/wind connected grid system. **31**(4), e12809 (2021)
12. George, M.A., Kamat, D.V., Kurian, C.P.: Electronically tunable ACO based fuzzy FOPID controller for effective speed control of electric vehicle. IEEE Access **9**, 73392–73412 (2021). https://doi.org/10.1109/ACCESS.2021.3080086
13. Farahani, G., Rahmani, K.: Speed control of a separately excited DC motor using new proposed fuzzy neural algorithm based on FOPID controller. J. Control Autom. Electric. Syst. **30**(5), 728–740 (2019). https://doi.org/10.1007/s40313-019-00485-8
14. Jovanović, A., Teodorović, D.: Pre-timed control for an under-saturated and over-saturated isolated intersection: a Bee Colony Optimization approach. Trans. Plan. Technol. **40**, 556–576 (2017). https://doi.org/10.1080/03081060.2017.1314498

15. Korkmaz, E., Akgüngör, A.P.: Comparison of artificial bee colony and flower pollination algorithms in vehicle delay models at signalized intersections. Neural Comput. Applic. **32**, 3581–3597 (2020). https://doi.org/10.1007/s00521-018-3670-3
16. Jones, A.: Particle Swarm Optimization (2023). https://doi.org/10.13140/RG.2.2.32162.20163
17. Celtek, S.A., Durdu, A., Alı, M.E.M.: Real-time traffic signal control with swarm optimization methods. Measurement **166**, 108206 (2020). https://doi.org/10.1016/j.measurement.2020.108206
18. Wang, F.L., Deng, H., Gao, Y., Lei, J. (eds.): AICI 2010. LNCS (LNAI), vol. 6320. Springer, Heidelberg (2010). https://doi.org/10.1007/978-3-642-16527-6
19. García-Nieto, J., Alba, E., Olivera, A.: Swarm intelligence for traffic light scheduling: application to real urban areas. Eng. Appl . AI **25**, 274-283 (2012). https://doi.org/10.1016/j.engappai.2011.04.011
20. García-Nieto, J., Olivera, A., Alba, E.: Optimal cycle program of traffic lights with particle swarm optimization. IEEE Trans. Evol. Comput. **17**, (2013). https://doi.org/10.1109/TEVC.2013.2260755
21. Yong, Z., Peng, W.: Optimal control for region of the city traffic signal based on selective particle swarm optimization algorithm. In: International Conference on Intelligent Transportation, Big Data and Smart City (2015). https://doi.org/10.1109/ICITBS.2015.140
22. Wijaya, I.G.P.S., Uchimura, K., Koutaki, G.: Traffic light signal parameters optimization using particle swarm optimization. In: 2015 International Seminar on Intelligent Technology and Its Applications (ISITIA), pp. 11–16. IEEE, Surabaya, Indonesia (2015). https://doi.org/10.1109/ISITIA.2015.7219945
23. Lo, K.-R., Tung, S.-L.: Traffic signal control based on particle swarm optimization. In: 21st World Congress on Intelligent Transport Systems, ITSWC 2014: Reinventing Transportation in Our Connected World (2014)
24. Han, K., Sun, Y., Liu, H., Friesz, T.L., Yao, T.: A bi-level model of dynamic traffic signal control with continuum approximation. Transp. Res. Part C Emerg. Technol. **55**, 409–431 (2015). https://doi.org/10.1016/j.trc.2015.03.037
25. Olivera, A.C., García-Nieto, J.M., Alba, E.: Reducing vehicle emissions and fuel consumption in the city by using particle swarm optimization. Appl. Intell. **42**, 389–405 (2015). https://doi.org/10.1007/s10489-014-0604-3
26. Gökçe, M., Oner, E., I k, G.: Traffic signal optimization with particle swarm optimization for signalized roundabouts. Simulation **91**, 456–466 (2015). https://doi.org/10.1177/0037549715581473
27. Jia, H., Lin, Y., Luo, Q., Li, Y., Miao, H.: Multi-objective optimization of urban road intersection signal timing based on particle swarm optimization algorithm. Adv. Mech. Eng. **11**, 168781401984249 (2019). https://doi.org/10.1177/1687814019842498
28. Abushehab, R.K., Abdalhaq, B.K., Sartawi, B.: Genetic vs. particle swarm optimization techniques for traffic light signals timing. In: 2014 6th International Conference on Computer Science and Information Technology (CSIT), pp. 27–35. IEEE, Amman, Jordan (2014). https://doi.org/10.1109/CSIT.2014.6805975
29. Angraeni, N., Muslim, M.A., Putra, A.T.: Traffic control optimization on isolated intersection using fuzzy neural network and modified particle swarm optimization. J. Phys.: Conf. Ser. **1321**, 032023 (2019). https://doi.org/10.1088/1742-6596/1321/3/032023
30. Huang, K., Zhu, J.: A novel method for reducing vehicle emissions utilizing IoT-based IS-APCPSO algorithm. Appl. Artif. Intell. **38**, 2344144 (2024). https://doi.org/10.1080/08839514.2024.2344144

31. Silva, G.O., Rocha, A.M.A.C., Witeck, G.R., Silva, A., Durães, D., Machado, J.: On tuning the particle swarm optimization for solving the traffic light problem. In: Gervasi, O., Murgante, B., Misra, S., Rocha, A.M.A.C., and Garau, C. (eds.) Computational Science and Its Applications - ICCSA 2022 Workshops, pp. 68–80. Springer International Publishing, Cham (2022). https://doi.org/10.1007/978-3-031-10562-3-6
32. Qin, H., Zhang, W., Zhai, H.: Cooperative control of multiple intersections combining agent and chaotic particle swarm optimization. Comput. Electr. Eng. **110**, 108875 (2023). https://doi.org/10.1016/j.compeleceng.2023.108875
33. Niu, S., Wang, J., Li, S., Zhang, X.: Optimization method of Intersection signal control scheme based on heuristic algorithm **12708**, 127080I (2023). https://doi.org/10.1117/12.2683878
34. Shao, J., Zhang, K., Wang, A., Li, S.: Dynamically signal timing optimization of isolated intersection traffic lights based on a dual-layer framework. J. Adv .Transp. **2024**, 1–13 (2024). https://doi.org/10.1155/2024/1260664
35. He, R., Zhao, H., Ye, Z., Yang, L.: Time optimization of urban arterial coordinated control based on multidimensional neighbourhood artificial bee colony algorithm. Transp. Plan. Technol. **47**, 113–145 (2024). https://doi.org/10.1080/03081060.2023.2246959
36. Zhao, H., He, R., Su, J.: Multi-objective optimization of traffic signal timing using non-dominated sorting artificial bee colony algorithm for unsaturated intersections. Archi. Transp. **46**, 85–97 (2018). https://doi.org/10.5604/01.3001.0012.2109
37. Ahmadu Bello University/Department of Electrical and Computer Engineering, Zaria, 234, Nigeria, O. Adebiyi, R.F., Ahmad Abubilal, K., Sunkary Tekanyi, A.M., Hadir Adebiyi, B.: Management of Vehicular Traffic System using Artificial Bee Colony Algorithm. IJIGSP. 9, 18-28 (2017). https://doi.org/10.5815/ijigsp.2017.11.03
38. Gao, K., Zhang, Y., Sadollah, A., Su, R.: Improved artificial bee colony algorithm for solving urban traffic light scheduling problem. In: 2017 IEEE Congress on Evolutionary Computation (CEC) (2017)
39. Dell'Orco, M., Başkan, Ö., Marinelli, M.: artificial bee colony-based algorithm for optimising traffic signal timings. In: Advances in Intelligent Systems and Computing (2013). https://doi.org/10.1007/978-3-319-00930-8-29
40. Chen, M., Zhang, H., Li, X.: Application of artificial bee colony algorithm for timing of road entrance signal. MATEC Web Conf. **232**, 03004 (2018). https://doi.org/10.1051/matecconf/201823203004
41. Hao, W., Ma, C., Moghimi, B., Fan, Y., Gao, Z.: robust optimization of signal control parameters for unsaturated intersection based on Tabu search-artificial bee colony algorithm. IEEE Access **6** (2018). https://doi.org/10.1109/ACCESS.2018.2845673
42. Dorigo, M., Gambardella, L.M.: Ant colony system: a cooperative learning approach to the traveling salesman problem. IEEE Trans. Evol. Comput. **1**, 53–66 (1997). https://doi.org/10.1109/4235.585892
43. Jabbarpour, M.R., Noor, R.M., Khokhar, R.H.: Green vehicle traffic routing system using ant-based algorithm. J. Netw. Comput. Appl. **58**, 294–308 (2015). https://doi.org/10.1016/j.jnca.2015.08.003
44. Nguyen, T.-H., Jung, J.J.: Ant colony optimization-based traffic routing with intersection negotiation for connected vehicles. Appl. Soft Comput. **112**, 107828 (2021). https://doi.org/10.1016/j.asoc.2021.107828

45. Dias, J.C., Machado, P., Silva, D.C., Abreu, P.H.: An inverted ant colony optimization approach to traffic. Eng. Appl. Artif. Intell. **36**, 122–133 (2014). https://doi.org/10.1016/j.engappai.2014.07.005
46. He, J., Hou, Z.: Ant colony algorithm for traffic signal timing optimization. Adv. Eng. Softw. **43**, 14–18 (2012). https://doi.org/10.1016/j.advengsoft.2011.09.002
47. Putha, R., Quadrifoglio, L., Zechman, E.: Comparing ant colony optimization and genetic algorithm approaches for solving traffic signal coordination under oversaturation conditions. Comput. Aided Civil Infrastruct. Eng. **27**, 14–28 (2012). https://doi.org/10.1111/j.1467-8667.2010.00715.x
48. Haldenbilen, S., Baskan, O., Oz, C.: An ant colony optimization algorithm for area traffic control. In: Barbosa, H. (ed.) Ant Colony Optimization - Techniques and Applications. InTech (2013). https://doi.org/10.5772/51695
49. Rida, N., Ouadoud, M., Hasbi, A.: Ant colony optimization for real time traffic lights control on a single intersection. Int. J. Interact. Mob. Technol. **14**, 196 (2020). https://doi.org/10.3991/ijim.v14i02.10332
50. Fu, X., Gao, H., Cai, H., Wang, Z., Chen, W.: How to improve urban intelligent traffic. a case study using traffic signal timing optimization model based on swarm intelligence algorithm. Sensors **21**, 2631 (2021). https://doi.org/10.3390/s21082631
51. Kouidri, C., Bachir Bouiadjra, R., Mahi, F.: ACO-DSOS hybrid approach to enhance traffic signal optimization. IJTDI **8**, 9–17 (2024). https://doi.org/10.18280/ijtdi.080102

Addressing Data Heterogeneity in Federated Learning: The FedWAQ Approach to Weighted Aggregation Based on Data Quality

Oussama Akram Bensiah(✉) and Rohallah Benaboud

Rela(CS)2 Laboratory, University of Oum El Bouaghi, Oum El Bouaghi, Algeria
{bensiah.oussama,benaboud.rohallah}@univ-oeb.dz

Abstract. Federated learning has surfaced as a promising paradigm for distributed machine learning, where multiple devices work collaboratively to train a model in a decentralized manner without shirring local data. Nonetheless, the issue of data heterogeneity posing a challenge within this paradigm due to variations in data distribution and quality across clients leads to less-than-optimal performance of the global model. The current aggregation methods, fail to inadequately tackle these disparities, leading to biased and inefficient models. This paper introduces Federated Weighted Aggregation grounded in data quality (FedWAQ), an approach that incorporates data quality metrics—specifically completeness, consistency, and timeliness into the federated learning paradigm. Each client calculates a data quality score (Q), which is used to adjust the weights of the global model based on proportionality factor (PF) in the process of the aggregation. Using data quality as a reference in weight aggregation enhances model performance and expedites convergence, especially in contexts characterized by substantial data heterogeneity. The experimental results indicate that FedWAQ outperforms traditional approaches like FedAvg, FedProx and SCAFFOLD. The model is also more accurate and robust across a variety of data scenarios.

Keywords: Federated Learning · Data Heterogeneity · Data Quality · Data privacy · Global model · Model accuracy

1 Introduction

Federated learning (FL) constitutes a distributed machine learning paradigm that enabling multiple decentralized devices to collaboratively train a shared model without the need to exchange local data. Federated learning (FL) approach offers significant benefits regarding data privacy, as it ensures that sensitives information's is retained on local devices, and reduces the communication overhead compared to traditional centralized training methods [1]. Federated learning (FL) is particularly beneficial in sectors such as healthcare, finance, and mobile applications, where data privacy is critical [2]. However, federated learning systems are often challenged by data heterogeneity, where clients contribute non-independent and identically distributed (non-IID) data due to differences in local data sources, collection methods, and storage environments. This variety

T. Marir et al. (Eds.): ICCSA 2024, ISEM 45, pp. 197–206, 2025.
https://doi.org/10.1007/978-3-031-90758-6_15

may impede the performance and convergence of the global model, leading to biased or suboptimal predictions [3]. For example, in a healthcare set-ting, patient data from different hospitals may vary significantly in completeness, consistency, and timeliness, impacting the predictive performance of the federated model [4].

Addressing data heterogeneity is crucial challenge in federated learning. Traditional FL algorithms, such as Federated Averaging (FedAvg), typically assume uniform data quality and distribution across clients, which is rarely the case in practical scenarios [5]. This assumption can result in inefficient learning and poor generalization of the global model. To tackle these challenges, this paper introduces the Federated Weighted Aggregation based on Data Quality (FedWAQ) approach. FedWAQ incorporates a quantitative assessment of data quality into the model aggregation process, aiming to mitigate the adverse effects of data heterogeneity and enhance overall model performance.

After introducing the concept of FedWAQ and the problem of data heterogeneity in federated learning, the remainder of this article is organized as follows: Sect. 2 provides a detailed review of relevant literature, focusing on existing approaches to address data heterogeneity in federated learning. In Sect. 3, the proposed FedWAQ approach is introduced and describing the client-side computation of data quality scores and the server-side aggregation process. Section 4 outlines the experimental design, including dataset descriptions and evaluation metrics. The results and discussion are presented in Sect. 5. Finally, Sect. 6 concludes the article by summarizing key findings, highlighting contributions, and suggesting potential directions for future research.

2 Background and Related Work

Federated learning (FL) represents a collaborative paradigm for model training in decentralized systems using a shared model by multiple devices or clients while preserving the confidentiality of raw data. This distributed paradigm effectively addresses privacy concerns and reduces communication overhead, especially in critical domains such as healthcare, finance, and the Internet of Things (IoT). However, federated learning (FL) encounters several challenges, with data heterogeneity standing out as particularly significant among the most challenging. Data heterogeneity arises when client devices have diverse data distributions, leading to non-IID (non-independent and identically distributed) data. This may decrease the performance of the global model, impede convergence rates, and produce biased outcomes as a result of varying data quality among clients.

The traditional federated learning approaches, such as Federated Averaging (FedAvg), fail to account for the variations in data quality across clients, leading to equal weighting of clients irrespective of the quality of their local data. Low-quality data can significantly impact the global model, which can reduce accuracy and efficiency. In order to address this issue, it is necessary to develop a mechanism that evaluates data quality and ensures that the contributions from clients with high-quality data are more prioritized during the global model aggregation process.

Federated learning was first popularized by [1] with the introduction of the Federated Averaging (FedAvg) algorithm. FedAvg aggregates locally computed model updates from clients to form a global model update. However, this approach assumes uniform data quality and distribution across clients, which can lead to inefficiencies in heterogeneous environments. FedAvg has become the foundation for many subsequent federated learning algorithms due to its simplicity and effectiveness in basic scenarios. Subsequent research has focused on addressing the limitations posed by statistical heterogeneity. Explored federated learning with non-IID (non-independent and identically distributed) data, proposing methods to handle statistical heterogeneity. Their work highlighted the importance of considering data distribution differences in federated learning and proposed solutions such as data sharing among clients to mitigate the impact of non-IID data on model performance.

To address the challenges of data heterogeneity, [2] proposed FedProx,a robust FL algorithm that tackles data heterogeneity by adding a proximal term to the local objective functions, thereby reducing client drift. FedProx ensures that updates from clients with di-verse data distributions do not deviate significantly from the global model, thus improving overall convergence. This approach addresses the challenge of varying data distributions by incorporating a regularization term that penalizes large deviations from the global model.

SCAFFOLD is an approach that mitigate the impact of data heterogeneity proposed by [6], SCAFFOLD is an algorithm that corrects for client drift by using control variates to reduce variance during training. SCAFFOLD introduces additional control parameters to track and correct the differences between local and global model updates, making it effective in scenarios with high data heterogeneity. The control variates act as correction factors that guide the local updates towards the global objective, thereby improving model accuracy and stability.

Recent advancements [7] and [5] have also focused on incorporating data quality into federated learning and discuss the implications of data quality on federated learning performance. They emphasize the need for adaptive strategies that account for the variability in data quality across clients. Their research highlights the potential benefits of weighting client contributions based on data quality scores, suggesting that high-quality data should have a greater influence on the global model.

The framework PrivAim developed by [8], is a quality-aware federated learning framework that presents solution to the dual challenges of privacy and incentivization in federated learning systems. It utilizes differential privacy to safeguard both local models and cost information from potentially untrustworthy servers. The proposed mechanism integrates a multi-dimensional reverse auction to incentivize high-quality, low-cost data contributions while maintaining bid confidentiality motivate high-quality, low-cost data contributions while safeguarding bid privacy. The importance of data quality in machine learning has been well documented in centralized systems and provided a comprehensive overview of data quality concepts, methodologies, and techniques [9], they emphasized the significance of data quality in ensuring the reliability and accuracy of analytical outcomes. Their work lays the foundation for understanding how data quality impacts machine learning models and the importance of maintaining high data standards. Several studies have developed frameworks for assessing data and information quality, emphasizing key dimensions such as accuracy, completeness, and consistency. For instance, frameworks discussed by [10]. These frameworks provide valuable insights into how data quality can be systematically evaluated and managed, which is crucial for effective machine learning model development. Similarly, [11] underscored the challenges of maintaining data quality in the big data era and proposed methods for data quality assessment. They argued that high-quality data is essential for deriving meaningful insights from large datasets. Their research is particularly relevant to federated learning, where the quality of local data can vary significantly across clients. Additionally, [12] introduced a methodology for data quality assessment, focusing on both subjective and objective measures of data quality. Their approach provides a structured way to evaluate data quality, which can be adapted for use in federated learning environments to ensure that only high-quality data contributes to the global model. In otherwise, [13] discussed the broader implications of data quality in the information age, stressing the need for systematic approaches to manage and improve data quality. His work highlights the impact of poor data quality on decision-making processes and underscores the importance of maintaining high data standards in all analytical endeavors.

Building on these foundations, the FedWAQ approach integrates data quality scores into the federated learning paradigm. By evaluating each client's data on completeness, consistency, and timeliness, FedWAQ ensures that higher-quality data client has a proportionally greater influence on the global model's aggregation. This approach aims to improve the robustness and performance of federated learning systems, particularly in environments with significant data heterogeneity. FedWAQ leverages established data quality assessment techniques to systematically evaluate the quality of local data. By incorporating these scores into the federated learning process, FedWAQ addresses the challenges posed by data heterogeneity and ensures a more equitable and effective model aggregation process. The FedWAQ approach takes into account the preservation of data privacy, which is one of the most prominent features of Federated learning (FL), and this is through the fact that the data quality scores used are an abstract presentation It does not contain detailed or sensitive individual data points. This abstraction prevents any direct link between the quality score and specific sensitive information.

The table (Table 1) presents a comparative analysis of the approaches were previously mentioned. The comparative based on key criteria such as data quality consideration, aggregation mechanisms, complexity, and robustness to data heterogeneity. The comparison highlights the strengths and limitations of each approach, with a particular focus on how they handle the varying of data across devices.

Table 1. A comparative analysis of the approaches aimed at addressing data heterogeneity in federated learning systems.

	FedWAQ	FedAvg	FedProx	FedNova	SCAFFOLD
Core Idea	Weighted aggregation based on data quality metrics	Average model updates from all clients	Adds a proximal term to the local objective function	Normalizes updates by client-specific scaling factors	Corrects client drift by using control variates
Data Quality Handling	Uses data quality scores (Completeness, Consistency, Timeliness) for weighting	Does not directly handle data quality	Handles data heterogeneity through proximal regularization	Normalizes updates to account for different data quantities	Mitigates drift by correcting model updates
Aggregation Method	Quality-based weighted aggregation	Simple average of local model weights	Proximal term added to aggregation	Normalized weighted aggregation	Uses control variates to correct for client drift
Complexity	Moderate; requires quality assessment and PF calculation	Low; straightforward averaging	High; involves proximal term tuning	Moderate; requires scaling factor estimation	Moderate to High; involves control variates and correction terms
Communication Overhead	Increased due to quality score transmission	Minimal; only model weights are transmitted	Increased due to proximal term requirements	Increased due to normalization factors	Increased; requires communication of control variates
Robustness to Data Heterogeneity	High; adjusts for quality variation among clients	Low; assumes similar data distributions	Moderate; improves robustness with proximal term	Moderate; addresses data quantity variations	High; corrects client drift to improve robustness

3 Methodology

The proposed FedWAQ approach includes two main aspects: client-side and server-side operations. On the client side, each device computes a data quality score (Q) using three normalized data quality metrics: Completeness, Consistency, and Timeliness. These scores are transmitted to the central server along with the local model updates. On the server side, these scores are used to calculate a Proportionality Factor (Pf) for each client, which then influences the weighted aggregation of the global model. This methodology ensures that data quality variations across clients are considered during aggregation, leading to more accurate and reliable global models.

3.1 Client Side

Each client device in the FedWAQ approach independently calculates a data quality score (Q) based on three key aspects:

Completeness: Measures the extent to which essential data fields are populated. Incomplete data can lead to biased or inaccurate model updates. For example, missing values in critical features such as patient information in a healthcare dataset can significantly impact model performance [12].

Consistency: Assesses the absence of contradictory information within the dataset. High consistency indicates that the data is reliable and free from conflicts. Consistency issues might arise from duplicated records or conflicting entries, which can distort model training [13].

Timeliness: Evaluates the relevance of data with respect to the time it is received. More recent data is often more valuable in dynamic environments, such as real-time traffic prediction systems. Timeliness ensures that the model adapts quickly to changing patterns [11].

The motivation for choosing completeness, consistency, and timeliness as key data quality metrics in the FedWAQ approach is the need to address critical challenge of data heterogeneity. Completeness ensures that data contributions clients are fully populated and reduce the risk of biased learning caused by missing data. Consistency highlights the reliability of the data by prioritizing contributions that are free from contradictions, therefore improving the robustness of the global model. Moreover, Timeliness ensures that the most recent and relevant data is weighted more considerably in the aggregation process, which is important in dynamic applications requiring real-time decision-making.

The overall data quality score $\boldsymbol{Q_i}$ for each client $\boldsymbol{i}$ is computed as the average of normalized scores of these three metrics:

$$Q_i = \frac{\boldsymbol{Normalized(completeness)} + \boldsymbol{Normalized(consistency)} + \boldsymbol{Normalized(Timeliness)}}{3} \quad (1)$$

These scores, along with the local model weights, are transmitted to the server.

3.2 Server Side

Upon receiving the local models and their corresponding quality scores, the server computes a Proportionality Factor ($\boldsymbol{Pf}$) for each set of weights. This factor determines the influence of each client's data on the final model based on the formula:

$$\boldsymbol{Pf}_i = \frac{\boldsymbol{Q_i}}{\sum_{j=1}^{N} \boldsymbol{Q_j}} \quad (2)$$

where $\boldsymbol{Q_i}$ is the quality score of the $\boldsymbol{i}$-th client and $\boldsymbol{N}$ is the total number of clients. The global model is updated by aggregating the local model updates weighted by their $\boldsymbol{Pf}$ values:

$$\boldsymbol{\Delta\theta_{global}} = \sum_{i=1}^{N} \boldsymbol{\Delta\theta_i} * \boldsymbol{Pf}_i \quad (3)$$

where $\boldsymbol{\Delta\theta_{global}}$ is the wights of the global model and $\boldsymbol{\Delta\theta_i}$ the weights of local model and $\boldsymbol{Pf}_i$ the Proportionality Factor for each set of weights.

This approach ensures that clients with higher-quality data exert a greater influence on the global model, thereby improving the overall robustness and performance of the federated learning system and considering maintaining data privacy by using abstract values.

Client-side and Server -side algorithms

Algorithm1: FedWAQ Client-side

1. **Output**: data quality score $\boldsymbol{Q}$;
2. **Calculate** Completeness count the number of missing values;
3. **Calculate** Consistency using duplicate detection;
4. **Calculate** Timeliness using data freshness;
5. $\boldsymbol{Q} = \frac{Normalized(completeness)+Normalized(consistency)+Normalized(Timeliness)}{3}$
6. **Send**: $\boldsymbol{Q}$, Local model weights $\boldsymbol{\theta}$;

Algorithm2: FedWAQ Server –side

1. **Input**: Model weights $\{ \Delta\theta_i \}$ from participating devices $\boldsymbol{i}$ =1,2... $\boldsymbol{N}$;
2. Data quality score Q_i ,where Q_i represent the quality of data from device $\boldsymbol{i}$
3. **Output**: Global Model weights $\Delta\theta_{global}$;
4. **For** each device $\boldsymbol{i}$ and in parallel **do**

   $$Pf_i = \frac{Q_i}{\sum_{j=1}^{N} Q_j}$$

   End For
5. $\Delta\theta_{global} = \sum_{i=1}^{N} \Delta\theta_i * Pf_i$
6. Update global model using $\Delta\theta_{global}$.

The client-side process is based on computing the Data Quality Score (***Q***) On each participating client device, the quality of the local dataset is evaluated using three key metrics Completeness, Consistency and Timeliness. The overall data quality score (***Q***) for each client is computed as a sum of the normalized value of these three metrics. Each client then sends its local model weights along with the computed data quality score Q_i to the central server.

The server receiving the local model updates and corresponding quality scores from the clients, the server performs a proportional aggregation. The Proportionality Factor (***Pf***) for each client is computed, this factor determines the influence of each client's data on the final model.

4 Implementation

In this study, we utilized EMNIST Balanced dataset to evaluate the effectiveness of the proposed FedWAQ approach in addressing data heterogeneity in federated learning. Dataset statistics are summarized in Table 2. This dataset is curated from prior work in federated learning [4].

Table 2. Used dataset characteristics.

Characteristics	Classes	Training Set	Test Set
EMNIST	47	112800	18800

The FedWAQ algorithm was implemented using Python and simulated on a distributed network of 50 virtual IoT devices, where each client is assigned a subset of data. The clients train their local models using convolutional neural networks (CNNs) and send only the model weights and quality scores (completeness, consistency, timeliness) to the central server. The FedWAQ approach was compared against a standard federated learning model without quality weighting. Tensor-Flow was used for model training and evaluation. Each client's dataset was artificially modified to simulate different levels of completeness, consistency, and timeliness. In each round, clients train their local models for 5 epochs and send their updated model weights along with their quality scores (completeness, consistency, timeliness) to the server. The total communication overhead is 7.5 MB for all clients per round. The Simulation took: 4 h and 15 min for 50 rounds motion that we use a Remote Desktop Protocol (RDP).

5 Results and Discussion

Table 3. Comparisons between the FedWAQ approach with FedAvg, FedProx, and SCAFFOLD based on key performance metric: Accuracy and Precision.

Metric	FedAvg	FedProx	SCAFFOLD	FedWAQ
Accuracy (%)	80.2	83.5	85.0	91.7
Precision (%)	77.5	80.5	83.2	89.3

Initial results as shown in the table above (Table 3) demonstrate that FedWAQ significantly improves model accuracy. In scenarios with high data heterogeneity, the overall accuracy of the FedWAQ approach was 91.7%, which is a 6.7% improvement over SCAFFOLD, the best alternative among the compared methods with an accuracy of 85.0%.

The precision of FedWAQ was 89.3%, otherwise the other methods with Fe-dAvg at 77.5%, FedProx at 80.5% and SCAFFOLD at 83.2%.

The superior performance of FedWAQ across the performance metrics men-tions above confirms the effectiveness of integrating data quality metrics into the global model aggregation in federated learning paradigm. FedWAQ approach ensures leading to more accurate and reliable global models while considered data quality variations across clients in aggregation process.

6 Conclusion

Our work has examined the impact of using data quality metric in the global model aggregation process in federated learning systems in order to address the challenge of data heterogeneity. The FedWAQ approach ensures that higher-quality data client has a proportionally greater influence on the global model. The FedWAQ approach mitigating the impact of low-quality contributions that can reduce the global model performance. On the other hand, The FedWAQ approach was evaluated with EMNIST dataset and the results proved the of superiority FedWAQ approach compared to traditional federated learning methods such as FedAvg, FedProx, and SCAFFOLD in heterogeneous environments. By continuing to improve the aggregation process and address data heterogeneity, FedWAQ has the potential to greatly enhance the power, robustness, and scalability of federal learning systems.

Acknowledgments. The authors appreciate support from the Research Laboratory on Computer Science's Complex Systems ReLaCS2 and would like to thank the Directorate General for Scientific Research and Technological Development (DGRSDT, MESRS).

References

1. McMahan, H.B., Moore, E., Ramage, D., Hampson, S., Arcas, B.A.: Communication-efficient learning of deep networks. In: Artificial Intelligence and Statistics (2017)
2. Li, T., Sahu, A.K., Talwalkar, A., Smith, V.: Federated learning: challenges, methods, and future directions. IEEE Sig. Process. Mag. **37**, 50–60 (2020)
3. Yue, Z., Meng, L., Liangzhen, L., Naveen, S., Damon, C., Vikas, H.: Federated learning with non-iid data. arXiv preprint arXiv:1806.00582 (2018)
4. Jie, X., Benjamin, G.S., Chang, S., Peter, W., Jiang, B., Fei, W.: Federated learning for healthcare informatics. J. Healthc. Inform. Res. **5**, 1–19 (2021)
5. Kairouz, P., McMahan, H.B., Avent, B., Bellet, A., Bennis, M., Bhagoji, A.N.: Advances and open problems in. Found. Trends® Mach. **14** (2021)
6. Karimireddy, S.P., Kale, S., Mohri, M.: Scaffold: stochastic controlled averaging for federated. In: International Conference on Machine (2020)
7. Harry, H.T.-M., Hang, Q., Matthe, B.: Measuring the effects of non-identical data distribution for federated visual classification. arXiv preprint arXiv:1909.06335 (2019)
8. Dan, W., Ju, R., Zhibo, W., Yichuan, W., Yaoxue, Z.: Privaim: a dual-privacy preserving and quality-aware incentive mechanism for federated learning. IEEE Trans. Comput. **72**(7), 1913–1927 (2022)
9. Batini, C., Scannapieco, M.: Data Quality: Concepts, Methodologies and Techniques. Springer, Cham (2006)

10. Stuart, M.E., Richard, W.Y., Yang, L.W., Hongwei, Z.: Overview and framework for data and information quality research. J. Data Inf. Qual. **1**(1), 22 (2009)
11. Cai, L., Zhu, Y.: The challenges of data quality and data quality assessment in the big data era. Data Sci. J. **14**, 2 (2015)
12. Pipino, L., Lee, Y.W., Wang, R.Y.: Data quality assessment. Commun. ACM **45**, 211–218 (2002)
13. Redman, T.C.: Data Quality for the Information Age (1996)

Dynamic Aspects of Multiagent Systems Organizations: A Survey

Narimane Sahel[1(✉)], Toufik Marir[1], Varun Gupta[2], and Maroua Bouzid[3]

[1] University of Oum El Bouaghi, Research Laboratory on Computer Science's Complex Systems, Oum El Bouaghi, Algeria
{narimane.sahel,marir.toufik}@univ-oeb.dz
[2] GISMA University of Applied Sciences, Potsdam, Germany
[3] University of Caen Normandie, GREYC Laboratory, 14000 Caen, France
maroua.bouzid-mouaddib@unicaen.fr

Abstract. Nowadays, the use of multiagent systems (MAS) has expanded significantly, particularly in complex and dynamic environments. This increased use of MAS is due to their openness, heterogeneity, and the complexity of the systems in which they are deployed. In fact, MAS allow users' requirements to be met better by including several social and organizational considerations. Hence, it is crucial to consider the dynamic organizational aspects to enhance stability and adaptability and optimize performance. By addressing both the static and dynamic aspects of MAS, their efficiency can be significantly improved. In this context, new approaches and models have been proposed to focus on the dynamic aspects of MAS' organizations. This paper synthesizes the dynamic aspects of MAS organization, highlighting innovative mechanisms and models that enable structural, functional, and normative changes.

Keywords: Multiagent systems · Organization · Self-Organization · Reorganization · Dynamic Reorganization

1 Introduction

In modern real-world systems, characterized by their high complexity and extensive distribution, non-linear processes, their operation in an open environment, heterogeneity, constant dynamic changes, and unpredictable interactions pose significant challenges. Faced with these challenges, new requirements are imposed on engineering technologies. They require an innovative engineering approach that integrates new styles, where concepts such as decomposition, autonomy, modularity, and adaptability can be seamlessly combined into a single system. In this regard, MAS emerged as a promising engineering style characterized by an architecture capable of addressing the increasing complexity due to the open, heterogeneous, and ever-evolving nature of these systems. MAS provides solutions for effectively managing these complex and and uncertain challenges, thus

T. Marir et al. (Eds.): ICCSA 2024, ISEM 45, pp. 207–221, 2025.
https://doi.org/10.1007/978-3-031-90758-6_16

paving the way for adaptive engineering in the real world. This new engineering style offers efficient system modelling through distributed autonomous agents that cooperate, act, and respond flexibly to achieve their individual objectives as well as the overall system goal.

Furthermore, they demonstrate an exceptional ability to dynamically reorganize themselves to adapt to dynamic changes in their environment. Initial research on MAS primarily emphasized the specific attributes of agents, such as their architectures, capabilities, and behaviours. However, current MAS research focuses more on the social and organizational aspects, covering both individual and collective facets. Furthermore, the new research highlights the dynamics and adaptability of systems through processes of change that involve both top-down and bottom-up approaches. Multiagent systems can dynamically change their organization, whether through self-organization or predefined changes. These changes result from the inadequate functioning of the system, whether due to the environment or the agent itself. These changes can occur on several dimensions, including the structure and functioning of the system as well as the normative dimension.

This survey paper synthesizes work on the dynamic aspects of multiagent system organizations, covering organizational dimensions, change techniques, and organizational models. The paper is structured as follows: Sect. 2 outlines the context and purpose of the study. Section 3 discusses the criteria for selecting change techniques. Section 4 reviews existing change techniques based on these criteria. Section 5 synthesizes existing organizational models, and Sect. 6 concludes the paper.

2 Multiagents Systems Organization

ACMAS (Agent-Centered Multiagent Systems) and OCMAS (Organization-Centered Multiagent Systems) represent two key approaches in multiagent systems engineering. ACMAS emphasizes agents' autonomy and individual decision-making, leading to unpredictable, emergent organizational structures. In contrast, OCMAS focuses on the organizational structure and communication between agents, treating the organization as the primary entity. Within the context of MASs, there are many definitions of organization, each offering a unique perspective that enhances our understanding of how these systems can be structured and managed effectively.

Dignum [8] proposed two definitions of organization [8]: organization as an entity and organization as a process. First, the organization is perceived as a group of agents with its own objectives. Second, organization is the process of organizing individual agents. Premm and Kirn [20] defined organization as an effective metaphor for the design of distributed software systems [20]. However, in the study conducted by Abbas et al. [3], an organization integrates both viewpoints (ACMAS and OCMAS) and is based on two essential characteristics, namely heterogeneity and openness. An organization can also provide effective system modulation through agent grouping, where each group represents a given context in which agents can freely interact, thus increasing system flexibility and maintainability [15].

The organizational mechanism of MAS mentioned previously addresses organizational aspects during design, requiring knowledge of the system's objectives from the outset. However, the growing requirements related to the openness, complexity, and heterogeneity of modern software systems impose new demands on agent-oriented software engineering (AOSE). This discipline aims to integrate dynamic reorganization, allowing the system to autonomously change in response to environmental conditions. Dynamic reorganization is a process of change initiated in response to inadequate system functioning or unexpected events in the dynamic environment.

Picard et al. [18] identified two types of dynamic reorganization in MAS based on whether agents are aware of the organizational structure. Reorganization occurs when agents, who are aware of the organizational level, explicitly manipulate the organization through specifications and constraints. This process can be initiated internally by the agents or externally by other entities, enabling the system to adapt to ensure adequate functioning when adjustments are needed. In contrast, self-organization is a bottom-up process where agents are unaware of the overall organizational state and rely only on local information. They adjust the system's configuration, such as topology and neighbourhood influences, through local interactions to adapt to environmental pressures without relying on predefined models. This approach enables the system to adapt dynamically by leveraging the emergent behaviour of agents.

Building on this understanding of dynamic reorganization, Picard et al. [18] also explored the various aspects of organizational dynamics, such as temporal and intentional aspects, that influence how and when changes occur within the system. These aspects include determining who initiates changes, how quickly the system can respond to environmental pressures and the intentions behind organizational adjustments. Understanding these factors is crucial for designing systems that can effectively manage and respond to dynamic conditions, ensuring flexibility and adaptability in dynamic environments. Figure 1 shows the main steps in the changing process in multiagent organizations.

Fig. 1. Process of Change in Multiagent Organizations

3 Dimensional Analysis of Organization in Multiagent Systems

Obviously, multiagent systems are sets of autonomous entities, called agents, that work together to achieve common goals. These agents are structured in organizations, allowing them to coordinate to achieve their common goal. Usually, we consider three dimensions when we study a community of agents: structural, functional and behavioural. Knowing that studying multiagent organizations considers agents as a black box, by omitting their structure and their behaviour, we think that studying and analyzing multiagent organizations can be done through three dimensions [15]: the structural dimension, which concerns the relationships and interactions between agents; the functional dimension, which focuses on the tasks performed by the roles played by agents; and the normative dimension, which discusses the rules governing agent interactions. Integrating these dimensions makes it possible to understand better and design efficient and adaptive MAS.

Structural Aspects. The structural aspect represents the way that agents are arranged and interconnected in a multiagent system. Agents are grouped based on their goals, resources, goals, ...etc. A MAS is, therefore, structurally organized according to various organizational paradigms [9,12]. Figure 2 shows some organizational structure.

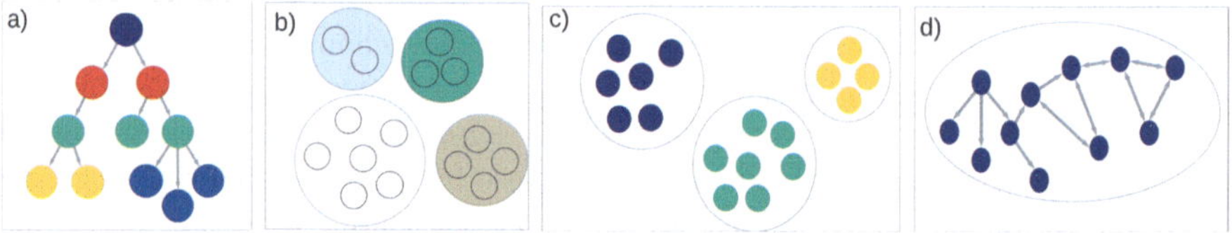

Fig. 2. Organizational Structures in MAS, namely: (a) Hierarchical, (b): Coalition, (c): Congregation, (d): Team

It is important to note that while some structures may look similar statically, how they work dynamically can make a difference. For example, consider the case of a coalition and congregation that have a similar static structure. However, the difference lies in their operational functioning. In a coalition, agents band together temporarily to pursue common goals, but they retain their autonomy and individual decision-making. In contrast, in a congregation, agents work together in a more closely coordinated manner, sharing tasks and responsibilities in a collaborative manner. Thus, the functional aspect, which determines how agents interact, cooperate and pursue their goals, helps to distinguish these two similar structures.

Functional Aspect. In multiagent organizations, the functional aspect refers to how the agents interact, coordinate their actions, and perform their tasks to achieve the organization's objectives. It encompasses coordination mechanisms, communication, collective decisions, and problem-solving processes. The functional aspect plays a key role in the effective collaboration between agents, ensuring the smooth functioning of the multiagent organization in achieving its objectives.

Normative Aspect. Norms play a crucial role in the organization of MAS. They define the expected behaviours of agents and can be pre-established by system designers or emerge in a distributed manner through interactions between agents. These norms guide agents' actions, promote cooperation, regulate conflicts and maintain the system.

4 Techniques for Organizational Change in Multi-Agent Systems

According to the dimensions of analysis of multiagent organizations mentioned above, we devote this section to present the techniques for changing organizations.

4.1 Structure-Based Techniques

A MAS can be initially organized using various organizational paradigms during the design phase. However, the chosen paradigm may need to be revised in dynamic environments to meet the system's evolving needs. Therefore, it becomes necessary to select another organizational paradigm dynamically. One effective strategy for adapting to such changes is through coalition formation, where agents collaborate and align their objectives to collectively transition to a new organizational structure.

Christianos and Chalkiaddakis [5] propose an approach to efficiently forming coalitions using hypergraphs. This approach represents agents by hypergraphs, where nodes represent agents and hyperedges represent qualities associated with given attributes. After this modelling, a hypergraph pruning algorithm is applied to reduce the search space, such that the new hypergraph represents only 1/20 of the initial hypergraph. This means that pruning allows the algorithms to maintain high performance while reducing the complexity of the representation. This technique is presented in Fig. 3.

Reducing the size of the hypergraph can be achieved using two methods. The first method in this work employs the Minimal Transversal Algorithm to create groups of agents of minimal size that cover all the hyperedges of the hypergraph. The second method employs a clustering algorithm called K-means, applied to eigenvectors. These eigenvectors are calculated from the Laplacian matrix representing the structure of the hypergraph. In the third method, a heuristic approach is employed based on sorting nodes by degree without requiring significant

Fig. 3. The change process proposed in [5].

exclusion of nodes from the hypergraph. This approach focuses on selecting high-quality nodes based on their degree to form effective coalitions.

Another work on coalition formation is proposed by Dutta et al. [10], which combines correlation clustering with a linear programming-based graph partitioning method and a region-growing strategy. The algorithm starts by constructing a graph where each node represents a task or agent, and edges represent their similarity. A binary classifier then labels the edges as "similar" or "dissimilar." The goal is to partition the graph into clusters with minimal disagreement between them. The algorithm determines the optimal number of clusters by minimizing these disagreements through a linear programming formulation without the need to pre-specify the number of clusters. Finally, a region-growing strategy is applied to iteratively merge clusters until no further mergers are possible. Figure 4 shows the process of coalition formation proposed in this work [10].

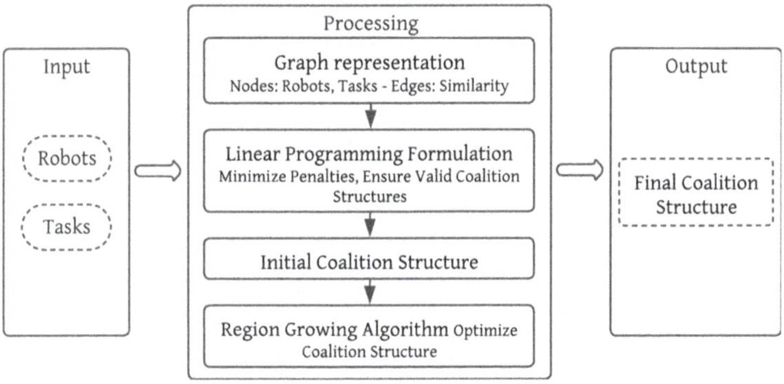

Fig. 4. Process Flow for Coalition Formation [10].

Building on these approaches, Wu et al. [23] extended the application of coalition formation by developing a method for forming coalitions in MAS that is designed for forest fire prevention and control. Their approach uses a model that considers the characteristics of the coalitions and optimization constraints. They proposed a hybrid algorithm combining K-means, Variable Neighborhood

Search (VNS), and Genetic Algorithm (GA) to form effective coalitions. This method helps improve resource management and optimize task execution in fire control systems.

In the previous approaches, the selection was limited to a single organizational paradigm, implying a configuration change with each iteration. It is important to note that these two coalition structure formation approaches exclusively focus on creating a coalition structure for each specific situation. However, another paradigm might exist that is more effective or optimal than the simple formation of coalitions. In this context, Rahmanzadeh et al. [21] introduced a dynamic model designed for self-organizing MAS that operates by selecting the optimal organizational paradigm at runtime. The model evaluates three paradigms: hierarchies, federation, and a hybrid, allowing the MAS to adapt its structure according to the current system state and requirements. The decision-making process within the model is based on system criteria such as the number of agents and overall utility, ensuring that the most appropriate paradigm is chosen to maximize efficiency. This approach was validated in an Internet of Things (IoT) scenario, specifically within a greenhouse sensor network. The results demonstrated the model's effectiveness in dynamically optimizing the organizational structure of the MAS, leading to improved system performance in alignment with real-time conditions. The process of selecting dynamically the adequate paradigm is presented in Fig. 5.

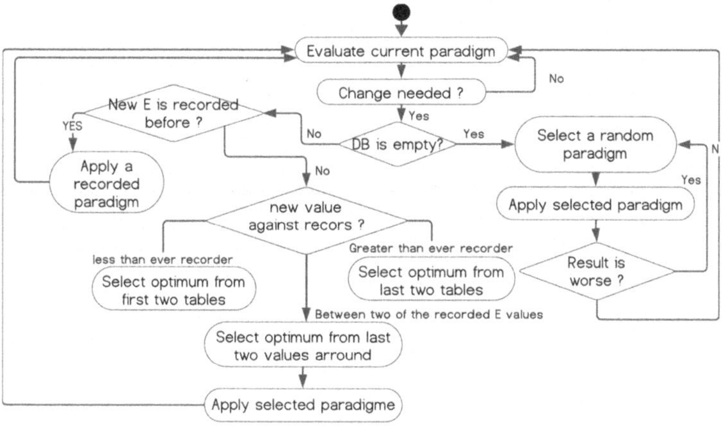

Fig. 5. Dynamic Paradigm Selection Process Proposed in [21].

4.2 Function-Based Techniques

Some works propose changing the functions realized by different components of the organization in addition to changing the organization's structure. It is well-known that the functions in multiagent organizations are represented abstractly by roles. Hence, the change in this category of work is based on manipulating agents' roles.

Picard et al. [19] proposed a role assignment technique that allows agents to modify their roles during execution based on the system's needs. This technique begins by checking the resources and prohibited actions for each role, then it eliminates agents who do not meet the criteria required by the role. In cases of agent unavailability, the algorithm evaluates the costs to reactivate the necessary agents [19]. This role-changing technique is applied at the design stage, providing significant flexibility in managing roles and resources within a multiagent system. However, its adaptability is limited by the designers' vision and may respond poorly to unexpected runtime issues. The process of the dynamic role assignment proposed in this technique is presented in Fig. 6.

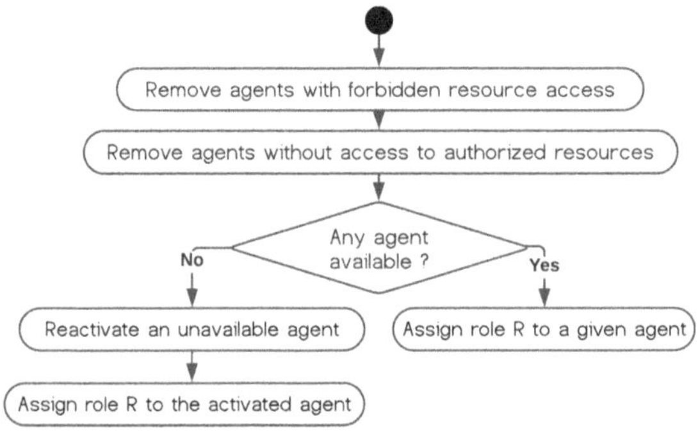

Fig. 6. The Process of Dynamic Role Assignment [19].

Picard et al. [19] proposed a second technique for changing interactions based on the AMAS theory. This technique involves analyzing and anticipating Non-Cooperative Situations (NCS) that may occur during system operation and defining the necessary actions to return to a cooperative state. NCS can include partnership conflicts, booking conflicts, unnecessary interactions, and more. System agents adapt their interactions based on their local environment and the detected NCS, resulting in a reorganization of the system and a change in its overall behaviour.

However, this approach introduces significant complexity and computational overhead during runtime, which can result in higher processing demands and slower system responses. It is important to note that changes made at the agent level can influence the system's overall structure, leading to a bottom-up reconfiguration.

4.3 Normes-Based Techniques

Dell'Anna et al. [6] introduce an execution supervision mechanism that automatically revises norms when their application appears ineffective. The decision to revise norms is based on a Bayesian network that provides information about the probability of achieving the system's overall goals by applying the norms. norms can be revised in three ways: relaxation, reinforcement, and modification. The supervision mechanism is evaluated using a smart urban traffic simulation.

Building on this, Dell'Anna et el. [7] proposed another mechanism for automating norm revision, which modifies the sanctions associated with norms based on real-time data collected during the system's operation and incorporates knowledge about agent preferences. This research also utilizes a Bayesian network to continuously learn the relationship between adherence to or violating a rule and achieving the system's goals. To facilitate the automatic updating of sanctions for enforced norms, the study presents two heuristic techniques that analyze the updated Bayesian network. Figure 7 shows the norms and sanctions revision framework proposed in the above works.

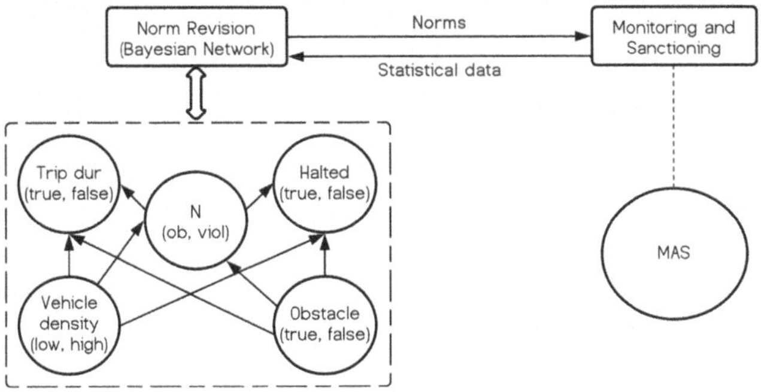

Fig. 7. Norm and Sanction Revision Framework [6,7].

Using Bayesian networks can improve decision-making but may lead to significant computational complexity and challenges in maintaining real-time performance, particularly in dynamic environments.

In the same context, Morris-Martin et al. [17] proposed an innovative conceptual framework for the evolution of norms in MAS, enabling real-time adaptation to the changing needs of participants and the environment. This framework employs specialized synthesizer agents with partial observability to revise or formulate new norms in response to agent requests. Validated using the InstAL specification language and the XHAIL symbolic learning system, this approach demonstrates the feasibility of decentralized, agent-directed norm revision to resolve institutional conflicts in real-time, paving the way for adaptable self-governance in socio-technical systems. Figure 8 shows the norms synhesis process proposed by [17].

Table 1 summarises the previous techniques.

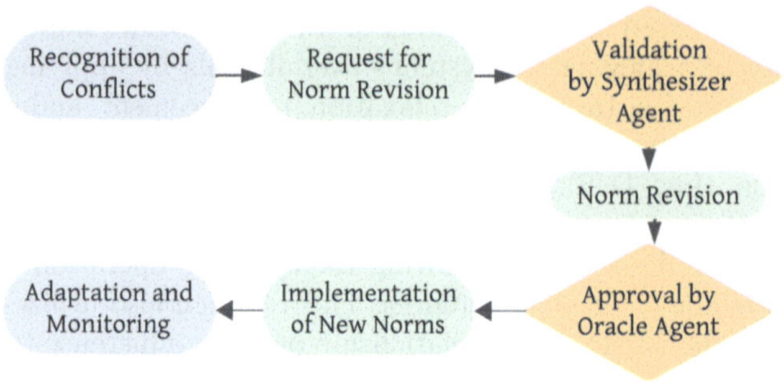

Fig. 8. Process Flow for Dynamic Norm Synthesis [17].

Table 1. Summary on Techniques for Changing Multiagent Organizations

Works	What	When	Who	Why	How
[5]	Structural Change (Top-Down)	Design Time and Runtime	Algorithmic Process	Resource Management	Hypergraph Prun- ing,Heuristic Selection, K-means Clustering Minimal transversal.
[10]	Structural Change (Top-Down)	Runtime	Algorithmic Process	Optimal Task Allocation	Correlation Clustering
[21]	Structural Change (Top-Down)	Runtime	Broker Agent	Adaptation to Environmental Changes	Contract-Net Protocol Database Utilization
[23]	Structural Change (Top-Down)	Runtime	Algorithmic Process	Resource Controland Task allocation	K-means, VNS, GA.
[19]	Structural Change (Top-Down) Functional Change	Design Time and Runtime	Algorithmic Process	Adaptation and Failure Recovery	Dynamic Role Assignment and Cooperative Behavior
[6]	Normative Change (Add, Remove, Update)	Runtime	Norm Updater	Ensure System Objectives Alignment	Bayesian Networks
[7]	Normative Change (Sanction Adjustment)	Runtime	Sanction Adjuster	Maximize System Objective Achievement	Bayesian Networks
[17]	Normative Change (Add, update)	Runtime	Synthesiser Agents	Adapting to environemental change and agent needs	InstAL specification language and XHail Symbolic Learning System

5 Organisational Models

An organizational model is a tool used within a development methodology to represent the system to be created from an organizational point of view. Generally, it is used from the analysis phase but can be extended to the design and implementation phases; in some cases, it can cover the entire development cycle. In addition to an organization's structure, a model can present its functionalities and even its normative rules [3]. This section is devoted to present some organizational models, focusing on their dynamic aspects.

5.1 AGR Model

Ferber et al. [11] proposed an organizational model called AGR, also known as Alaadin, which is based on three fundamental concepts: agent, group, and role. An agent is an active entity capable of adopting various architectures and playing multiple roles within groups. A group consists of agents sharing common characteristics and allows communication only among its members. A role represents the functionalities and services an agent can provide within a group. In addition, several notations, including first-order logic, are proposed to define the model precisely. Also, the authors introduced introduced the *"cheeseboard diagram"* to describe the organizational structure and an interaction diagram to represent the dynamic aspect of the organization. Figure 9(a) present an example of an organization presented by both the cheeseboard and interaction diagrams.

The dynamic aspect is based on creating agent groups, including partitioning and forming hierarchies. Additionally, the dynamic creation of groups allows agents to play new roles and leave roles, leading to agents taking on multiple roles, potentially resulting in an agent-level overload.

This model is implemented using the Madkit framework, which provides a platform for developing and managing multi-agent systems based on the AGR concepts [16].

5.2 Moise+

MOISE+ is an extension of the MOISE model for specifying the organization of a multiagent system, enhanced by role compatibility, role cardinalities, group recursivity, and role inheritance. It specifies three key aspects: structural, functional, and deontic. The structural aspect defines the relationship between agents through roles and groups. The functional aspect is represented by a social scheme (SCH) that decomposes goals into plans within a tree structure, with the root as the system's global goal. The deontic aspect connects the structural and functional aspects by describing roles' permissions and obligations toward missions with time constraints [14]. Indeed, Hübner et al. [13] introduced an innovative dimension to the MOISE+ model, integrated with the deontic aspect, that specifies which protocols should or can be used to achieve the goals within the roles' missions.

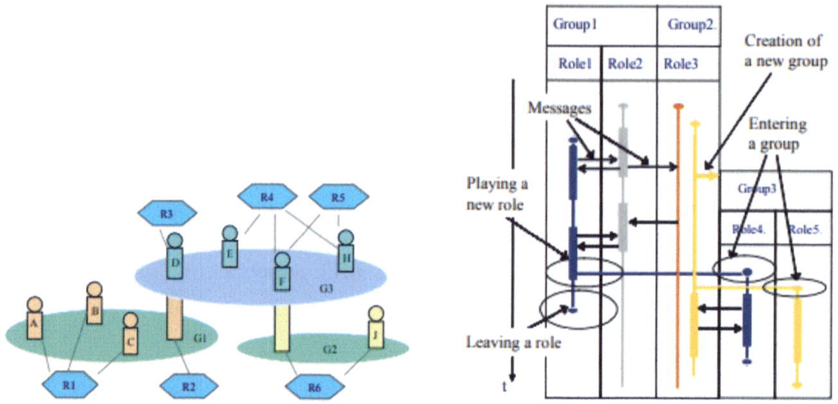

(a) Cheeseboard Diagram. (b) Interaction Diagram.

Fig. 9. Illustrations of the AGR Model: (a) Organizational Structure and (b) Dynamic Interactions [11].

Regarding the dynamic aspect of MOISE, organizational changes are predefined at design time and executed by a monitor agent. These changes include role reassignment, introducing new roles, and adding new obligations. However, because these changes are predefined, they may not address all potential issues, which can limit the system's ability to respond to unexpected situations. Figure 10 presents the whole aspects used to model a multiagent organization.

The JaCaMo framework, which supports the implementation of the organizational dimension, provides a comprehensive platform for developing multiagent system organizations [4].

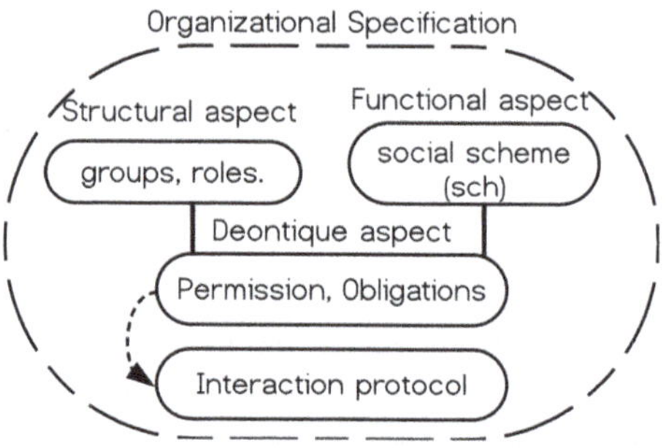

Fig. 10. Aspects Used to Specify an Organization according to MOISE+ [13,14].

5.3 Macodo

Weyns et al. [22] propose an organizational model for modelling dynamic organizations based on the context of the environment. This model is part of an integrated approach called MACODO (Middleware Architecture for COntext-driven Dynamic agent Organizations). The authors have applied this organizational model within a traffic monitoring application context. In this model, the organization is dynamically created according to predefined specifications in response to changes in context, and it evolves with each modification of this context, following predefined laws that govern operations such as merging organizations, exiting agents, joining agents, and splitting organizations. It is important to note that MACODO can be considered less general as it primarily applies to specific domains of collaboration.

5.4 Noshape

Noshape [1,2] is a bioinspired model that aims to combine the two viewpoints, the ACMAS and OCMAS viewpoints from MAS engineering. Noshape makes three levels of abstraction: universe level, which contains an infinite number of worlds; world level, which contains an infinite number of organizations; and an organization level, which contains an infinite number of static and dynamic roles. Roles, which are considered static, are managers of the organization's organizational structure that cannot be exchanged or moved. However, dynamic roles can be created, removed, configured dynamically, and moved from one organization to another. According to this model, the interaction between dynamic and static roles leads to the reorganization of a level higher. In other words, dynamic roles send functional triggers to static roles to bring them into interaction with other dynamic roles. The restructuring of higher level levels uses organizational dynamic behaviours inspired by amoeba, such as expand, overlap, partition, merge, shrink, home, and retire.

Finally, Table 2 presents a comparative study between these models where we focused on the dynamic aspects.

Table 2. Comparative Overview of Multi-Agent System Models.

Models	Engineering View Point	Modeled Aspect	Dynamics Aspect	Timing	Framework
AGR	OCMAS	Structure Function	Dynamic creation of agent groups and hierarchy of groups Dynamic role: entering a group, playing a new role, leaving a role.	Runtime	MADKIT
Moise+	OCMAS	Structure Function Norms	Role changes, give new obligations, new role defined.	Design time	JACAMO
Macodo	ACMAS-OCMAS	Structure Function	Merging organization, exiting agents, joining agents, and splitting organization.	Design time	MACODO
Noshape	ACMAS-OCMAS	Structure Function	Creation, suppression, and configuration of roles dynamically, role movement from one organization to another.	Runtime	Noshape

6 Conclusion

Organizations play a crucial role in multiagent systems (MAS) by enabling the natural modeling of complex systems. Key aspects such as agent relationships, task execution, and governing rules can be effectively studied through an organizational lens. In this paper, we have focused on the dynamic aspects of multiagent organizations, synthesizing key works on this topic. These works were analyzed from two complementary perspectives: the techniques used to ensure dynamism and the models that support it.

In conclusion, the dynamic nature of organizations allows MAS to adapt flexibly to unpredictable situations. However, the process of change in many techniques or models can be resource-intensive, potentially affecting overall system performance. Therefore, we suggest further research on the impact of the change process to identify the most appropriate techniques for different situations.

Acknowledgments. The authors appreciate support from the Research Laboratory on Computer Science's Complex Systems ReLa(CS)[2] and would like to thank the Directorate General for Scientific Research and Technological Development (DGRSDT, MESRS).

References

1. Abbas, H.A., Shaheen, S.: Towards a hybrid mas organizational model: combining the ACMAS and OCMAS viewpoints. Int. J. Org. Coll. Intell. (IJOCI) **7**(4), 18–50 (2017)
2. Abbas, H.A.: Realizing the noshape mas organizational model: an operational view. Int. J. Agent Technol. Syst. (IJATS) **7**(2), 75–104 (2015)
3. Abbas, H.A., Shaheen, S.I., Amin, M.H.: Organization of multi-agent systems: an overview. arXiv preprint arXiv:1506.09032 (2015)
4. Boissier, O., Hübner, J.F., Ricci, A.: The JaCaMo framework. In: Social coordination frameworks for social technical systems, pp. 125–151 (2016)
5. Christianos, F., Chalkiadakis, G.: Efficient multi-criteria coalition formation using hypergraphs (with application to the V2G problem). In: Multi-Agent Systems and Agreement Technologies: 14th European Conference, EUMAS 2016, and 4th International Conference, AT 2016, Valencia, Spain, December 15-16, 2016, Revised Selected Papers, pp. 92–108. Springer (2017)
6. Dell'Anna, D., Dastani, M., Dalpiaz, F.: Runtime norm revision using Bayesian networks. In: Miller, T., Oren, N., Sakurai, Y., Noda, I., Savarimuthu, B., Cao Son, T. (eds.) PRIMA 2018. LNCS (LNAI), vol. 11224, pp. 279–295. Springer, Cham (2018). https://doi.org/10.1007/978-3-030-03098-8_17
7. Dell'Anna, D., Dastani, M., Dalpiaz, F.: Runtime revision of sanctions in normative multi-agent systems. Auton. Agent. Multi Agent Syst. **34**(2), 1–54 (2020). https://doi.org/10.1007/s10458-020-09465-8
8. Dignum, V.: Handbook of research on multi-agent systems: Semantics and dynamics of organizational models: Semantics and dynamics of organizational models. IGI Global (2009)

9. Dorri, A., Kanhere, S.S., Jurdak, R.: Multi-agent systems: A Survey. Access **6**, 28573–28593 (2018)
10. Dutta, A., Ufimtsev, V., Asaithambi, A.: Correlation clustering based coalition formation for multi-robot task allocation. In: Proceedings of the 34th ACM/SIGAPP Symposium on Applied Computing, pp. 906–913 (2019)
11. Ferber, J., Gutknecht, O., Michel, F.: From agents to organizations: an organizational view of multi-agent systems. In: International Workshop on Agent-oriented Software Engineering, pp. 214–230. Springer (2003)
12. Horling, B., Lesser, V.: A survey of multi-agent organizational paradigms. Knowl. Eng. Rev. **19**(4), 281–316 (2004)
13. Hübner, A., Dimuro, G.P., da Rocha Costa, A.C., Mattos, V.: A dialogic dimension for the Moise+ organizational model. In: MALLOW (2010)
14. Hübner, J.F., Sichman, J.S., Boissier, O.: MOISE+ towards a structural, functional, and deontic model for mas organization. In: Proceedings of the First International Joint Conference on Autonomous Agents and Multiagent Systems: part 1, pp. 501–502 (2002)
15. Maestro-Prieto, J.A., Rodríguez, S., Casado Vara, R.C., Corchado, J.M., et al.: Agent organisations: from independent agents to virtual organisations and societies of agents. ADCAIJ Adv. Distrib. Comput. Artif. Intell. J. **9**(4), 55–70 (2020)
16. Michel, F.: MaDKit (2021). https://hal-lirmm.ccsd.cnrs.fr/lirmm-03833158
17. Morris-Martin, A., De Vos, M., Padget, J., Ray, O.: Agent-directed runtime norm synthesis. In: Proceedings of the 2023 International Conference on Autonomous Agents and Multiagent Systems, pp. 2271–2279 (2023)
18. Picard, G., Hübner, J.F., Boissier, O., Gleizes, M.P.: Réorganisation et auto-organisation dans les systèmes multi-agents. In: Journées Francophones sur les Systèmes Multi-Agents (JFSMA'09), pp. 89 (2009)
19. Picard, G., Mellouli, S., Gleizes, M.P.: Techniques for multi-agent system reorganization. In: Engineering Societies in the Agents World VI: 6th International Workshop, ESAW 2005, Kuşadasi, Turkey, October 26-28, 2005, Revised Selected and Invited Papers 6, pp. 142–152. Springer (2006)
20. Premm, M., Kirn, S.: A multiagent systems perspective on industry 4.0 supply networks. In: Müller, J.P., Ketter, W., Kaminka, G., Wagner, G., Bulling, N. (eds.) MATES 2015. LNCS (LNAI), vol. 9433, pp. 101–118. Springer, Cham (2015). https://doi.org/10.1007/978-3-319-27343-3_6
21. Rahmanzadeh, A., Rahmani, M., Nazemi, E.: A model for selecting the best organizational paradigm at run time for self-organizing multi-agent systems; an IoT case study. Int. J. Inf. Commun. Technol. Res. **11**(4), 29–39 (2019)
22. Weyns, D., Haesevoets, R., Helleboogh, A.: The MaCoDo organization model for context-driven dynamic agent organizations. ACM Trans. Auton. Adapt. Syst. (TAAS) **5**(4), 1–29 (2010)
23. Wu, R., Zhang, J., Xin, B.: Research on the coalition formation of multi-agent systems for forest fire prevention and control tasks. In: 2024 IEEE 18th International Conference on Control and Automation (ICCA), pp. 880–885. IEEE (2024)

Cosinus-Based vs Sinus-Based Transfer Functions Effect on Modified Binary Rat Swarm Optimizer

Hichem Rahab[1(✉)], Dalal Bardou[2], Karima Saidi[3], Hichem Haouassi[1], and Said Kouachi[4]

[1] ICOSI Lab, Department of Computer Science, Abbes Laghrour University, 40004 Khenchela, Algeria
{rahab.hichem,haouassi.hichem}@univ-khenchela.dz

[2] Department of Computer Science, Abbes Laghrour University, 40004 Khenchela, Algeria
dalal.bardou@univ-khenchela.dz

[3] Mathematics and Computer Science Faculty Batna2 University, Batna, Algeria
karima.saidi@univ-batna2.dz

[4] Department of mathematics Abbes Laghrour University, 40004 Khenchela, Algeria
Kouachi@univ-khenchela.dz

Abstract. A recent swarm algorithm, Rat Swarm Optimizer (RSO), has demonstrated its efficiency in solving various mathematical problems. The Modified Binary RSO (MBRSO) is a binary version of RSO designed to address binary problems with enhanced performance. Transfer functions, used in MBRSO, serve as the medium for mapping an algorithm from continuous to discrete binary space. Despite the widespread use of swarm algorithms, studies focusing on transfer functions are still limited in both number and diversity. This study proposes eight new transfer functions, grouped into two families: sinus-based and cosinus-based, for MBRSO. Two standard benchmark datasets for Arabic sentiment analysis were used to evaluate the proposed transfer functions in terms of fitness, accuracy and feature reduction rate. Compared with state-of-the-art transfer functions, the newly introduced functions demonstrate their competitiveness and efficiency with MBRSO algorithm. Additionally, parameter fine-tuning appears to be a crucial factor in improving results.

Keywords: RSO · metaheuristic · Feature selection · Transfer functions · Swarm-based optimization

1 Introduction

Sentiment analysis can be considered as the most emergent field in natural language processing domain. In morphology rich languages such Arabic, the feature space is extremely important. Thus, a feature selection step is crucial in both

T. Marir et al. (Eds.): ICCSA 2024, ISEM 45, pp. 222–233, 2025.
https://doi.org/10.1007/978-3-031-90758-6_17

noise elimination and processing cost optimization purposes [1]. Opinion mining seeks to analyze opinionated texts, such as reviews, on social media networks and discussion forums to determine whether they contain positive or negative sentiments. Sentiment analysis is considered both a summarization and a classification task regarding on its results [2].

Feature selection the step in machine learning process that involves identifying and selecting the most relevant features from a dataset to improve the performance of predictive models. By reducing the number of input variables, feature selection helps in minimizing the computational cost and complexity of the model, while also enhancing its accuracy and interpretability [3]. There are various techniques for feature selection, including filter methods, wrapper methods, and hybrid methods. Filter methods use statistical measures to evaluate the relevance of each feature independently of the model, while wrapper methods involve training and evaluating a model to select the best subset of features. Hybrid methods perform feature selection by combining wrapper and filter methods to achieve the best possible performance during the model training process. Effective feature selection can lead to better generalization, faster training times, and improved model performance [4].

Swarm intelligence employs nature-inspired metaheuristic algorithms to solve NP-hard problems. Recently, the Rat Swarm Optimizer (RSO) algorithm was introduced with the aim of solving continuous hard optimization problems [5] with enhanced performance. Subsequently, in the work by [6], a Modified Binary version of RSO (MBRSO) was introduced to address binary problems. The RSO binary versions are primarily applied in feature selection [6] and [7], demonstrating their efficiency in this domain. Similar to continuous versions of swarm algorithms, the binary versions suffer from the common issue of local optima, leading to limited results and slow convergence. This necessitates further improvements to overcome these challenges.

Transfer functions distinguish between the continuous and binary versions of swarm intelligence-based algorithms. These functions enable the mapping from continuous to binary dimensions, with only 1 and 0 values, using a probability function. This is particularly important in swarm intelligence-based algorithms, where transfer functions help convert continuous position vectors into binary ones, facilitating the optimization process. Common transfer functions include the sigmoid and hyperbolic tangent functions, which are widely used due to their smooth and differentiable properties. By effectively transforming data, transfer functions enhance the performance and applicability of machine learning models in diverse problem domains [8].

Following the aforementioned reasons, transfer functions can be considered as the most crucial part of the Binary RSO algorithm variants. Despite this apparent importance, there appears to be insufficient work on the transfer function in the literature. In this paper, two new different families of transfer functions, sinus-based and cosinus-based, consisting of eight transfer functions, are introduced and evaluated. The effectiveness of employing these new transfer functions is investigated in terms of avoiding local minima and convergence speed. Additionally, in order to validate their performance, a comparative study is provided to compare the newly proposed functions with four well-known transfer functions.

The remainder of this paper is organized as follows, In the second section, related works on transfer function with metaheuristic algrithmes were presented and discussed. Then, in Sect. 3 the MBRSO algorithm, which is the version used in this study, is described. Then, Sect. 4 introduces the newly proposed sinus-based and cosinus-based transfer functions. Section 5 is dedicated to the experimental study design, followed by a discussion of the results in Sect. 6. Conclusions and some perspectives are provided in the seventh section.

2 Related Works

Several transfer functions have been proposed in the literature for mapping different metaheuristic algorithms to binary space. In the following, an overview of state-of-the-art work on transfer functions in different algorithms, with a focus on the application in RSO algorithm. RSO with S-shaped transfer function is enhanced and explored for FS problems in [7]. Three new versions of RSO were produced in the base of this enhancements, Binary RSO (BRSO), Binary Enhanced RSO (BERSO), and Binary Enhanced RSO with Crossover operators (BERSOC). A set of 24 benchmark datasets were employed to asses the effectiveness of these versions. The employed dataset include, biology, artificial, politics, physics, game and chemistry domains which prove the effectiveness of RSO versions on different domains.

In [9] the authors employed a sigmoid transfer function to maps Arithmetic Optimization Algorithm (AOA) to Binary AOA (BAOA). To prove the perfromance of BAOA, 18 benchmark datasets from University of California, Irvine(UCI) repository were used. BAOA out performs a set of well known meta-heuristic algorithms like Binary Particle Swarm Optimization (BPSO) and Binary Genetic Algorithm (BGA).

S-shaped transfer functions, presented in Eqs. 1, 2, 3, and 4, were used in [10] for Binary PSO together with four V-shaped functions. Experimental results with several mathematical function shows superior performance of V-shaped functions. A Binary Grasshopper Optimization Algorithm with a new time-varying Gaussian transfer function (BGOA-TVG) is introduced in [11]. BGOA-TVG is evaluated against S-shaped and V-shaped transfer functions cited above on UCI, DEAP, and EPILEPSY datasets. According to the authors of BGOA-TVG, it achieves better results outperformaing S-shaped and V-shaped transfer functions on the different experiments. Different transfer functions from the family of S-shaped, V-shaped, U-shaped[12], Z-shaped, and quadratic [13] transfer functions were used to binarize the Greater Cane Rat Algorithm (GCRA) [14]. Then, another variant BGCRA using a threshold that maps a variable to 0 or 1 is introduced. For experiments, 12 datasets were exploited to evaluate the performance of the six variants. The study concludes that the new threshold version known as BGCRA showed better performance over the five other version using S-shaped, V-shaped, U-shaped, Z-shaped, and quadratic transfer functions.

$$S1(x) = \frac{1}{1 + e^{-2x}} \tag{1}$$

$$S2(x) = \frac{1}{1 + e^{-x}} \tag{2}$$

$$S3(x) = \frac{1}{1 + e^{-\frac{x}{2}}} \tag{3}$$

$$S4(x) = \frac{1}{1 + e^{-\frac{x}{3}}} \tag{4}$$

From this review of the literature, a common conclusion is that the experimental results depend mostly on the algorithm and the domain of the application. The most used functions in the previous works belong to th S-shaped family.

3 Modified Binary RSO

The RSO is a novel nature-inspired algorithm that has proven its efficiency in solving numerous continuous problems. Chasing and fighting the prey are the main phases in the RSO algorithm. Modified Binary RSO (MBRSO) is an enhanced version that improves the RSO's performance, private thinking in essence, by considering the best value so far for each agent when updating its position [6]. Equation 5 models the chess behavior of the rats in MBRSO as follows:

$$\boldsymbol{P} = A.\boldsymbol{P}_{i(best)}(x) + C.\left(\boldsymbol{P_r}(x) - \boldsymbol{P_i}(x)\right) \tag{5}$$

where A and C are defined in 6 and 7 respectively.

$$A = R - x\left(\frac{R}{Max_{iteration}}\right) \tag{6}$$

R is a randum number in the range [1, 5].

$$C = 2 \times rand() \tag{7}$$

$$\boldsymbol{P_i}(x+1) = |\boldsymbol{P_r}(x) - \boldsymbol{P}| \tag{8}$$

In the binary space, since only the numbers '0' and '1' are used, the position cannot be updated using Eq. 8. The method for changing the rat's position is described below. The sigmoid function, given in Eq. 9, is used.

$$S(x) = \frac{1}{1 + e^{-x}} \tag{9}$$

For rat's position mapping from continuous to binary spaces we use Eq. 10. Where *rand* is a random number in [0, 1].

$$x_{ij}^b(t+1) = \begin{cases} 1 \ \ if \ \ S(x_{ij}^b(t+1)) > rand \\ 0 \ else \end{cases} \tag{10}$$

4 Our Proposed Sinus-Based and Cosinus-Based Functions

In the case of continuous RSO, the rat agents navigate the search space using direction vectors, with their positions represented in the continuous real domain. However, the binary search space is conceptualized as a hypercube, where an agent can navigate to closer or more distant corners by flipping different numbers of bits. A transfer function determines the probability of changing the elements of a position vector from 0 to 1 and vice versa. These functions push rat agents to navigate within a binary space [15]. In the next subsections eight newly proposed transfer function intended to replace the sigmoid function (Eq. 9) for the MBRSO algorithm in Sect. 3.

4.1 Sinus-Based Functions

The sinus-based functions has the form give in Eq. 11.

$$Sin\_f(x) = sin(\beta \ log(x)) \ x^{\alpha} \tag{11}$$

The arguments β and α are varied to obtain the four Sinus-based functions 12, 13, 14 and 15 (depicted in Fig. 1) as follows.

$$Sin1(x) = sin(4 \ \log(x))x^{0.2} \tag{12}$$

$$Sin2(x) = sin(4 \ \log(x)) \ x^{0.4} \tag{13}$$

$$Sin3(x) = sin(2 \ \log(x)) \ x^{0.2} \tag{14}$$

$$Sin4(x) = sin(2 \ \log(x)) \ x^{0.4} \tag{15}$$

4.2 Cosinus-Based Functions

The cosinus-based functions has the form in Eq. 16.

$$Cos\_f(x) = cos(\beta \ \log(x)) \ x^{\alpha} \tag{16}$$

As the sinus-based functions, we have fixed the arguments α and β for cosinus-based functions to obtain functions 17, 18, 19 and 20 as follows. Figure 2 shows the graphs of the different new introduced cosinus-based transfer functions.

$$Cos1(x) = cos(4 \ \log(x)) \ x^{0.2} \tag{17}$$

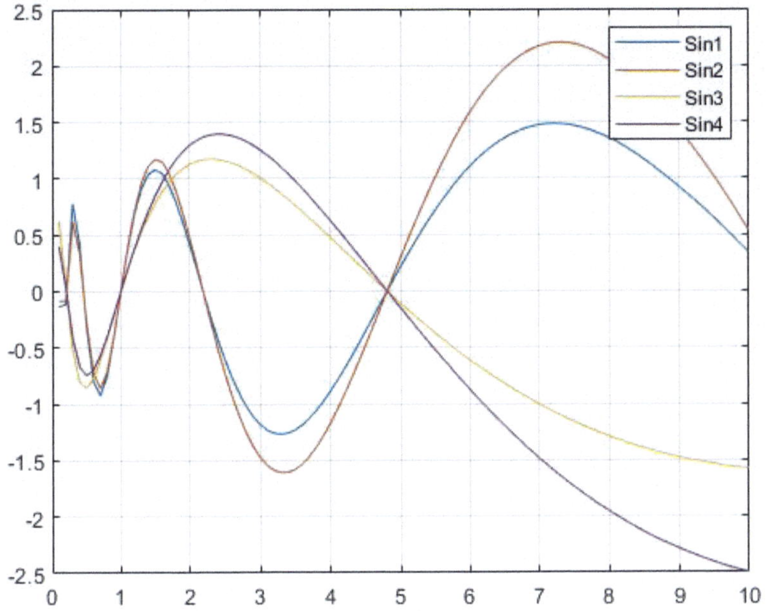

Fig. 1. The newly introduced Sinus-based transfer functions.

$$Cos2(x) = cos(4\ \log(x))\ x^{0.4} \tag{18}$$

$$Cos3(x) = cos(2\ \log(x))\ x^{0.2} \tag{19}$$

$$Cos4(x) = cos(2\ \log(x))\ x^{0.4} \tag{20}$$

5 Experimental Study

In this study, two standard benchmark datasets were employed, i.e. Sana [16] and OCA [17]. Sana corpus is collected from reviews of Algerian newspaper readers and manually annotated at the document level into positive and negative by two experts. On the other hand, the OCA dataset domain is related to Arabic movie review and it is labeled automatically using the websites' rating systems. Table 1 and Table 2 show the statistics of Sana and OCA datasets, respectively. Both OCA and Sana datasets are publicly available for sentiment analysis scientific community.

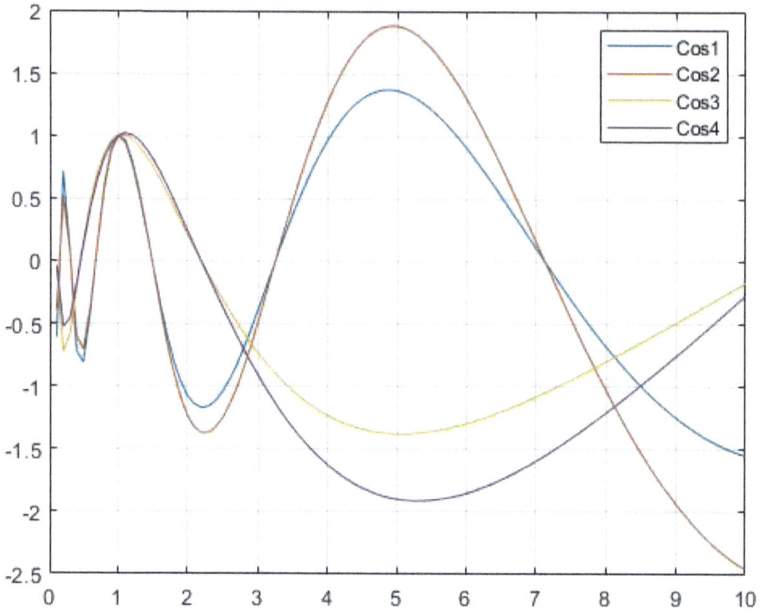

Fig. 2. The newly introduced Cosinus-based transfer functions.

Table 1. Sana dataset statistics.

Nbre instances	positive	negative	# Features
388	194	194	2 860

The focus of this work is on feature selection, thus, two simultaneous objectives, maximization and minimization, were targeted. The maximization problem is related to accuracy, where a higher value of accuracy means better classification performance. On the other hand, minimizing the number of selected features is desired due to its impact on processing time and resources optimization. The fitness function is defined in Eq. 21 to achieve this balance.

$$\uparrow Fitness = Acc + \alpha \left(1 - \frac{|R|}{|C|}\right) \tag{21}$$

Table 2. Oca dataset statistics.

Nbre instances	positive	negative	# Features
500	250	250	11 701

The parameter α is fixed at 0.8, $|R|$ is the number of selected features and $|C|$ the total vocabulary size. The number of agents (rats) is set to 30 and the maximum number of iterations to 300.

Experimental results on SANA and OCA datasets are depicted in Table 3 and table 4 respectively. The best performance results were shown in **bold**. For comparison sake, the performance of the newly introduced transfer functions is compared with four standard benchmark transfer function in the occurrence the S-shaped functions described earlier in this paper (Eqs. 1, 2, 3 and 2).

6 Results and Discussion

Table 3. Sana experimental results.

function	Fitness	Accuracy	# Selected Features / 2860
S1	1.3591	0.9487	1393(48.71%)
S2	1.3481	0.9744	1524(53.29%)
S3	1.3451	0.9487	1443(50.45%)
S4	1.3888	1.0000	1470(51.40%)
Sin1	1.5645	0.9231	567(19.830%)
Sin2	1.5831	0.8718	317(11.08%)
Sin3	1.5786	0.8718	333(11.64%)
Sin4	**1.6125**	0.8718	212(7.41%)
Cos1	1.5883	0.9231	482(16.85%)
Cos2	**1.6454**	0.8974	186(6.50%)
Cos3	1.5693	0.8974	458(16.01%)
Cos4	**1.6100**	0.8718	221(7.73%)

From Table 3, the Cos2 function achieves the best results on Sana dataset. In second place, the functions Sin4 and Cos 4 reaches roughly the same results in this dataset. The Sana dataset, as described by its authors, is related to sentiment analysis of Algerian content on the web, thus the size of its vocabulary is limited, 2860. Achieving good results with such dataset improves the ability of these functions for application with morphologically poor dialects and languages. It can be seen from Fig. 4 that there is a large gap between the new introduced functions both sinus-based and cosinus-based and the state of the art of literature S-shaped functions i.e., S1, S2, S3, and S4 which proves the effectiveness of the propositions in this work.

Moreover, the Fig. 3 shows a good convergence rate of the Cos2 function. We can see that before 150 iterations, around 125, the Cos2 function has more than 1.6246 fitness outperforming all other functions even on 300 iterations. It is worth-mentioning at this stage that, even the S-shaped functions, S1-4, reach a

Table 4. OCA experimental results.

Function	Fitness	Accuracy	# Selected Features/11 701
S1	1.3628	0.9600	5809(49.65%)
S2	1.3696	0.9800	6002(51.29%)
S3	1.3500	0.9600	5997(51.25%)
S4	1.3516	0.9600	5974(51.06%)
Sin1	1.5960	0.9200	1814(15.50%)
Sin2	**1.6259**	0.8800	792(6.77%)
Sin3	**1.6316**	0.9200	1293(11.05%)
Sin4	**1.6537**	0.9000	677(5.79%)
Cos1	**1.6259**	0.9000	1084 (9.26%)
Cos2	1.5934	0.8800	1267(10.83%)
Cos3	1.5991	0.9600	2354.00 (20.12%)
Cos4	**1.6386**	0.8800	605(5.17%)

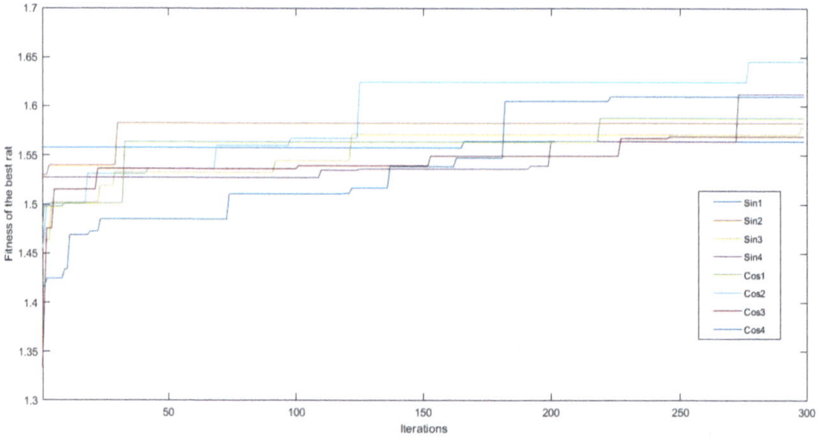

Fig. 3. Sana Fitness using the newly introduced transfer functions.

good accuracy, their fitness is very low due to the limited feature reduction. For this reason, we didn't consider this performance since we search for a combination of high accuracy with important reduction in the feature space.

The results of the OCA dataset, depicted in Table 4, continue to prove the effectiveness of both the Sin4 and Cos4 functions, with Sin4 achieving the highest score in all experiments. Sin2, Sin3, and Cos1 also achieve competitive results in this dataset, as shown in Fig. 5. Like the Sana dataset, Fig. 6 indicates the high performance of the newly introduced functions over the S-shaped functions used in the literature on the OCA dataset. OCA dataset is one of the widely used resources for Arabic sentiment analysis. It is characterized by its rich content

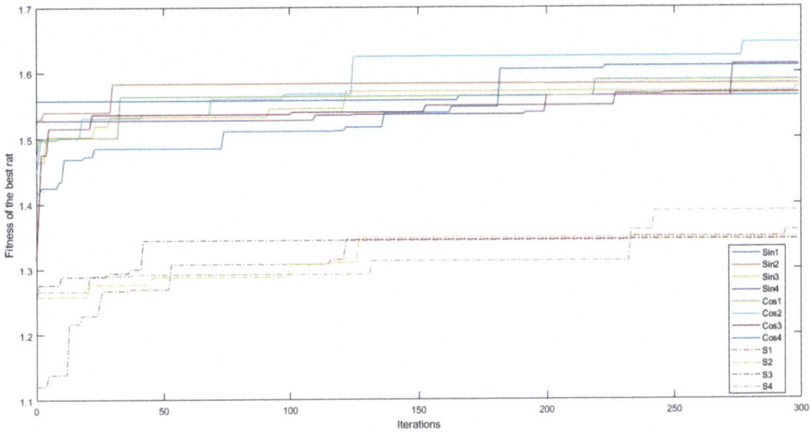

Fig. 4. A comparison between the newly introduced transfer functions and other transfer functions on fitness value for Sana dataset.

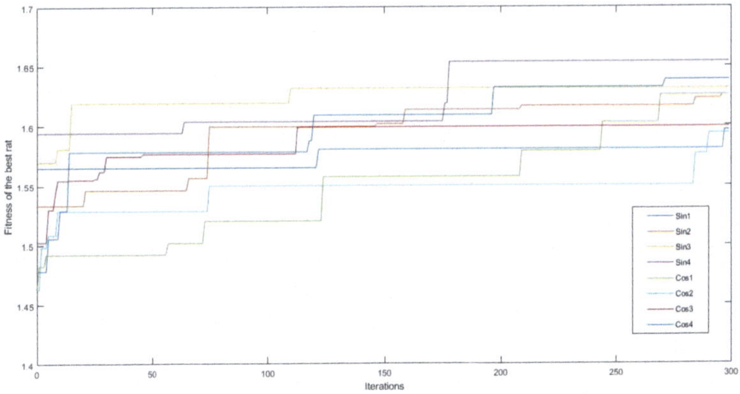

Fig. 5. OCA Fitness using the newly introduced transfer functions.

focusing on Modern Standard Arabic (MSA). The obtained results indicate the importance of these functions for rich morphological languages and different Arabic dialects.

In summary, this experimental study proves that although the earlier proposed S-shaped transfer functions are suitable for other swarm algorithms, their performance with RSO is very limited. On the other hand, both sinus-based and cosinus-based transfer function families prove their efficiency in mapping RSO to Binary space, especially in sentiment analysis domain. Beside achieving satisfying results with almost all newly introduced functions, Sin4 and Cos4 appear as the leading functions in these experiments. Recall that these two functions have the same arguments β and α values of 2 and 0.4 respectively. This conclusion sheds light on the importance of efficient parameter fine-tuning for these novel transfer functions.

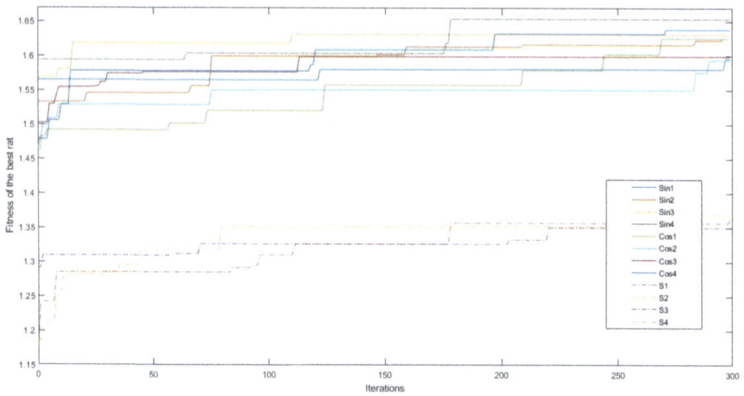

Fig. 6. A comparison between the newly introduced transfer functions and other transfer functions on fitness value for OCA dataset.

7 Conclusions and Perspectives

The current study introduced a set of eight new transfer functions grouped into two families,i.e. sinus-based and cosinus-based. The new functions were evaluated on Arabic sentiment analysis using two standard benchmark datasets. To prove the efficiency of the proposed transfer functions, their results on both datasets were compared with four standard state-of-the-art functions in the occurrence the S-shaped functions.

Obtained results of the newly introduced functions prove their competitive performance against other functions. The last two functions, Cos4 and Sin4 outperform the other functions, and knowing that these function have the same parameters sheds light on the critical importance of parameter fine-tuning for these functions. These results demonstrate the suitability of these functions for metaheuristic algorithms binarization and in essence the RSO algorithm.

For future work, we recommend parameter fine-tuning with different values of β and α. Use these functions on binary version of other swarm algorithms may lead to additional important results.

References

1. Rahab, H., Djoudi, M., Zitouni, A.: Sentiment analysis of Arabic documents: main challenges and recent advances. In: Natural Language Processing For Global and Local Business, pp. 307–331 (2021)
2. Bekhouche, M., Haouassi, H., Bakhouche, A., Rahab, H., Mahdaoui, R.: Improved binary crocodiles hunting strategy optimization for feature selection in sentiment analysis. J. Intell. Fuzzy Syst. **45**, 369–389 (2023)
3. Theng, D., Bhoyar, K.: Feature selection techniques for machine learning: a survey of more than two decades of research. Knowledge Inf. Syst. **66**, 1575–1637 (2024)

4. Yu, L., Liu, H.: Feature selection for high-dimensional data: A fast correlation-based filter solution. In: Proceedings Of The 20th International Conference On Machine Learning (ICML-03), pp. 856–863 (2003)
5. Dhiman, G., Garg, M., Nagar, A., Kumar, V., Dehghani, M.: A novel algorithm for global optimization: rat swarm optimizer. J. Amb. Intell. Hum. Comput. **12**, 8457–8482 (2021)
6. Rahab, H., Haouassi, H., Souidi, M., Bakhouche, A., Mahdaoui, R., Bekhouche, M.: A modified binary rat swarm optimization algorithm for feature selection in Arabic sentiment analysis. Arabian J. Sci. Eng. **48**, 10125–10152 (2023)
7. Awadallah, M., Al-Betar, M., Braik, M., Hammouri, A., Doush, I., Zitar, R.: An enhanced binary Rat Swarm Optimizer based on local-best concepts of PSO and collaborative crossover operators for feature selection. Comput. Biol. Med. **147**, 105675 (2022)
8. Qasim, O., Algamal, Z.: Feature selection using different transfer functions for binary bat algorithm. Int. J. Math. Eng. Manag. Sci. **5**, 697 (2020)
9. Khodadadi, N., et al.: BAOA: binary arithmetic optimization algorithm with K-nearest neighbor classifier for feature selection. IEEE Access. **11**, 94094–94115 (2023)
10. Mirjalili, S., Lewis, A.: S-shaped versus V-shaped transfer functions for binary particle swarm optimization. Swarm Evol. Comput. **9**, 1–14 (2013)
11. Li, M., Luo, Q., Zhou, Y.: BGOA-TVG: binary grasshopper optimization algorithm with time-varying gaussian transfer functions for feature selection. Biomimetics **9**, 187 (2024)
12. Mirjalili, S., Zhang, H., Mirjalili, S., Chalup, S., Noman, N.: A novel U-shaped transfer function for binary particle swarm optimisation. In: Soft Computing for Problem Solving 2019: Proceedings of SocProS 2019, vol. 1, pp. 241–259 (2020)
13. Jordehi, A.: Binary particle swarm optimisation with quadratic transfer function: a new binary optimisation algorithm for optimal scheduling of appliances in smart homes. Appl. Soft Comput. **78**, 465–480 (2019). https://www.sciencedirect.com/science/article/pii/S1568494619301127
14. Agushaka, J., Akinola, O., Ezugwu, A., Oyelade, O.: A novel binary greater cane rat algorithm for feature selection. Results Control Optim. **11**, 100225 (2023)
15. Rashedi, E., Nezamabadi-Pour, H., Saryazdi, S.: BGSA: binary gravitational search algorithm. Nat. Comput. **9**, 727–745 (2010)
16. Rahab, H., Zitouni, A., Djoudi, M.: SANA: sentiment analysis on newspapers comments in Algeria. J. King Saud Univ.-Comput. Inf. Sci. **33**, 899–907 (2021)
17. Rushdi-Saleh, M., Martin-Valdivia, M., Urena Lopez, L., Perea Ortega, J.: OCA: opinion corpus for Arabic. J. Am. Soc. Inf. Sci. Technol. **62**, 2045–2054 (2011)

Design, Specification, and Verification of a Cooperative Multi-robots System for Delivery Service

Mahdi Kahoul(✉), Leyla Belaiche, Laid Kahloul, and Samir Tigane

LINFI Laboratory, University of Biskra, Biskra, Algeria
mahdikahoul07@gmail.com, {l.kahloul,s.tigane}@univ-biskra.dz

Abstract. This paper presents the specification and design of a cooperative multi-robot system (CMRS) tailored for delivery services. In the context of increasing demand for efficient logistics solutions, especially in underdeveloped regions like Algeria, the deployment of mobile robots offers promising avenues for enhancing delivery operations. Leveraging advancements in robotics, this paper outlines a comprehensive approach toward developing a fleet of collaborative robots capable of navigating diverse environments to facilitate delivery tasks. The proposed system addresses the unique challenges faced in delivery services, such as communication and seamless coordination among multiple robots. In this work, a prototype of a mobile robot for delivery service is provided to test the proposed cooperative multi-robot system at the University of Biskra in Algeria as a case study. Afterward, a formal model for the proposed system is prepared based on the Buchi automaton. Under the UPPAAL model-checker tool, the behavior of the proposed CMRS is verified in terms of two criteria: the system's safety and the system's liveness which are formulated in a set of computational tree logic (CTL) properties. The UPPAAL model-checker tool results show that the proposed CMRS represents a safe and robust delivery mobile system.

Keywords: Delivery operations · Mobile robotics · Cooperative multi-robot system · Buchi automata · CTL logic · Model-checking · UPPAAL tool

1 Introduction

In recent years, the robotics field has seen incredible developments that have revolutionized several industries and redefined the possibilities of task automation. Mobile robots are advantageous because they can be used in a wide range of environments and are highly versatile [20]. Mobile robots have proven their effectiveness in streamlining workflows and increasing productivity in a variety of settings, including industry [18], hospitals, warehouses, delivery services [10,12], and exploration and mission missions [14].

T. Marir et al. (Eds.): ICCSA 2024, ISEM 45, pp. 234–247, 2025.
https://doi.org/10.1007/978-3-031-90758-6_18

In the context of delivery services, mobile robots offer a promising solution to address the growing demands of logistics. With the emergence of e-commerce and the increasing need for last-mile delivery solutions, mobile robots present an efficient service to optimize the delivery process, reduce costs, and improve customer satisfaction [3]. Examples of such applications include Amazon's delivery drones [1] and Star-Ship Technologies' autonomous delivery robots [2]. These examples produced in the United States of America illustrate the potential of mobile robots in transforming the delivery landscape. Therefore, developed countries have exploited the integration of mobile robot technologies into their logistics infrastructure, however, underdeveloped regions such as Algeria still face challenges in adopting such technologies. With limited resources and infrastructure, Algeria is poised to take advantage of deploying mobile robots and robotic solutions in a variety of sectors, including delivery, industrial, and hospital.

Several research studies have substantially contributed to the advancement of mobile robotics, particularly in navigation, localization, and multi-robot coordination. These studies have focused on refining localization techniques via integrating sensor fusion and probabilistic methods [16]. This has resulted in an enhanced accuracy of position estimation in dynamic environments. Navigation algorithms have undergone extensive development to address challenges posed by complex terrains and dynamic obstacles, enabling robots to traverse indoor [16,17,25] and outdoor [15,24] spaces with increased agility and precision. Additionally, research efforts have investigated communication protocols, task allocation strategies, and collaborative decision-making mechanisms for multi-robot systems [7], paving the way for developing cooperative robots capable of tackling diverse challenges.

Formal methods use mathematical description and analysis to insure systems reliability and to study their performance [5,6,8,9,11,21–23,26,27]. In the field of multi-robotics systems, numerous researches have exploited formal methods for specification and verification of multi-robots-system to ensure simulation and evaluate the behavior of some complex multi-robot systems [13] by verifying some properties such as (system's safety, system's liveness) [19]. Collectively, these research endeavors highlight the interdisciplinary nature of mobile robotics and underscore the ongoing quest for innovative solutions to real-world problems in autonomous systems.

This work addresses the requisite for delivery robotics solutions in Algeria by proposing a mobile delivery robot. The proposed solution is made of three stages: (i) mobile robot design and prototype, (ii) formal specification of a cooperative multi-robot system for delivery service applications via timed automata, and (iii) based on model-checking adopted in UPPAAL tool, safety and liveness properties are verified for the cooperative multi-robot system. The proposed system aims to optimize the delivery process by leveraging the collective capabilities of multiple robots by working in coordination.

The remainder of this paper is organized as follows. Section 2 describes the hardware setup and the experimental environments for the proposed mobile robot and illustrates its final prototype and robust localization scheme. The specification and verification of the proposed system's behavior, are analyzed in Sect. 3. Finally, conclusions are given in Sect. 4.

2 Mobile Robot for Delivery Service

To implement the proposed multi-robot system for the delivery service, we started by designing and implementing a mobile robot that can move across the university campus, because it offers wide open space and a well-prepared infrastructure for testing the robot. This work takes the University of Biskra in Algeria as a case study. Figure 1 shows three maps of the university campus, streets, satellites, and topology maps.

(a) streets Map (b) satellites Map (c) topology Map

Fig. 1. Maps of the University of Mohamed Khider Biskra, Algeria

2.1 Hardware and System Design

We have designed and implemented a prototype to test it on our proposed system. The prototype is a Delivery four-wheeled mobile robot that has a safety box to carry products. Figure 2 illustrates the final prototype of the proposed mobile robot for delivery service. The following set of required devices including sensors, the battery, and the motor are employed to accomplish the final prototype and design the proposed mobile robot.

- **Raspberry Pi 4**: The main processing unit of the robot is the Raspberry Pi 4 4G RAM micro-controller board, which is responsible for communicating with the coordinator and receiving the specified path to the target through web platform using the TCP/IP protocol, which in turn sends the path data to the Arduino unit through UART Serial communication protocol.
- **Arduino Mega micro-controller board**: processes all the input GPS and compass data and generates pulse width modulation (PWM) signals to control the motors. The operating voltage of the board is 5 V.

- **Power Supply**: We used a 14.8 V 3300 mAh (milliampere per hour) lithium polymer battery to power the whole robot. The battery provides sufficient power to run the robot for 45–60 min.
- **Ublox Neo-M8N GPS module** The Ublox Neo-M8N module was utilized which comes with an HMC5883L compass. Mounting the Ublox series compact GPS compass module away from sources of interference within the system's confines is easy thanks to the latest Ublox series compact GPS compass module. Under good reception conditions, the receiver can provide a high level of location tracking accuracy. This sensor component allows us to obtain precise data on latitude, longitude, altitude, time, and direction angle from a single device.
- **Motor driver**: The L298N motor driver board was utilized to power the 4 DC geared motors employed in our robot's wheels. A lithium polymer battery with a capacity of 14.8 V 3300 mAh powers the motor driver. The motor driver can handle the load of 2 geared motors with a peak output current of 2 Amp per channel.
- **Servo motor**: The container's lid is opened by a Tower Pro MG996R servo motor. The stall torque of this servo motor is 9.4 kg/cm and it operates at an operating voltage of 4.8 V.

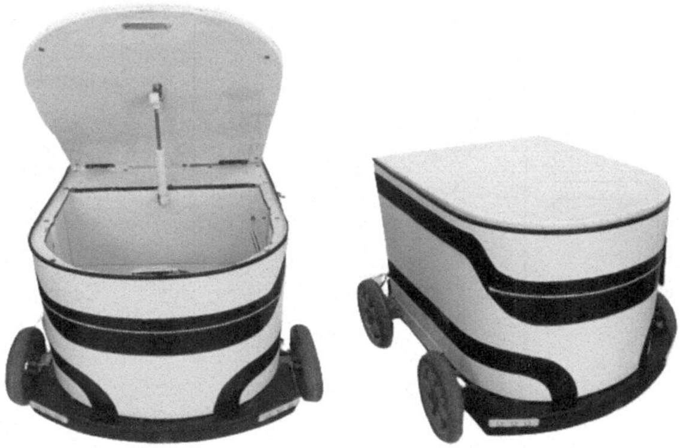

(a) front side (b) left side

Fig. 2. The final prototype of the proposed mobile robot

After selecting all the required components for prototyping the proposed mobile robot, we propose an architecture of the system composed of four subsystems: motion, localization, communication, and perception systems. Each of these has a specific role in the overall robot behavior. Figure 3 illustrates the system architecture of the proposed mobile robot with its sub-systems equipped with hardware components.

- **Motion System**: is responsible for the robot's movement in various directions by operating the motors through signals issued by the monitoring unit.

- **Localization System**: represents a system that answers the question "Where am I ?" concerning the robot, by determining the robot's coordinates and the steering angle concerning the desired path.
- **Communication System**: Communication system: This system provides the robot with the ability to communicate with the command station to receive and send all the necessary data, for example (path coordinates, robot coordinates, etc.).
- **Perception System**: allows the robot to recognize its surrounding environment through the camera and sensors to overcome obstacles that hinder the navigation process.

The robot was programmed using the Arduino Integrated Development Environment (Arduino IDE) and C++ and Python programming languages.

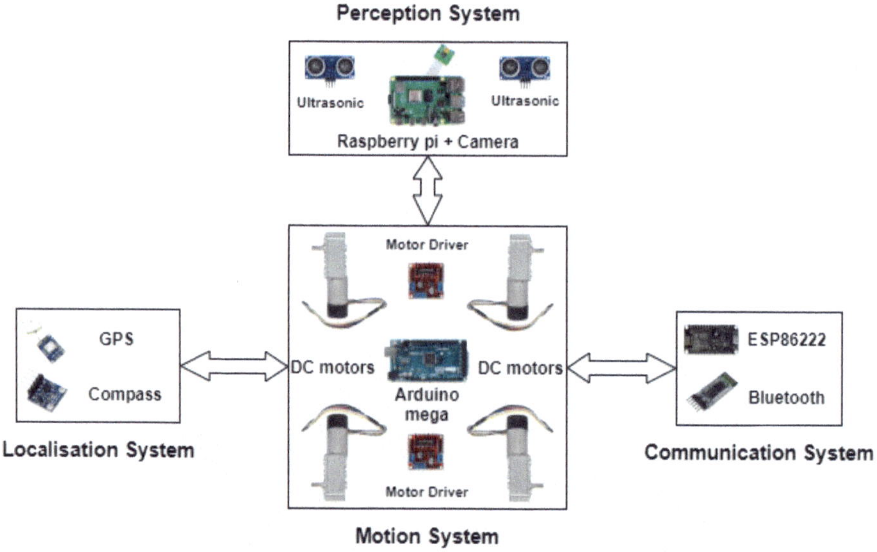

Fig. 3. System Architecture

2.2 Working Principle

The user can use the proposed mobile robot via a web application, starting by selecting the desired destination, also he can show the path from the robot's position to the selected destination in the satellite map by pressing the "Show" button. Figure 4 depicts the designed web application user interface for the proposed mobile robot and an example of a path to be passed by the robot.

The robot software begins by initializing the GPS sensors and compass. The working principle of the robot follows N steps. These steps are described in the

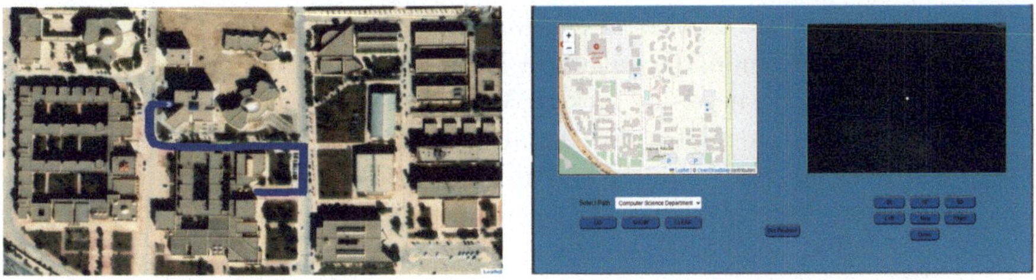

(a) Sample of a path (b) Web application interface

Fig. 4. The web application interface for the proposed mobile robot with an example a path

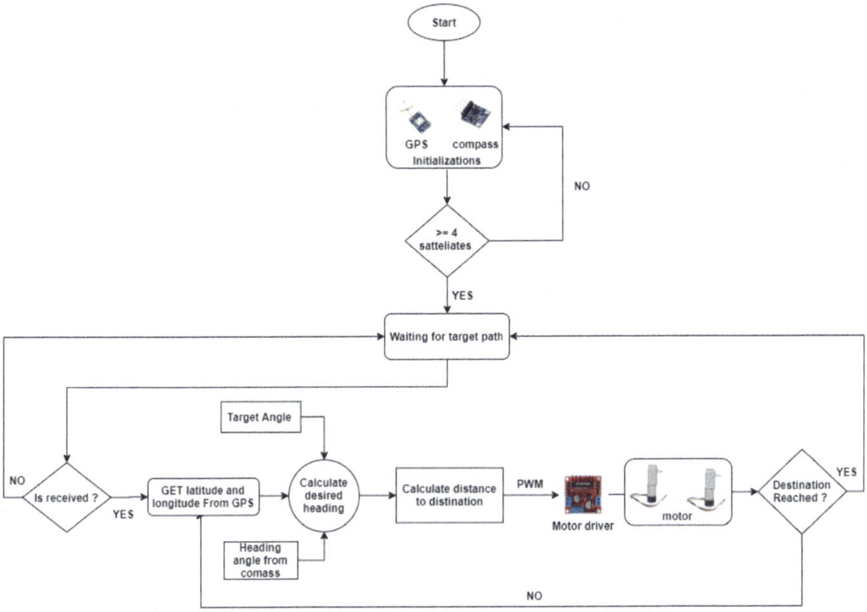

Fig. 5. Program flowchart of the robot

following. Figure 5 shows the flowchart of the robot program, which summarizes all the steps of the robot program equipped with their required hardware.

1. **The user** determines the target path through the website interface shown in Fig. 4, based on a map that shows the path from the starting point to the destination point in the form of a drawn line, then sends the command by pressing the Go button.
2. **Raspberry Pi unit** receives the path data and converts it to a JSON object, then sends the path in an array format of latitude and longitude coordinates to the Arduino module through the UARTs Serial communication protocol.

3. **The GPS module** Looks for satellites and examines the data on current latitude and longitude. To calculate latitude and accurately determine the longitude, the GPS needs be locked With at least 4 satellites.
4. Then **the robot** calculates the Distance to the target and GPS Cours from the latitude and longitude data. We have used the haversine formula [4]to calculate the great-circle distance between two points on the Earth's surface:

$$d = 2R \cdot \arcsin\left(\sqrt{\sin^2\left(\frac{\Delta\phi}{2}\right) + \cos(\phi_1)\cos(\phi_2)\sin^2\left(\frac{\Delta\lambda}{2}\right)}\right)$$

where:

$$\Delta\phi = \phi_2 - \phi_1$$
$$\Delta\lambda = \lambda_2 - \lambda_1$$

- **d** represents the distance between two points on the surface of a sphere (in this case, the Earth), measured along the sphere's surface.
- **R** is the radius of the sphere. In the context of the haversine formula, it represents the radius of the Earth. The R-value is approximately 6371 km or 3959 mi, depending on the units you're using.
- $\Delta\phi$ is the difference in latitude between two points, measured in radians. Where ϕ_1 and ϕ_2 are the latitudes of the two points, respectively.
- $\Delta\lambda$ is the difference in longitude between the two points, measured in radians. Where λ_1 and λ_2 are the longitudes of the two points, respectively.

5. Then **the robot** calculates the target angle from the latitude and longitude data. The angles must be measured in radians.
6. The current heading angle is received from the compass unit and the error angle is measured from the difference between the current heading angle and the target's angle.
7. After knowing the angle of error, we know whether it is a robot It needs to turn left or right. The four motors get pulses according to the Error angle of a micro-controller. Engine L298N The driver operates the motors to align the robot with the target angle.
8. **The robot** checks if it has reached the threshold value from its destination (in our robot, we set the threshold value to 1 m). If the robot does not reach the destination, it will recalculate two angles (the target and the current heading) and the distance to the destination. This loop continues till the robot reaches its destination.
9. When **the robot** reaches its destination, it sends a signal to the control unit, then stops and waits for the user to type the entered password to receive a delivery.

3 Formal Modeling and Verification of the System

To prove the efficiency of the proposed cooperative multi-robots system, we decided to use formal modeling tools such as Timed Automaton and UPPAAL tool to specify, simulate, and model-check the proposed system. In this section, we will present (i) the basics of the automata in the UPPAAL tool, (ii) the specification of the system using timed automata, and (iii) the model-checking of some basic properties.

3.1 Automata and Temporal Logic in UPPAAL Tool

In the context of the UPPAAL tool, an automaton (or timed automaton) is a mathematical model used to describe the behavior of real-time systems. It consists of a finite set of states and transitions between those states, where the transitions are triggered by events or conditions. Each state can have an associated clock constraint, and transitions can be guarded by conditions on clocks, making the automaton suitable for modeling systems with timing requirements. In UPPAAL, automata are used to represent individual components or processes within a system, and these automata can be composed to model the entire system.

A **timed automaton** in UPPAAL is formally defined as a tuple $A = (L, l_0, C, \Sigma, E, I)$, where:

- L is a finite set of locations (or states).
- $l_0 \in L$ is the initial location.
- C is a finite set of clocks.
- Σ is a finite set of actions (which includes input, output, and internal actions).
- $E \subseteq L \times \Sigma \times \mathbb{B}(C) \times 2^C \times L$ is a set of edges (or transitions). Each edge $e = (l, a, g, r, l')$ consists of:
  - l and l' are the source and destination locations.
  - $a \in \Sigma$ is the action labeling the transition.
  - $g \in \mathbb{B}(C)$ is a guard, a condition on the clocks that must be true for the transition to occur. $\mathbb{B}(C)$ is the set of formulas built upon the set of clocks C.
  - $r \subseteq C$ is a set of clocks to be reset when the transition is taken.
- $I : L \rightarrow \mathbb{B}(C)$ is an invariant condition that must hold as long as the system is in a given location.

In UPPAAL, a system is typically modeled as a network of such automata, where each automaton interacts with others through synchronization on shared actions or through shared variables. The tool then analyzes the possible behaviors of this network to verify properties such as safety (something bad never happens) and liveness (something good eventually happens).

Temporal Logic in UPPAAL: *Timed Computation Tree Logic (TCTL)* is an extension of Computation Tree Logic (CTL) that incorporates timing constraints, making it particularly useful for verifying real-time systems in tools like

UPPAAL. TCTL allows the specification of temporal properties that involve not only the sequence of states but also the timing between them. In TCTL, the usual CTL operators such as X (next), F (eventually), G (globally), and U (until) are extended with timing constraints. For example, $F < 5$ means "eventually within 5 time units", and $G \leq 10$ means "globally, within the next 10 time units". The Uppaal tool provides only two operators: F denoted as $<>$ and G denoted as [ ].

In UPPAAL, TCTL is used to express properties like "a certain state must be reached within a specified amount of time" or "a condition must hold continuously for a certain duration". This allows for the verification of timing-specific requirements in the system, ensuring that not only does the system behave correctly, but it also meets the necessary timing constraints. UPPAAL's model checker can evaluate these TCTL queries against the model, providing a powerful way to verify the correctness and timeliness of real-time systems.

3.2 Modeling of the System Using Automata

The model of the system consists of two timed automata templates: the coordinator automaton (see Fig. 6) and the robot automaton (see Fig. 7). By instantiating these two templates and combining them using the UPPAAL tool, we can simulate and verify the whole system's behavior which may be composed of several cooperative robots.

Coordinator Automaton: The coordinator automaton represents the main states of the coordinator system with all possible basic transitions from selecting the given target path to reaching the desired destination.

"`start`" state is the starting point of the entire control system. After starting the system, it switches to "`Wait`" state to communicate with the robots. In the "`Wait`" state, the system waits to receive the "`ID`" of each robot through the channel "`Connect_to_robot`" and it saves all received robots identifiers in an array which will be used for direct communication with each robot individually.

When the coordinator ensures communication with all robots ("`all_robots_connect = true`"), it moves to the "`wait_for_set_destination`" state that is determined by the user through the website interface. After receiving the path, the system sends the path to all connected robots via the path transmission channel "`send_path`".

Hence, each robot is expected to send its distance from the target to allow the coordinator to select the closest robot to the target. Following selecting the nearest robot, the system sends the selection command to the selected robot and then waits for the robot to start moving.

During the waiting phase to reach the target "`wait_for_destination_reached`", the system receives the target arrival signal from the robot, switches to "`Destination_reached`" state to inform the user, and then returns to the path determination waiting state again.

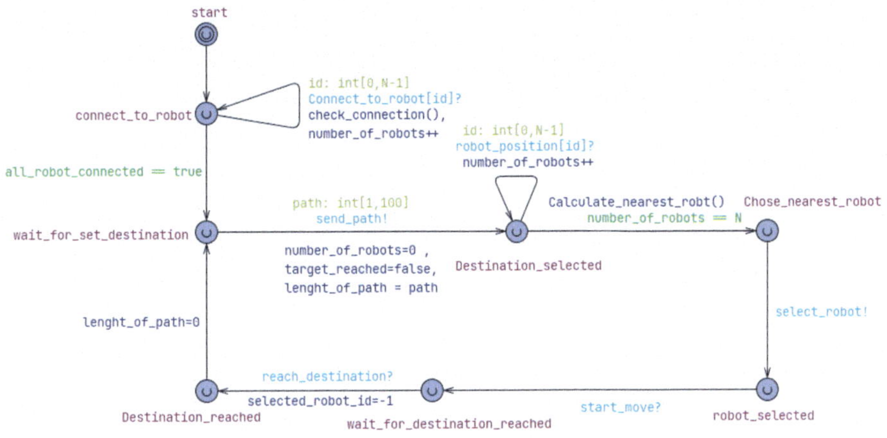

Fig. 6. Coordinator Automaton

Robot Automaton: The robot automaton represents the behavior of the robot and the main states that the robot visits, with all possible basic transformations during the navigation process.

When the robot starts working, it starts from the "`start`" state and sends its "`ID`" to the coordinator to establish direct communication with that coordinator.

After sending the "`ID`", the robot automaton reaches the "`Wait`" state to receive the destination from the coordinator. Upon receiving the destination of the target, the robot "`distance_from_goal`" and sends this distance to the coordinator, which in turn performs the process of electing the nearest robot from the target, hence, the robot moves to "`waiting_to_be selected`" state.

After the election process, all robots receive the decision of the selected robot. Every robot compares its "`ID`" with "`selected_robot_id`" and if its "`ID`" is not equal to "`selected_robot_id`" then it moves to "`not_selected`" state, hence this robot returns again to "`Wait`" state, waiting for any new path (thus, a new delivery mission).

In a case where the "`ID`" of the robot is equal to the select "`selected_robot_id`" by the coordinator, the robot receives permission to navigate. The process of navigating begins as we have explained in the previous section about the principle of work. In this case, this robot sends the move signal to the coordinator.

Finally, when the robot reaches the destination, it sends the "`reach_destination`" command to the coordinator, then this robot returns to the "`Wait`" state, waiting for any new path (thus, a new delivery mission).

3.3 Model-Checking and Evaluation

The model-checking based on automata using UPPAAL consists of checking CTL (computational tree logic) properties against the composed automata of

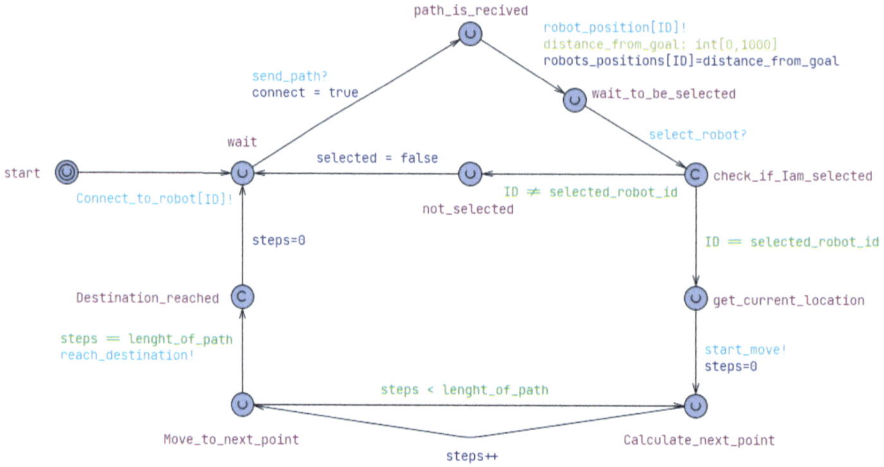

Fig. 7. Robot Automaton

the entire system. CTL represents a logic-based method for describing properties related to the system's behavior. It provides a set of operators that allow us to express various types of properties. Two main categories of properties are to be checked: **safety properties** and **liveness properties**. The following paragraphs present several formulas belonging to each of the previous categories and show if these formulas are fulfilled or not in the proposed system.

Safety Properties: "something Bad Does Not Happen". Safety properties express that "nothing bad will happen" when the system's running. These properties focus on preventing undesirable states or behaviors.

- *Deadlock Freedom*: This property ensures that the system does not stop working unexpectedly throughout its operation, and is considered one of the most important features to check. We use formula 1 to ensure this property.

$$A[\,]\ not\ Deadlock \tag{1}$$

- *Resource Availability*: This property allows us to verify the availability of system components. For example, in our case, we can use 2 to verify that at least one robot is in service.

$$A <> \ (coordinator.all\_robot\_connected = true) \tag{2}$$

Liveness: "Something Good Eventually Happens". Liveness properties state that "something good will happen." They ensure that specific conditions eventually hold while the system is running.

- *Response to Requests*: by ensuring this property we ensure that every delivery request issued by the coordinator must be completed by at most one robot.in CTL we can verify this property using formula 3:

$$cor.Chose\_nearest\_robot \longrightarrow cor.robot\_selected \tag{3}$$

- *Task Completion*: Every delivery task lunched by the coordinator must be completed. Formula 4 defines the property of task completion.

$$(cor.wait\_for\_distination\_reached \longrightarrow cor.Distination\_reached) \tag{4}$$

- *Fairness*: Tasks must be distributed among the robots fairly for, in CTL we can verify this property using formula 5:

$$E <> (r1.ID = selected\_robot\_id \text{ or } r2.ID \ = \ selected\_robot\_id) \tag{5}$$

We have modeled the model using the UPPAAL Version 5.0 and the model was simulated and properties are checked on a computer of Intel(R) Core(TM) i5-6300U CPU @ 2.40 GHzU and 8 GB RAM, running Windows 10 Pro 64-bit (Table 1).

Table 1. Results of model-checker in UPPAAL

Property name	CTL formula	Result
Deadlock Freedom	A[ ] not Deadlock	Satisfied
Resource Availability	A<> (coordinator.all_robot_connected = true)	satisfied
Response to Requests	(cor.Chose_nearest_robot ⟶ cor.robot_selected)	satisfied
Task Completion	(cor.wait_for_distination_reached ⟶ cor.Distination_reached)	satisfied
Fairness	E<> (r1.ID = selected_robot_id or r2.ID = selected_robot_id)	satisfied

4 Discussion and Conclusion

In this work we have developed a prototype of an autonomous mobile delivery robot with a password-protected container that can deliver up to 10kg of products to a certain GPS location.

The proposed mobile robot can be useful in delivering food, grocery, medical equipment in hospitals, and daily useful products to any desired destination.

The proposed mobile robot provides a simple user interface for customers, making it smart and easy to use. The lightweight construction of our robot makes it safe to use in an environment containing humans.

Additionally, we have proposed a cooperative multi-robot system to accomplish the delivery task. The results of the formal model-checker UPPAAL show that the proposed system ensures cooperation between robots.

The proposed cooperative multi-robot system can play an important role in accomplishing transporting tasks. Our system can be a smart logistics system that solves significant logistics problems in the supply chain, specifically by minimizing last-mile delivery cost and time.

References

1. Amazon releases updates on drone delivery, robots, and packaging. https://www.aboutamazon.com/news/operations/amazon-delivering-the-future-2023-announcements. Accessed 09 Apr 2024
2. Starship technologies: Autonomous robot delivery - the future of delivery - today!. https://starship.co/. Accessed 17 Apr 2024
3. Alverhed, E., Hellgren, S., Isaksson, H., Olsson, L., Palmqvist, H., Flodén, J.: Autonomous last-mile delivery robots: a literature review. Eur. Transp. Res. Rev. **16**(1), 4 (2024)
4. Azdy, R., Darnis, F.: Use of haversine formula in finding distance between temporary shelter and waste end processing sites. J. Phys. Conf. Ser. 1500, 012104 (2020)
5. Ben Attia, H., Kahloul, L., Benhazrallah, S., Bourekkache, S.: Using hierarchical timed coloured petri nets in the formal study of TRBAC security policies. Int. J. Inf. Secur. **19**(2), 163–187 (2020)
6. Bouzegag, W., Belaiche, L., Kahloul, L., Bennoui, H.: Leveraging formal methods to blockchain consensus protocols: a scoping literature review. In: 2022 International Symposium on iNnovative Informatics of Biskra (ISNIB), pp. 1–6. IEEE (2022)
7. Gielis, J., Shankar, A., Prorok, A.: A critical review of communications in multi-robot systems. Current Robot. Rep. **3**(4), 213–225 (2022)
8. Hmidi, Z., Kahloul, L., Benharzallah, S.: A new mobility and energy harvesting aware medium access control (meh-mac) protocol: modelling and performance evaluation. Ad Hoc Netw. **142**, 103108 (2023)
9. Houimli, M., Kahloul, L., Benaoun, S.: Formal specification, verification and evaluation of the MQTT protocol in the internet of things. In: 2017 International Conference on Mathematics and Information Technology (ICMIT), pp. 214–221. IEEE (2017)
10. Ivanov, S., Webster, C.: Robots, Artificial Intelligence, and Service Automation in Travel, Tourism and Hospitality. Emerald Publishing Limited (2019)
11. Kahloul, L., Bourekkache, S., Djouani, K.: Designing reconfigurable manufacturing systems using reconfigurable object petri nets. Int. J. Comput. Integr. Manuf. **29**(8), 889–906 (2016)
12. Khaliq, A., Waqas, A., Nisar, Q.A., Haider, S., Asghar, Z.: Application of AI and robotics in hospitality sector: a resource gain and resource loss perspective. Technol. Soc. **68**, 101807 (2022)
13. Luckcuck, M.: Using formal methods for autonomous systems: five recipes for formal verification. Proc. Inst. Mech. Eng. Part O J. Risk Reliab. **237**(2), 278–292 (2023)
14. Nesnas, I., Fesq, L.M., Volpe, R.A.: Autonomy for space robots: past, present, and future. Curr. Robot. Rep. **2**(3), 251–263 (2021)
15. Nilwong, S., Capi, G.; Reinforcement learning based outdoor navigation system for mobile robots. In: 2020 IEEE 18th World Symposium on Applied Machine Intelligence and Informatics (SAMI), pp. 219–224. IEEE (2020)
16. Novitsky, A., Yukhimets, D.: The navigation method of wheeled mobile robot based on data fusion obtained from onboard sensors and camera. In: 2015 15th International Conference on Control, Automation and Systems (ICCAS), pp. 574–579. IEEE (2015)

17. Oleiwi, B.K., Mahfuz, A., Roth, H.: Application of fuzzy logic for collision avoidance of mobile robots in dynamic-indoor environments. In: 2021 2nd International Conference on Robotics, Electrical and Signal Processing Techniques (ICREST), pp. 131–136. IEEE (2021)
18. Ross, L.T., Fardo, S.W., Walach, M.F.: Industrial Robotics Fundamentals: Theory and Applications. Goodheart-Willcox Co. (2017)
19. Schlingloff, B.-H.: Specification and verification of collaborative transport robots. In: 2018 4th International Workshop on Emerging Ideas and Trends in the Engineering of Cyber-Physical Systems (EITEC), pp. 3–8. IEEE (2018)
20. Sharma, N., Pandey, J.K., Mondal, S.: A review of mobile robots: applications and future prospect. Int. J. Precis. Eng. Manuf. **24**(9), 1695–1706 (2023)
21. Tigane, S., Guerrouf, F., Hamani, N., Kahloul, L., Khalgui, M., Ali, M.A.: Dynamic timed automata for reconfigurable system modeling and verification. Axioms **12**(3), 230 (2023)
22. Tigane, S., Guerrouf, F., Kahloul, L.: A GSPN-based formalism under infinite-server semantics for reconfigurable WANETs. Computing **105**(10), 2263–2291 (2023)
23. Tigane, S., Kahloul, L., Hamani, N., Khalgui, M., Ali, M.A.: On quantitative properties preservation in reconfigurable generalized stochastic petri nets. IEEE Trans. Syst. Man Cybernet. Syst. **53**(6), 3311–3323 (2022)
24. Velagic, J., Osmic, N., Hodzic, F., Siljak, H.: Outdoor navigation of a mobile robot using GPS and GPRs communication system. In: Proceedings ELMAR-2011, pp. 173–177. IEEE (2011)
25. Yao, J.: Path planning algorithm of indoor mobile robot based on ROS system. In: 2023 IEEE International Conference on Image Processing and Computer Applications (ICIPCA), pp. 523–529. IEEE (2023)
26. Zroug, S., Kahloul, L., Tigane, S., Hamani, N., Benharzallah, S.: A new formal approach for performance evaluation of green mac protocol in energy harvesting WSNs. Telecommun. Syst. **84**(1), 53–67 (2023)
27. Zroug, S., Remadna, I., Kahloul, L., Terrissa, S.L.: Simulation and formal verification in WSNs empowered by machine learning: a review. In: 2022 International Symposium on iNnovative Informatics of Biskra (ISNIB), pp. 1–6. IEEE (2022)

A New Approach for Checking the Consistency and Coherence of Self-adapting Software

Saadi Abdelfetah[1(✉)], Hammal Youcef[1], and Mourad Oussalah[2]

[1] University of Science and Technology Houari Boumediene, Algiers, Algeria
{afsaadi,yhammal}@usthb.dz
[2] University of Nantes, Nantes, France
mourad.oussalah@univ-nantes.fr

Abstract. A self-adaptive software architecture can alter its behavior and structure during run-time in order to achieve its objectives. It may change its requirements according to its environment. It adapts itself in response to changes that are difficult to predict before deployment. This paper presents a new approach that maintains the coherence, adequacy, and consistency of self-adaptive software architecture. It ensures that the software architecture conforms to the requirements specification according to functional and non-functional properties. The proposed approach uses mainly the meta-model, CSP, graph, and automata concepts to specify the behavior, structure, functional, and non-functional properties of a software architecture. After each adaptation or change in the software, a verification and validation process is performed. It is ensured by the use of the UPPAAL model checker tool, which aims at checking the coherence of the software architecture before adaptation. Finally, we have proposed a framework that implements all the proposed concepts.

Keywords: Software · self-adaptation · architecture · verification · run-time change

1 Introduction

In recent years, self-adaptive software systems have advanced considerably [16]. Self-adaptive software is able of changing and evaluating its own behavior and structure whenever the checking process shows that the software is not accomplishing what it was intended to do or when better functionality or performance may be possible [44]. Cheng et al. [9] define a self-adaptive system as a system that "is able to adjust its behavior in response to their perception of the environment and the system itself". Run-time changes are generally unpredictable or difficult to predict and check at design time and before software deployment. Unpredictable situations such as modifications in the environment of the system, reconfiguration, evolution, system problems, and new requirements are some of the reasons for triggering adaptation in a self-adaptive system [32]. To deal with these unpredictable situations, self-adaptive software must be designed and implemented with rigorous methods and strategies. It must provide specific features to model dynamic changes in the running environment and have the ability

T. Marir et al. (Eds.): ICCSA 2024, ISEM 45, pp. 248–263, 2025.
https://doi.org/10.1007/978-3-031-90758-6_19

to describe the structure and the behavior of dynamic software architectures, it must also handle run-time changes in the software and its elements [34].

Self-adaptive software must be flexible and able to autonomously readjust itself in the presence of any change in its requirements [17]. As the requirements' changes may occur at run-time, the running software has to adapt itself without stopping, which is important in critical systems. At that point, traditionally, the software can encounter a set of problems (incoherence, inconsistency, etc.). To avoid such a situation, the designed architecture must include solutions that allow the system to dynamically adapt its own structure and/or behavior at run-time. This means monitoring its own behavior and structure with respect to its current goals and ensuring their consistency and fulfillment of safety properties before adaptation [46].

To achieve this goal, we propose in this paper a new approach called *CC-SelfArchi* (Checking the Consistency of SELF-Adapting software Architecture). The aim is to maintain the coherence, consistency, and adequacy of self-adaptive software architecture before, during, and/or after adaptation. In *CC-SelfArchi*, a self-adaptive software architecture has a set of non-functional properties (denoted with $NFPs$) that express some constraints and help to overcome deviations from expected goals after adaptation. The *CC-SelfArchi* approach mainly deals with the non-functional specification, given that the structural and behavioral specification have already been covered in our previous work. The main contribution in this paper is to provide a solution that will be integrated with our previous proposals in order to allow the self-adaptation in software architecture and to preserve the coherence, consistency, and adequacy of the software specification before, during, and after adaptation.

The *CC-SelfArchi* approach has three objectives: (i) allow software architecture to adapt itself without halting its execution, (ii) present a set of concepts for describing and handling functional and non-functional properties (denoted with $F\&NFPs$), and (iii) verify the consistency of dynamic updates of the software architecture to keep $F\&NFPs$ fulfilled. To reach these objectives, the *CC-SelfArchi* approach uses the meta-model concept for the static part specification and the CSP (*Communicating Sequential Processes*) language for the dynamic part specification, especially behaviors and interactions between components. The idea is to combine the CSP specification and a set of adaptation actions (add component, delete connector, etc.) in order to generate a graph representing all possible configurations of the initial architecture, including a set of $NFPs$ (time constraints, memory size, energy, etc.). The resulting graph is then transformed into a set of timed automata that will be executed in parallel in order to check (using the UPPAAL model checker tool) their consistency before any software architecture adaptation. This makes it possible to verify that the specification and the software are correctly adapted and enable software-safe adaptations according to the consistent specification of requirements and $F\&NFPs$. It is worth noting that the *CC-SelfArchi* approach extends and revises our work proposed in [40–42]. Unlike the existing work and approach, the *CC-SelfArchi* approach formally verifies the coherence, adequacy, and consistency of self-adaptive software architectures against their $F\&NFPs$ and requirements. It combines three concepts (CSP, meta model, and graph) for the modeling, specification, and verification of self-adaptive software.

The rest of this paper is structured as follows: Sect. 2 presents the related work on self-adapting software. Section 3 describes the *CC-SelfArchi* approach. Section 4 presents the adaptation and self-adaptation process in the *CC-SelfArchi* approach. Section 5 describes the verification and validation process of self-adaptive software architecture. Section 6 presents the evaluation and the implementation of the proposed approach. Finally, the conclusions are drawn in Sect. 7 with some directions for future work.

2 Related Work

There are many works that are proposed about the topic of self-adaptive software architecture [42]. We can classify existing works into five main categories: works on the state of the art and survey, works that propose solutions for self-adaptive systems, works in the context of component-based software development, works that use properties verification in self-adaptive systems, and models of reference for engineering self-adaptive software systems. In the *state of the art and survey category*, there exist some work, the main works are: Cheng et al. [9], Macias-Escriva et al. [13], Mahdavi et al. [32], Raibulet et al. [39], Wong et al. [50], Weyns [49], Mazeiar and Ladan [33], and Krupitzer et al. [28]. In the *solutions for self-adaptive systems category*, we cite: Angelopoulos et al. [2] that consider the scenario in which the environment changes dynamically over time, and Sabatucci et al. [43] that provide a unified meta-model that describes many adaptation categories. In the *context of component-based software development category*, we cite: certain ADL, like Fractal and ACME, deal with run-time adaptation [7]. Other ADLs propose to specify changes with the use of languages like FPath/FScript [11] or Plastik [5], or by defining *Event Condition Actions (ECA)* rules [1]. Krupitzer et al. [28] classified research and work into several classes: [42]: *Architecture-based approaches* (Rainbow Framework [17], Kramer and Magee [24], Georgas et al. [18], Carmo et al. [8]), *model-based approaches* (Vogel and Giese [45], Giese and Schafer [20]), *agent-based approaches* (Ballagny et al. [4], De Wolf and Holvoet [12]), and *formal methods based approaches* (Weyns et al. [48], Gudemann et al. [21], Nafz et al. [35], Franco et al. [15], Filieri and Tamburrell [14]). In the *properties verification in self-adaptive systems category*, there exist some works, the principal are: Manzoor et al. [31], Benghazi et al. [6], Apvrille et al. [3], Laleau et al. [29], Ghezzi and Sharifloo [19], Ramnath et al. [38], Lee et al. [30], Paraiba and Martins [37]. Finally, in the *models of reference for engineering self-adaptive software systems category*, and as cited in [32], a number of models of reference have been proposed for engineering self-adaptive software systems, we cite: the feedback loop model of control theory [23,36], the mape-k model [23,27], kramer and magee's self-management reference model [24,26], the dynamico reference mode [47], the autonomic computing reference architecture [25].

As a conclusion, we can say that most of the existing work was focused on the management of requirements and $F\&NFPs$ in self-adaptive systems. Therefore, it is necessary to propose a solution that formally verifies the coherence, consistency, and adequacy of self-adaptive software architectures against their $F\&NFPs$, and requirements before, during, and after adaptation. The aim is to adjust the software's behavior or structure at run-time and ensure the satisfaction of requirements.

3 Approach

This section presents our approach, called *CC-SelfArchi*, to verify the coherence, consistency, and adequacy of self-adapting software. This approach uses the meta-model concept and the CSP (*Communicating Sequential Processes*) language to specify a self-adaptive software architecture.

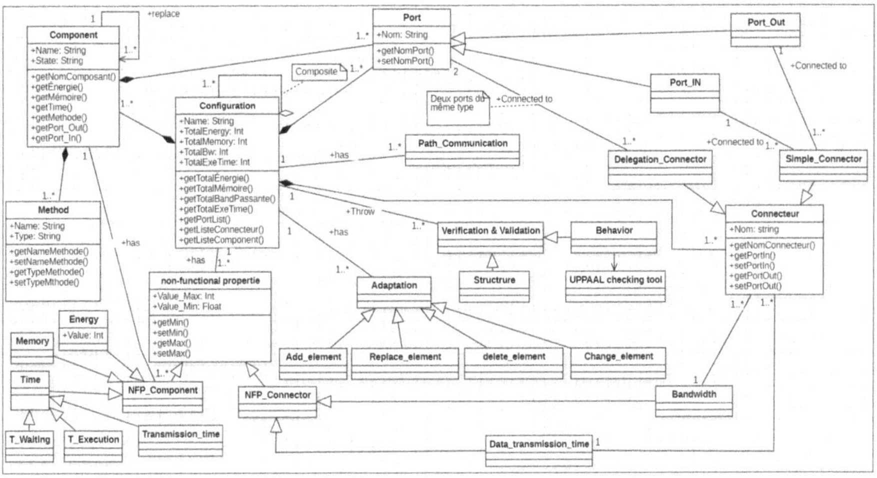

Fig. 1. CC-SelfArchi syntax.

Figure 1 shows the syntax of *CC-SelfArchi*, it is composed of two parts:

- The static part: it represents a port, connector, component, and configuration.
- The dynamic part: it represents interactions of architectural elements, behaviors, verification and validation, and adaptation.

CSP Specification: In *CC-SelfArchi*, ports define how a component can communicate and interact with other components. We discern two types of ports [41]:

- The required port: used to express a required functionality that may be provided by another component. It receives and sends data, but it always starts

by receiving data as the first action. The CSP expression that we have proposed to specify the behavior of a required port is given by [41]:
$In\_behavior = In?a : integer- > In!b : integer- > In\_behavior$. In this expression, In represents the name of the required port, it receives a (which is an integer), then, after a treatment executed on a, it sends the result b.

- The provided port: expresses a provided functionality that might be used by other components. It sends and receives data, it always starts by sending data. The CSP expression proposed to specify the behavior of a provided port is given by the following expression [41]:
$Out\_behavior = Out!a : integer- > Out?b : integer- > Out\_behavior$. This expression describes the behavior of a provided port (named Out).

A component [41] is identified by a name, it has ports. It implements one or more methods for its internal processing, and it is associated with a set of $NFPs$. In our approach, the CSP expression of a component defines communication and interactions between component ports. For example, in Fig. 2 a CSP expression of the component Cmp_1 (that has two ports: a provided port ($Port_1$) and a required port ($Port_2$)) is given by the following expression: $Cmp_1\_beh = Port_2?a : integer- > Port_1!b : integer- > Cmp_1\_beh$. In this expression, $Port_2$ receives a value stored in a, then the component Cmp_1 executes a task that outputs b which will be sent using $Port_1$.

A connector [41] depicts the interactions of data exchange between components. It has a set of $NFPs$. The behavior of a connector $Connector_1$ (Fig. 2) that connects two component instances (Cmp_1 and Cmp_2) is given by:
$Connector_1 = Cmp_1.Port_1?a : integer- > Cmp_2.Port_2!a : integer- >$
$Connector_1$. This means that the connector $Connector_1$ begins the exchange by receiving data a incoming from the output port $Port_1$ of the component Cmp_1. Then, the received data will be sent to the input port $Port_2$ of the component Cmp_2. The behavior of the connector $Connector_1$ specifies the order of operations between $Port_1$ (Cmp_1) and $Port_2$ (Cmp_2). Such a behavior states which common actions are shared by Cmp_1 and Cmp_2 components.

A configuration [41] encompasses a set of ports, connectors, components, and a set of $NFPs$. The behavior of a configuration is created from the behaviors of components, connectors, and ports. It represents the concurrent combination of the configuration instances, along with synchronization on common actions of the configuration elements. The CSP expression of a configuration is produced according to the next formula: $(|||_{i\in\{1...n\}}(Cmp_i))|S|(|||_{i\in\{1...n-1\}}(Connector_i))$. $S = \cup(S_{ij}(Cmp_i, Cmp_j))$ Cmp_i and Cmp_j are components connected by a connector. $S_{ij}(Cmp_i, Cmp_j)$ represents the shared actions between Cmp_i and Cmp_j and their connectors, respectively.

Non-functional Properties: *CC-SelfArchi* approach considers five $NFPs$:

- The execution time ($Time$): a component implements methods for its internal computation. An *execution time* is associated with each method.

- The memory consumption (Mem): each component of the software architecture requires *memory space* to perform its tasks.
- The energy (Egy): components require *energy* (which is measured in joules) for their functioning. Energy will decrease over time, any change in energy values implies the need to make changes to these components in order to satisfy constraints on energy consumption.
- The data transmission time ($Trans$): sending and receiving data between components requires timing constraints that should not be overlooked in order to ensure a proper synchronization between components.
- The bandwidth (Bdw): connectors have a *bandwidth* used for the data transmission and data exchange.

All $NFPs$ are boolean-valued expressions formed from predicate symbols, constants, and variables.

Modeling with Graphs: The software architecture structure is modeled in the form of a graph. Each node represents a configuration, and edges represent adaptation actions (add component, delete connector, etc.) that can be performed on a configuration to switch to another configuration. Each node is associated with a set of properties that remain true in the corresponding configuration.

Let NF be a set of non-functional constraints of a software architecture. As in Clarke et al. [10] we define the graph Gph as a 5-tuple: $Gph = (S, I, Rt, Lf, A)$ such that:

- S denotes a finite set of states. States represent configurations.
- $I \subseteq S$ denotes the set of initial states.
- $Rt \subseteq S$ x S denotes a relation of transition: $\forall s \in S$, $\exists s' \in S$ such that $(s, s') \in Rt$.
- $Lf : S \rightarrow 2^{NF}$ denotes a labeling function.
- A denotes a set of adaptation actions.

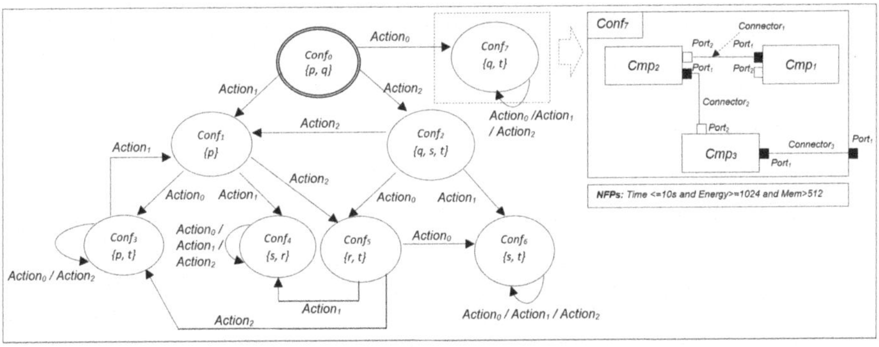

Fig. 2. The generated graph.

Let us consider $NF = \{p, q, s, r, t\}$ such that p, q, s, r and t express boolean properties of a configuration ($p = Time > 3s, q = Bdw < 1024$, $r = Mem >= 512$, etc.). The Fig. 2 illustrates an example of a graph $Gph = (S, I, Rt, Lf, A)$, where:

- $S = \{Conf_0, Conf_1, Conf_2, Conf_3, Conf_4, Conf_5, Conf_6, Conf_7\}$.
- $I = \{Conf_0\}$.
- $Rt = \{(Conf_0, Conf_1), (Conf_0, Conf_2), (Conf_0, Conf_7), (Conf_1, Conf_3), (Conf_1, Conf_4), (Conf_1, Conf_5), (Conf_2, Conf_1), (Conf_2, Conf_5), (Conf_2, Conf_6), (Conf_3, Conf_1), (Conf_5, Conf_3), (Conf_5, Conf_4), (Conf_5, Conf_6)\}$.
- $Lf = \{(Conf_0, \{p, q\}), (Conf_1, \{p\}), (Conf_2, \{q, s, t\}), (Conf_3, \{p, t\}), (Conf_4, \{s, r\}), (Conf_5, \{r, t\}), (Conf_6, \{s, t\}), (Conf_7, \{q, t\})\}$.
- $A = \{Action_0, Action_1, Action_2\}$.

The creation of the graph Gph is done at design-time and before deploying the software architecture. This creation is achieved by applying all adaptation actions ($a \in A$) on the initial configuration $Conf_0$. The application of each adaptation action generates a new configuration that will be processed in the same process. This means applying the action $Action_0$ to $Conf_0$ generates $Conf_7$, applying the action $Action_1$ to $Conf_0$ generates $Conf_1$, applying the action $Action_2$ to $Conf_0$ generates $Conf_2$. In the same way, we apply for each configuration ($cf \in S$) all actions ($\forall a \in A$).

An adaptation action that does not affect the status of the configuration can be modeled by a single edge outgoing back to the original state. The labeling function Lf assigns to each state ($s \in S$) a set denoted $Lf(s)$ containing all non-functional constraints that are valid in s. It is possible to construct an infinite path through the graph Gph. A path ph of the graph Gph is defined as a set of states $ph = \{s_1, s_2, s_3, ...\}$ and adaptation actions $ad = \{Action_0, Action_1, Action_3, ...\}$, such that for each $i > 0, Rt(s_i, s_{i+1})$ holds. The word (w) on the path ph is represented by a set of the labeling functions of its states ($w = \{Lf(s_1), Lf(s_2), Lf(s_3), ...,\}$) and adaptation actions. With this definition, a graph Gph that has unique initial state ($i \in I$) could be identified with a unique input word w, with the output function being its labeling function.

4 Adaptation and Self-adaptation

An adaptation is usually initiated by an event. Events can occur during the run-time or during deployment. Event occurrences are unpredictable before running the software. Events are classified into two types [42]: *internal events* and *external events*. *Events* can be triggered by another event. Events may concern any modification in the internal structure or behavior of the software or environment. Moreover, the software specification and requirements may be changed following the receipt of an event at run-time.

In *CC-SelfArchi*, the software architecture interactions, behaviors, and $NFPs$ are specified at design-time by an architect. An adaptation aims to modify the structure, the behavior, or even the properties and constraints of the architecture. In such a situation, we examine the generated graph *Gph* to find and determine the configuration that satisfies the new requirements for adaptation. After determining the desired configuration, we generate a script called *adaptation script* (denoted *ads*) that represents the shortest path to follow in the graph *Gph* to reach the node containing the new configuration. Then, we apply the *adaptation script* to the initial software architecture to obtain the final configuration. *The adaptation script* guides the adaptation actions (adding or removing instances of components and connectors) to be performed according to the new configuration specification. After applying each adaptation action, a verification and validation process (see Sect. 5) is performed to ensure the validity of the architecture at each stage of the adaptation. This confirms that the configuration architecture remains correctly defined and consistent with the meta-model and its constraints.

For further details, we consider the software architecture of Fig. 2. We want to make an adaptation of the initial configuration $Conf_0$ that satisfies the properties $\{p, q\}$ to a new configuration that satisfies the properties $\{s, t\}$. After an examination and a search in the graph *Gph*, we find that there exists the configuration $Conf_6$ that satisfies the new properties $\{s, t\}$. Hence, there are two paths (ph_1 and ph_2) to make the transition from configuration $Conf_0$ to configuration $Conf_6$: ph_1=$\{Conf_0$, $Conf_2$, $Conf_5$, $Conf_6\}$ with $ads_1 = \{Action_2, Action_0, Action_0\}$, and ph_2=$\{Conf_0$, $Conf_2$, $Conf_6\}$ with $ads_2 = \{Action_2, Action_1\}$. ph_2 is the shortest path in which the new configuration $Conf_6$ satisfies the proposition $\{s, t\}$. Therefore, the *adaptation script* (ads_2) associated with the path ph_2 will be applied on the $Conf_0$ to obtain $Conf_6$.

5 Verification and Validation Process

The rationale under the process of verification and validation is to preserve the coherence and adequacy of the software architecture. Like in [40], the verification and validation process is divided into two steps:

1. **The structural verification and validation step**: it consists of checking the structural modifications of the software architecture. The process consists of two sub-steps:
   - Verification step: aims at checking the coherence of the software architecture. It guarantees that [40]: all required ports are linked to provided ports and component ports are linked to the composite ports (using connectors of delegation).
   - Validation step: consists of checking if the software architecture model (specified in XML language) conforms to its meta-model (specified in XML schema).
2. **The behavioral verification and validation step**: it consists of checking the behavioral modification of the software architecture. The process consists of two sub-steps:

- Verification step: it consists of analyzing CSP expressions and adaptation actions in order to check whether they are valid or not.
- Validation step: it is composed of three parts:
  (a) Transformation of the graph into an automaton using SAS2A approach [42].
  (b) Validation of CSP expression. Using FDR tool, we have to ensure that [41]:
    - A component's CSP expression must satisfy the formula: $(|||_{i} \in_{I} (Port_i)) \sqsubseteq_F Cmp\_beh$. This indicates that a component's CSP expression refines the concurrent execution (without synchronizing ports of the component).
    - A connector's CSP expression must satisfy the formula: $(Cmp_2.Port_2|||Cmp_1.Port_1) \sqsubseteq_F (Connector_1|||...|||Connector_n)$. n represents the number of connectors between Cmp_1 and Cmp_2.
    - Verify if the formula: $(|||_{i\in\{1...n\}}(Cmp_i))|S|(|||_{i\in\{1...n-1\}} (Connector_i))$ does not display a deadlock error.
  (c) Using UPPAAL, we verify that the deadlock situation is not attainable within the architecture and its $NFPs$. That means we verify the software architecture against the $NFPs$ that are expected from this architecture in order to ensure that it preserves software architecture constraints and properties.

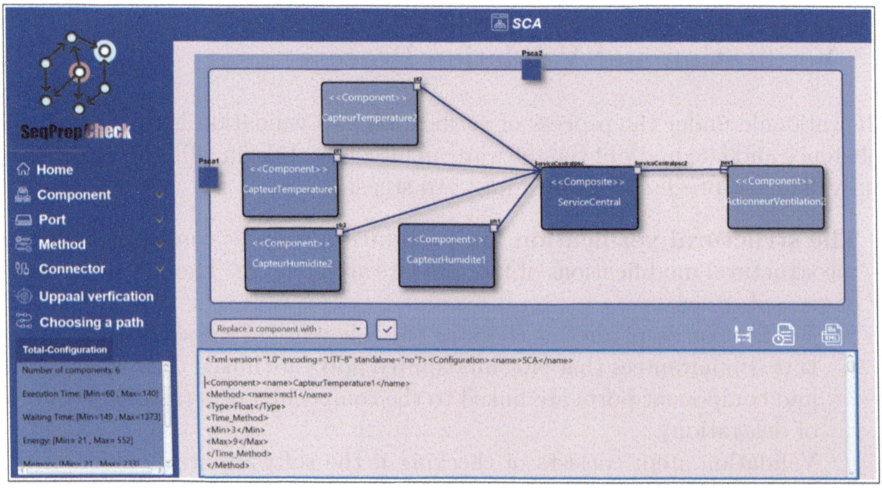

Fig. 3. The developed tool (*SeqPropCheck*).

6 Evaluation and Prototyping

To validate the proposed approach, we have considered the *Agricultural Monitoring and Control System (AMCS).* This system is composed of components that are deployed on sensors and actuators. The aim is to monitor and control different aspects of the agricultural environment (such as temperature, humidity, etc.). The AMCS system collects data incoming from various sensors and then sends them to the central service for treatment. Then, the latter sends the commands to the actuators in the farm.

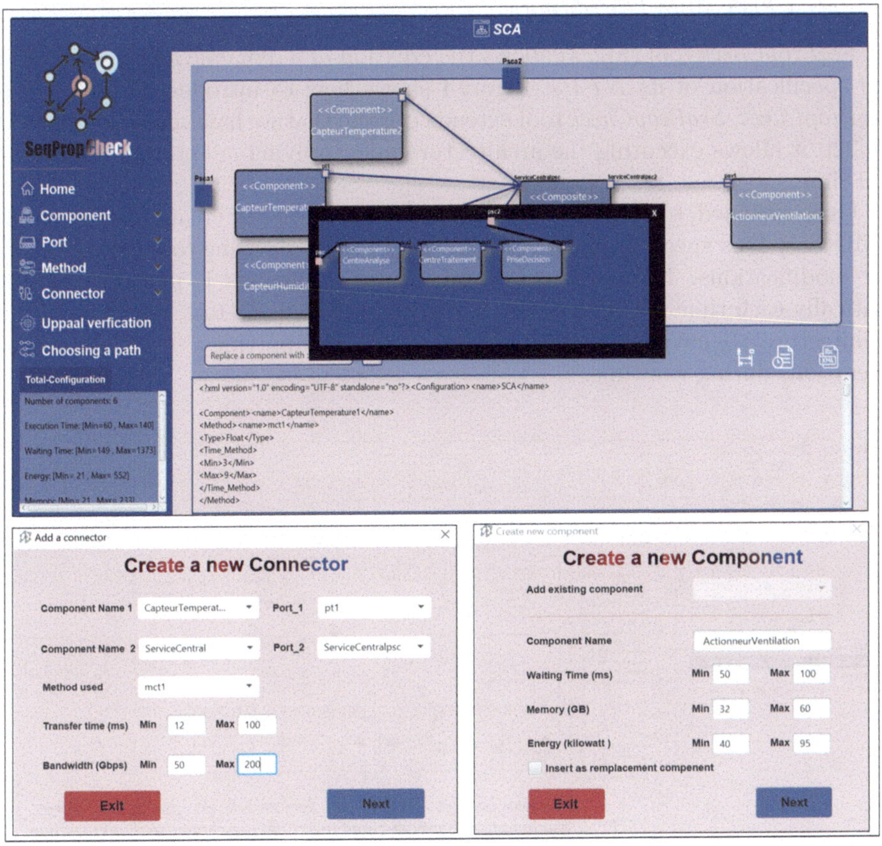

Fig. 4. Specification of *NFPs* of components and connectors.

AMCS software structure is composed of a set of components (*SensorTemperature, SensorHumidity, CentralService, ActuatorVentilation, etc.*). Figure 3 shows the AMCS system architecture and its main components created using our developed tool (named *SeqPropCheck*).

AMCS considers the following constraints:

- *SensorTemperature, SensorHumidity, CentralService, and ActuatorVentilation* component must be able to manage energy properly (*Energy*).
- All components of the AMCS system must be able to store and process the collected data efficiently (*Memory*).
- Control commands must be processed within a minimum delay to ensure a rapid response to changes in the agricultural environment (*Response time*).
- Some actions (such as activating the ventilation actuator in response to a drop in humidity) must respect strict time constraints (*Time Constraints*).
- The communication between components uses a bandwidth that must not exceed 2048 Mbps (*Bandwidth*).

SeqPropCheck tool (Fig. 3) allows the creation of a dynamic architecture and the specification of its *NFPs*. Figure 4 shows how to introduce *NFPs* using *SeqPropCheck*. *SeqPropCheck* tool extends the one that we have already proposed in [22], it allows executing the architecture and applying adaptation actions at run-time.

Once specified, software specifications are automatically created and saved in XML files. This specification is verified by UPPAAL checking tool before enacting modifications. The verification and validation process is performed automatically each time the software architecture is altered. To test, we changed the architecture's behavior, structure, and/or *NFPs* by applying a set of adaptation operations during run-time.

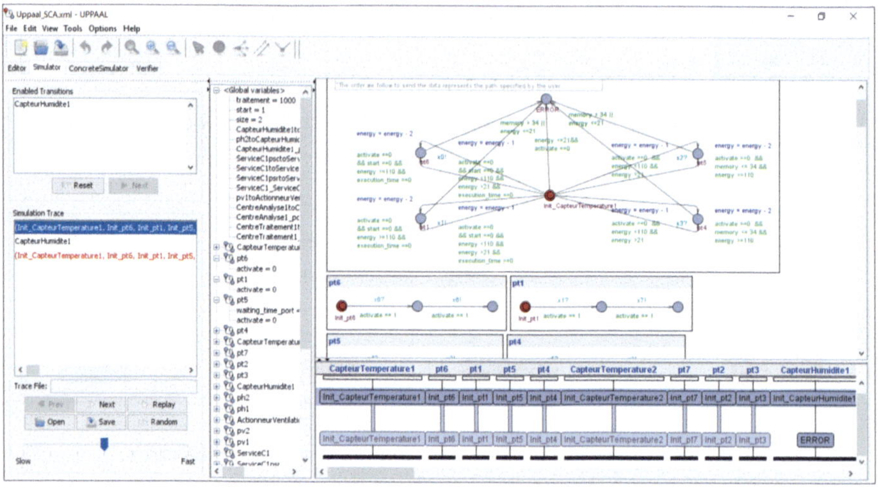

Fig. 5. The verification and validation of the AMCS system using UPPAAL tool.

Figure 5 shows the generated automata relating to the AMCS system, including all *NFPs*. *NFPs* are translated to the UPPAAL query language for execution. Figure 6 shows the execution and the validation of these queries.

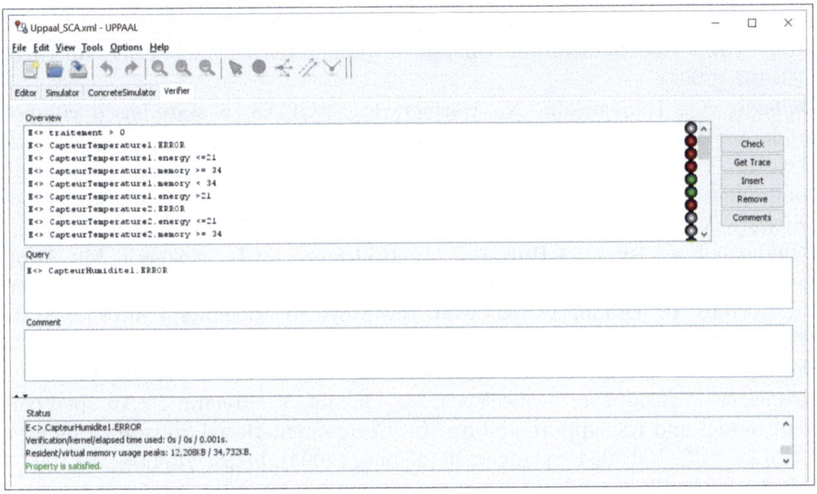

Fig. 6. The verification and validation of requests using UPPAAL tool.

7 Conclusion

In this work, we presented a new approach for checking the coherence and adequacy of self-adapting software. This approach offers the necessary steps to specify and control at run-time the adaptation of the architecture. The *CC-SelfArchi* approach uses a set of concepts to specify the requirements, the behavioral part, the structural part, and the $F\&NFPs$ of a given self-adapting software, it satisfies goals we have established for this paper. *CC-SelfArchi* guarantees that dynamic aspects are treated separately from the remaining parts of the system, it proposes a monitoring system for its adaptation equipped with a process of verification and validation. In the future, we plan to use artificial intelligence methods and techniques in *CC-SelfArchi* to reduce complexity and optimize performance. We plan also to generalize the *CC-SelfArchi* approach in order to use it in complex systems, IoT (internet of things) systems, and service-oriented computing.

References

1. Alvares, F., Rutten, E., Seinturier, L.: A domain-specific language for the control of self-adaptive component-based architecture. J. Syst. Software (2017)
2. Angelopoulos, K., Papadopoulos, A., Silva Souza, V., Mylopoulos, J.: Engineering self-adaptive software systems: from requirements to model predictive control. ACM Trans. Auton. Adapt. Syst. **13**(1), 27 (2018). Article 1, https://doi.org/10.1145/3105748

3. Apvrille, L., Courtiat, J., Lohr, C., Saqui-Sannes, P.D.: TURTLE: a real-time UML profile supported by a formal validation toolkit. IEEE Trans. Software Eng. **30**, 473–487 (2004)
4. Ballagny, C., Hameurlain, N., Barbier, F.: MOCAS: a state-based component model for self-adaptation. In: Proceedings of the SASO, pp. 206–215. IEEE (2009)
5. Batista, T., Joolia, A., Coulson, G.: Managing dynamic reconfiguration in component-based systems. In: Proceedings of the 2nd European Conference on Software Architecture, EWSA'05, pp. 1–17. Springer, Berlin (2005)
6. Benghazi, K., Visitación Hurtado, M., Rodríguez, M.L., Noguera, M.: Applying formal verification techniques to ambient assisted living systems. In: Meersman, R., Herrero, P., Dillon, T. (eds.) On the Move to Meaningful Internet Systems: OTM 2009 Workshops. OTM 2009. Lecture Notes in Computer Science, vol. 5872. Springer, Heidelberg (2009). https://doi.org/10.1007/978-3-642-05290-3_50
7. Bruneton, E., Coupaye, T., Leclercq, M., Quema, V., Stefani, J.: An open component model and its support in Java. In: Component-Based Software Engineering. CBSE. LNCS, Vol 3054. Springer, Heidelberg (2004). https://doi.org/10.1007/978-3-540-24774-6_3
8. Carmo, D., Carvalho, S., Murta, L., Loques, O.: Run-time monitoring and auditing of self-adaptive systems. In: Proceedings of the International Conference on Software Engineering and Knowledge Engineering, SEKE 2013, pp. 731–736 (2013)
9. Cheng, B.H.C., et al.: Software engineering for self-adaptive systems: a research roadmap. In: Cheng, B.H.C., de Lemos, R., Giese, H., Inverardi, P., Magee, J. (eds.) Software Engineering for Self-Adaptive Systems. LNCS, vol. 5525. Springer, Heidelberg (2009). https://doi.org/10.1007/978-3-642-02161-9_1
10. Clarke Edmund Jr., M., Grumberg, O., Peled, D.: Model Checking. Cyber Physical Systems Series, p. 14. MIT Press (1999). ISBN 978-0-262-03270-4
11. David, P., Ledoux, T., Léger, M., Coupaye, T.: FPath and FScript: language support for navigation and reliable reconfiguration of Fractal architectures. Ann. Telecomm. **64**, 45–63 (2009)
12. De Wolf, T., Holvoet, T.: Design patterns for decentralized coordination in self-organizing emergent systems. In: Engineering Self-Organizing Systems. LNCS, vol. 4335, pp. 28–49. Springer (2007)
13. Macias-Escriva, F., Haber, R., del Toro, R., Hernandez, V.: Self-adaptive systems: a survey of current approaches, research challenges and applications, Expert Syst. Appl. **40**(18), 7267–7279 (2013). ISSN 0957-4174, https://doi.org/10.1016/j.eswa.2013.07.033
14. Filieri, A., Tamburrelli, G.: Probabilistic verification at run-time for self-adaptive systems. In: Cámara, J., de Lemos, R., Ghezzi, C., Lopes, A. (eds.) Assurances for Self-Adaptive Systems. LNCS, vol. 7740. Springer, Heidelberg (2013). https://doi.org/10.1007/978-3-642-36249-1_2
15. Franco, J., Correia, F., Barbosa, R., Zenha-Rela, M., Schmerl, B., Garlan, D.: Improving self-adaptation planning through software architecture-based stochastic modeling. J. Syst. Softw. **115**(2016), 42–60 (2016)
16. Faniyi, F., Bahsoon, R.: Economics-driven software architecting for cloud. In: Mistrik, I., Bahsoon, R., Kazman, R., Zhang, Y. (eds.) Economics-Driven Software Architecture, pp. 83–103. Morgan Kaufmann (2014). ISBN 9780124104648, https://doi.org/10.1016/B978-0-12-410464-8.00005-2
17. Garlan, D., Cheng, S., Huang, A., Schmerl, B., Steenkiste, P.: Rainbow: architecture-based self-adaptation with reusable infrastructure. IEEE Comput. **37**(10), 46–54 (2004)

18. Georgas, J., Van der Hoek, A., Taylor, R.: Using architectural models to manage and visualize run-time adaptation. Computer **42**(10), 52–60 (2009). https://doi.org/10.1109/MC.2009.335
19. Ghezzi, C., Sharifloo, A.: Dealing with non-functional requirements for adaptive systems via dynamic software product-lines. In: de Lemos, R., Giese, H., Müller, H.A., Shaw, M. (eds.) Software Engineering for Self-Adaptive Systems II. Lecture Notes in Computer Science, vol. 7475. Springer, Heidelberg (2013). https://doi.org/10.1007/978-3-642-35813-5_8
20. Giese, H., Schäfer, W.: Model-driven development of safe self-optimizing mechatronic systems with MechatronicUML. In: Cámara, J., de Lemos, R., Ghezzi, C., Lopes, A. (eds.) Assurances for Self-Adaptive Systems. LNCS, vol. 7740, pp. 152–186. Springer, Heidelberg (2013). https://doi.org/10.1007/978-3-642-36249-1_6
21. Güdemann, M., Ortmeier, F., Reif, W.: Formal Modeling and Verification of Systems with Self-x Properties. In: Yang, L.T., Jin, H., Ma, J., Ungerer, T. (eds.) ATC 2006. LNCS, vol. 4158, pp. 38–47. Springer, Heidelberg (2006). https://doi.org/10.1007/11839569_4
22. Guenatri, L., Djebbara, F.: Techniques de génération des scripts de reconfiguration des architectures logicielles. Master thesis, June 2023, USTHB University, Algiers, Algeria (2023)
23. Hellerstein, J.L., Diao, Y., Parekh, S., Tilbury, D.M.: Feedback Control of Computing Systems. Wiley, Chichester (2004)
24. Kramer, J., Magee, J.: Self-managed systems: an architectural challenge. In: Proceedings of the 2007 Workshop on the Future of Software Engineering (FOSE 2007), pp. 259–268. IEEE Computer Society (2007)
25. I.B.M. Corporation: An Architectural Blueprint for Autonomic Computing, IBM Corporation (2006)
26. Kramer, J., Magee, J.: Dynamic structure in software architectures. SEN **21**(6), 3–14 (1996)
27. Kephart, J.O., Chess, D.M.: The vision of autonomic computing. Computer **36**(1), 41–50 (2003)
28. Krupitzer, C., Roth, F., VanSyckel, S., Schiele, G., Becker, C.: A survey on engineering approaches for self-adaptive systems. Perv. Mob. Comput. J. **17**(B), 184206 (2015). https://doi.org/10.1016/j.pmcj.2014.09.009
29. Laleau, R., Semmak, F., Matoussi, A., et al.: A first attempt to combine SysML requirements diagrams and B. Innovations Syst. Softw. Eng. **6**, 47–54 (2010). https://doi.org/10.1007/s11334-009-0119-y
30. Lee, S., Park, S., Park, Y.: Self-adaptive system verification based on SysML. In: 2019 International Conference on Electronics, Information, and Communication (ICEIC), Auckland, New Zealand, pp. 1–3 (2019). https://doi.org/10.23919/ELINFOCOM.2019.8706383
31. Manzoor, A., Nicolas, B., Jean-Michel, B.: Modeling and verification of functional and non-functional requirements of ambient self-adaptive systems. J. Syst. Software **107**, 50–70 (2015). ISSN 0164-1212. https://doi.org/10.1016/j.jss.2015.05.028
32. Mahdavi-Hezavehi, S., Avgeriou, P., Weyns, D.: A classification framework of uncertainty in architecture-based self-adaptive systems with multiple quality requirements. In: Mistrik, I., Ali, N., Kazman, R., Grundy, J., Schmerl, B. (eds.) Managing Trade-Offs in Adaptable Software Architectures, pp. 45–77. Elsevier, Berlin (2017). ISBN 9780128028551, https://doi.org/10.1016/B978-0-12-802855-1.00003-4
33. Mazeiar, S., Ladan, T.: Self-adaptive software: landscape and research challenges. TAAS. 4. ACM Trans. Auton. Adapt. Syst. (2009)

34. Medvidovic, N., Taylor, R.: A classification and comparison framework for software architecture description languages. Trans. Software Eng. **26**(1), 70–93 (2000)
35. Nafz, F., Seebach, H., Steghöfer, J.-P., Bäumler, S., Reif, W.: A Formal framework for compositional verification of organic computing systems. In: Xie, B., Branke, J., Sadjadi, S.M., Zhang, D., Zhou, X. (eds.) ATC 2010. LNCS, vol. 6407, pp. 17–31. Springer, Heidelberg (2010). https://doi.org/10.1007/978-3-642-16576-4_2
36. Ogata, K.: Modern Control Engineering, 3rd edn. Prentice Hall, Upper Saddle River (1996)
37. Paraiba, J., Martins, L.: PERSA: a requirements specification process for self-adaptive systems based on fuzzy logic and NFR-framework. Internat. J. Uncertain. Fuzziness Knowl.-Based Syst. **25**(01), 145–178 (2017). https://doi.org/10.1142/S0218488517500064
38. Ramnath, R., Gupta, V., Ramanathan, J.: RED-transaction and goal-model based analysis of layered security of physical spaces. In: Computer Software and Applications. COMPSAC'08. 32nd Annual IEEE International, pp. 679–685. IEEE (2008)
39. Raibulet, C., Arcelli Fontana, F., Capilla, R., Carrillo, C.: Chapter 13 - An overview on quality evaluation of self-adaptive systems. Mistrik, I., Ali, N., Kazman, R., Grundy, J., Schmerl, B. (eds.) Managing Trade-Offs in Adaptable Software Architectures, pp. 325–352. Morgan Kaufmann (2017). ISBN 9780128028551, doi: https://doi.org/10.1016/B978-0-12-802855-1.00013-7
40. Saadi, A., Oussalah, M., Henni, A.: An approach for managing the dynamic reconfiguration of software architectures. Int. J. Comput. Sci. Eng. **17**(2) (2018)
41. Saadi, A., Hammal, Y., Oussalah, M.: A CSP-based approach for managing the dynamic reconfiguration of software architecture. Int. J. Inf. Technol. Syst. Approach (IJITSA) **14**(1), 156–173 (2021). https://doi.org/10.4018/IJITSA.2021010109
42. Saadi, A., Hammal, Y., Oussalah, M.: Automata based approach to manage self-adaptive component-based architectures. Int. J. Software Innovat. (IJSI) **10**(1), 1–22 (2022). https://doi.org/10.4018/IJSI.297623
43. Sabatucci, L., Seidita, V., Cossentino, M.: The Four Types of Self-adaptive Systems: A Met-Model, p. 440–450 (2018). https://doi.org/10.1007/978-3-319-59480-4_44
44. Salehie, M., Tahvildari, L.: Towards a goal-driven approach to action selection in self-adaptive. Software Pract. Exp. (2011)
45. Vogel, T., Giese, H.: A language for feedback loops in self-adaptive systems: executable run-time mega-models. In: Proceedings of the SEAMS, pp. 129–138. IEEE (2012)
46. Villegas, N.M., Tamura, G., Müller, H.A.: Architecting software systems for run-time self-adaptation: concepts, models, and challenges. In: Mistrik, I., Ali, N., Kazman, R., Grundy, J., Schmerl, B. (eds.) Managing Trade-Offs in Adaptable Software Architectures, pp. 17–43. Morgan Kaufmann (2017). ISBN 9780128028551, https://doi.org/10.1016/B978-0-12-802855-1.00002-2
47. Villegas, N., Tamura, G., Müller, H., Laurence D., Rubby, C.: DYNAMICO: a reference model for governing control objectives and context relevance in self-adaptive software systems (2012). https://doi.org/10.1007/978-3-642-35813-5_11
48. Weyns, D., Malek, S., Andersson, J.: FORMS: unifying reference model for formal specification of distributed self-adaptive systems. ACM Trans. Auton. Adapt. Syst. **7**(1) (2012)

49. Weyns, D.: Software engineering of self-adaptive systems. In: Cha, S., Taylor, R., Kang, K. (eds.) Handbook of Software Engineering, pp. 399–443. Springer, Cham (2019). https://doi.org/10.1007/978-3-030-00262-6_11
50. Wong, T., Wagner, M., Treude, C.: Self-adaptive systems: a systematic literature review across categories and domains. Inf. Software Technol. **148**, 106934 (2022). ISSN 0950-5849, https://doi.org/10.1016/j.infsof.2022.106934

Enhancing Software Architecture Modernization with Mixed Oriented Aspect and Fuzzy Logic: A Novel Approach

Djezar Sarra(✉) and Aouag Sofiane

University of Batna 2 Pole Fesdis RN, 05110 Batna, Fesdis, Algeria
djezarsara@gmail.com

Abstract. Model-driven engineering (MDE) has brought various advancements to software development, yet it has also resulted in greater complexity of applications in terms of distribution, security, and performance. The Object Management Group (OMG) works to promote Model Driven Architecture (MDA), as well as the Architecture Driven Model (ADM), to combat this rising complexity. Unfortunately, modifications of these systems often result in issues with the design, leading to the degradation of their quality properties. In order to control the complexity of the modernization process, our research suggests combining fuzzy logic with aspect-oriented modeling. By enabling a more logical progression of inference, we recommend using fuzzy logic to convert the Knowledge Discovery Metamodel (KDM) into an oriented aspect version. This method enables conversion from the old system to the new one and provides abstraction to support modularization of crosscutting concerns and detect violations, while allowing the conversion from the old system to a new one. Our development approach, which is based on fuzzy logic, includes a few sets of rules that have been established as the rule basis in order to automate complexity measurement. By measuring the complexity of the architecture using our model, it will be possible to assess other software attributes utilized in the context of our case study, such as security and performance.

Keywords: Modernization · Knowledge Discovery Meta-Model (KDM) · Model Transformation · Fuzzy Logic · Software Architecture Quality

1 Introduction

Defining the architecture of a software system is an imperative step for its development. The first phase of this process involves making design decisions, along with considering the non-functional requirements, including quality requirements [1].

Introducing new technologies can sometimes lead to having to adapt the existing architecture of the software system to meet certain functional requirements, which may then result in pattern violations and a subsequent loss of architectural choices from the original design. This impedes the maintainability and validity of the quality attributes that were previously implemented. Rather than completely replacing a system, legacy

T. Marir et al. (Eds.): ICCSA 2024, ISEM 45, pp. 264–278, 2025.
https://doi.org/10.1007/978-3-031-90758-6_20

applications can instead incorporate modernization and be adapted to meet the business needs of the digital age, thereby supplying greater business value.

The Object Management Group, in order to lessen this process, has introduced modernization standards, as well as representations, such as the Knowledge Discovery Meta Model (KDM) tool, to reflect all the elements of the associated architecture. Modernization requires an understanding of the system before it can be modernized [2]. In other words, this route requires a thorough understanding of the current system through reverse engineering techniques before the journey of modernization can begin.

Aspect-oriented (AO) programming is an approach that allows one to solve preexisting problems through separation at design and implementation levels [3]. Thereby increasing modularity and allowing for the analysis of non-functional aspects and attributes representing quality. As a result, KDM can be reformed into an Aspect-Oriented version, the KDM-OA.

This paper focuses on describing the model transformation approach used to convert KDM into KDM-OA, which is applied according to a set of rules that will be applied to the source model, which conforms to the source meta-model, to obtain the target model, which conforms to the target meta-model. Generally, these rules are defined by the designers themselves in a manual way. In fact, those rules must be generated precisely to preserve the initial architecture and avoid any data loss. Therefore, we have chosen to generate them automatically based on fuzzy logic.

This work presents a methodology aimed at enhancing the KDM abstraction layer and generating a destination model while reducing analysis costs. By separating the model into functional and non-functional aspects, our proposal ensures that the resulting model maintains its quality properties. The contribution presented in this work is the methodology we have conducted whose goal is to answer the following research questions:

- How can we preserve the quality properties of legacy systems when using KDM?
- Is the integration of fuzzy logic a potential means to improve the KDM abstraction layer?

To answer all these questions, the paper is structured around three main points, which are then followed by a presentation of related works, an overview of our proposal, a case study, and, finally, a conclusion.

2 Related Works

Our research focuses on improving the MDA process in managing legacy system architecture. This covers a broad spectrum of analysis levels, from business objective analysis to architecture evaluation. It is worth mentioning that diverse approaches are being explored as a substitute for KDM through various studies.

Feliu et al. [4] suggested a method founded on the principles of Architecture-Driven Modernization (ADM), which automates the migration of Web applications to content management system-based Web applications. It is based on model-to-model transformation rules that generate a KDM model automatically from an anciennement American Society for Testing and Materials (ASTM) model. Vasilecas et al. in [5] have presented a model-driven process for extracting business rules from legacy systems. An ASTM model at the Platform Specific Model (PSM) level describes the legacy source code, and the code model is described at the Platform Independent Model (PIM) level. They define Text to Model transformations to extract the ASTM model, and then they specify Model-to-Model transformations to auto-generate the code model. This approach enables the extraction of business rules from any existing information system. The SDL KDM toolkit supports editing, querying, importing, and exporting KDM models in XMI format.

Alvaro et al. in [6] introduce a model-to-model transformations process that holds trace generation named MeTAGeM-Trace; it can be used to develop model transformations that produce the associated target models, and thus a trace model between the source and target model elements, which are involved in the transformation process.

Whittle et al. [7] have presented a transformation approach to obtain an Aspect-Oriented Model from an Object-Oriented Model using the Modeling Aspects Using a Transformation Approach tool (MATA) that allows model reuse, the dependency between aspects, conflict detection, and has a concrete syntax.

In recent years, Khan et al. [8] have developed a CMMI compliant modernization framework to transform legacy systems into reliable enterprise solutions. This framework consists of eight stages that ensure quality and reliability, validated through an industrial case study.

In the literature, a few other works exist, in contrast to our research that provides model transformations between KDM and some EA concepts. For example, Perez et al. in [9] presented an approach to transform KDM into an ArchiMate model (exogenous transformation).

While the authors in [10] focus on Architecture-Conformance Checking (ACC) to identify architectural drifts between the Current and Planned Architectures, our work goes a step further by enabling transformations that ensure the preservation of quality attributes within the Planned Architecture. This approach not only detects drifts but also facilitates the alignment of the Current Architecture with the planned one, ensuring that the desired architectural qualities are maintained.

The work proposed in [11] emphasizes the alignment of architectural elements across abstraction layers to aid in management, it does so without making decisions regarding the necessary transformations. In contrast, our work goes further by not only aligning these elements but also enabling transformations that make the system more adaptable, ultimately ensuring a more performant and efficient system. In this paper, we have proposed a new method for software modernization while preserving the architectural qualities and ensuring that there is no data loss. Most of the mentioned works do not follow an automatic approach to generate transformation rules contrary to our proposal in this paper, where we adopted a model transformation where the generation of rules is based on fuzzy logic proposed by Zadeh. L in [12].

3 Overview of Model Transformation Based on Fuzzy Logic

In Model-Driven Engineering (MDE), model transformation (Fig. 1) is a crucial process that enables the conversion of a source model, conforming to one metamodel, into a target model that conforms to another. In our approach to software system modernization, the

Knowledge Discovery Metamodel (KDM) [13] serves as our source metamodel. Developed by the Object Management Group (OMG), KDM is a standardized framework designed for managing and modernizing software architectures. It provides a comprehensive structure for representing various elements within a software system, including code, data, and runtime behavior, which facilitates the analysis and transformation of legacy systems.

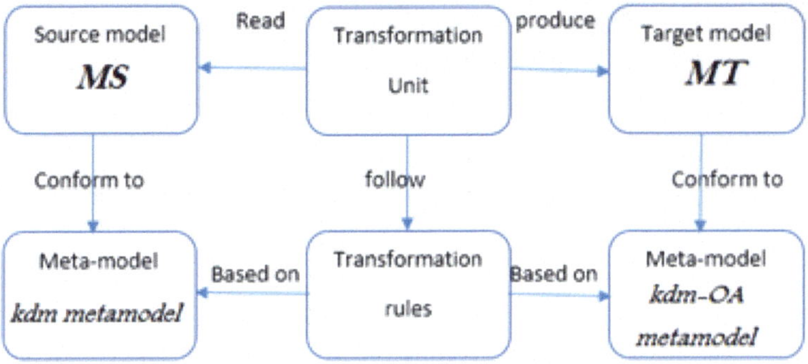

Fig. 1. Model transformation process.

The Knowledge Discovery Metamodel (KDM) is our source metamodel, serving as a standardized framework developed by the Object Management Group (OMG) for managing and modernizing software architectures. KDM provides a comprehensive structure for representing various elements and relationships within a software system, including code, data, and runtime behavior. This enables organizations to understand, analyze, and transform legacy systems effectively, facilitating their adaptation to new technologies and requirements. By leveraging KDM, software engineers can gain deep insights into complex systems, making it easier to plan and execute modernization efforts with precision. Within KDM, there are four layers: Infrastructure, Program Elment, Runtime Resource, and Abstraction. The Abstraction Layer is crucial in defining high-level concepts, but focusing exclusively on this layer can introduce security vulnerabilities if critical details are obscured or poorly managed. Additionally, abstracion layers may omit essential details, limiting the system's ability to dynamically adjust and evolve, which could lead to an inaccurate assessment of vital system attributes. Since KDM is a holistic framework, it is essential to consider all elements linked to the Abstraction Layer to ensure comprehensive system modernization. To harness the full potential of KDM, we configured the metamodel within the Eclipse environment by importing the KDM machine-readable file provided by OMG and installing the necessary plugins.

To further enhance the transformation process, we integrated fuzzy logic across different abstraction layers (Fig. 2), which provides significant advantages. In the Design Layer, fuzzy logic enhances the flexibility and precision of abstraction, allowing for more accurate modeling of system attributes and cross-cutting concerns, leading to a more robust design foundation. In the Implementation Layer, it supports the translation of design models into code, resulting in higher code quality and improved maintainability. In the Integration Layer, the abstraction enhancements provided by fuzzy logic

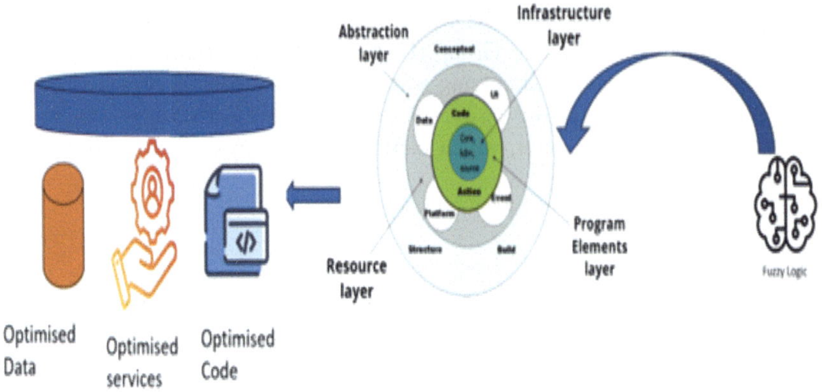

Fig. 2. Optimizing Software System Components through KDM Transformation and Fuzzy Logic Integration.

clarify integration points between components, ensuring a smoother and more reliable integration of various system parts. During the Testing Phase, fuzzy logic allows for more targeted and effective testing, ensuring that the system's performance and security are comprehensively evaluated under various conditions. Additionally, in the Maintenance Layer, it simplifies the system's complexity, making maintenance and updates less complicated and more cost-effective. This multi-layer enhancement is particularly crucial in the Security Layer, where it improves the identification and management of vulnerabilities, thereby strengthening the system's security framework. Finally, in the User Layer, the optimization of data and processes enabled by fuzzy logic enhances the user experience by providing more efficient and responsive interactions.

The source model was extracted by generating the KDM from existing Java code using the MoDisco tool, an Eclipse-based framework designed for model-driven reverse engineering [14]. MoDisco automatically discovers models from various software artifacts, such as source code, by parsing and analyzing the underlying structure of the system. In this case, MoDisco scanned the Java code and generated a corresponding KDM model that accurately represents the system's architecture, components, and interrelationships. This KDM model serves as a foundational representation, facilitating subsequent transformation and modernization tasks.

The target metamodel for our transformation is an extension of KDM (Fig. 3), incorporating aspect-oriented concepts. We introduced an additional package dedicated to aspects, including essential elements such as aspects, pointcuts, and advice—key concepts in aspect-oriented programming [15]. Aspects modularize cross-cutting concerns, such as transaction management or security, by separating them from the main code, thereby facilitating their management and reuse. Pointcuts define the points in the code where aspects should be applied, while advice specify the actions to be executed when these points are reached. By integrating these elements into KDM, we create an enriched version that offers greater flexibility for modeling and transforming cross-cutting concerns in the software system. Following this transformation, an aspect-oriented target model is generated, enabling more effective management of cross-cutting concerns in the modernized system architecture.

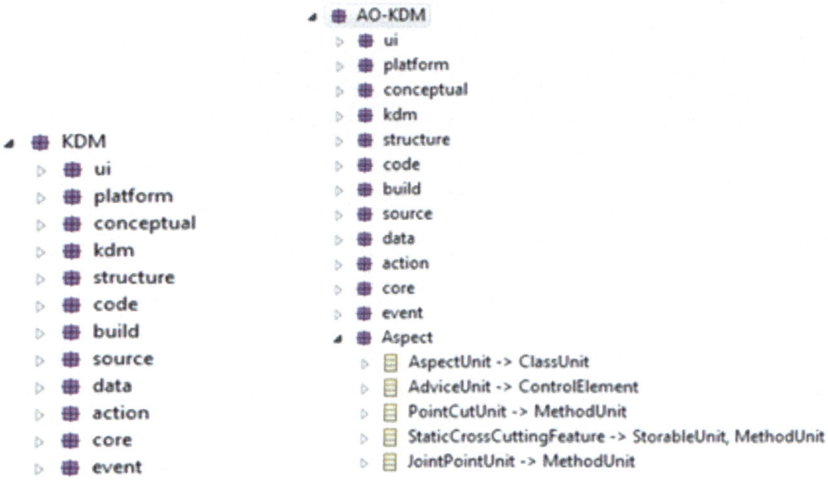

Fig. 3. KDM meta-model provided by OMG vs proposed one.

For generating transformation rules using fuzzy logic, a mathematical tool designed to deal with vagueness and uncertainty, we address the critical aspect of managing uncertainties inherent in non-functional criteria and their metrics [16]. Non-functional criteria, such as performance and security, are often challenging to assess precisely due to their complex and variable nature. Fuzzy logic is particularly well-suited for handling this uncertainty by providing a method to evaluate these criteria through degrees of truth rather than a simple binary approach.

In our approach, we developed a fuzzy logic engine (Fig. 4) based on Mamdani's four-step method to model these uncertainties [17]. The first step, fuzzification, involves transforming the measured values of non-functional criteria into fuzzy sets using appropriate membership functions. Next, we evaluate the defined fuzzy transformation rules by combining the fuzzy values obtained for each criterion through conditional rules, which express complex relationships between criteria in a more nuanced manner. After evaluating the rules, we proceed to the aggregation step, where the results of different rules are combined to obtain a consolidated representation of quality, using operations such as maximum calculation to aggregate the results. Finally, defuzzification converts this aggregated representation into a precise value using the center of gravity method, providing a final estimate of the non-functional qualities of the transformed model. This fuzzy logic-based transformation approach effectively manages the uncertainties associated with non-functional criteria, offering a more robust and adaptable evaluation in the model transformation process.

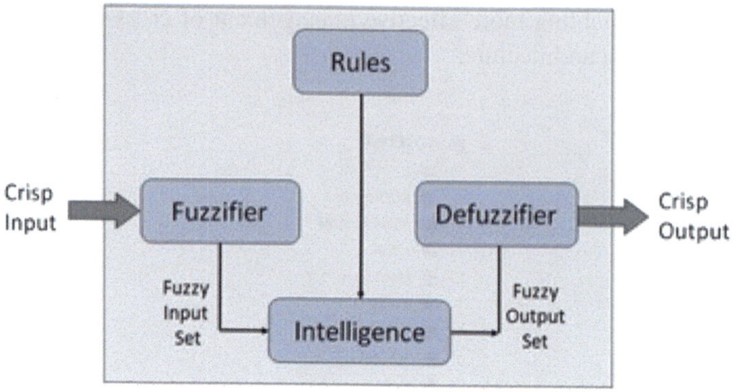

Fig. 4. Architecture of fuzzy engine [18].

Given the complexity of the proposed work, an initial implementation has been developed to validate our approach within the context of model transformation. This preliminary implementation demonstrates the feasibility and effectiveness of our method, laying the groundwork for further refinement and application.

4 A Case Study

Let us take the example of a pharmacy management system used by a pharmacy to manage and oversee all of its pharmaceutical operations. This system may have been installed several years ago and may have outdated features or limitations that hinder its efficiency. To illustrate the transformation process, we consider the following source meta-model and model (Fig. 5).

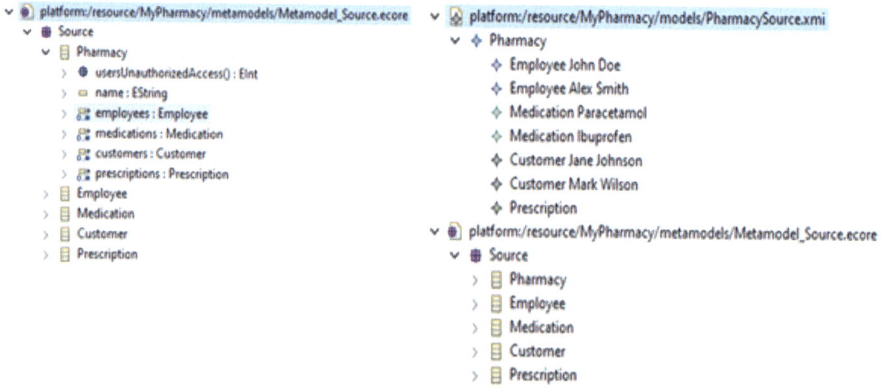

Fig. 5. Source meta-model and model.

By modernizing this pharmacy management system, it is assumed that sensitive information such as usernames and passwords are stored in plaintext, like this:

```
xml
name="MyPharmacy" username="admin" password="admin123"
```

In the transformation to the target model, this information might be inadvertently exposed or mismanaged. For example, if the target model is accessible to unauthorized parties, it could compromise the security of sensitive information, such as usernames and passwords of system users. Also, prescriptions are referenced by their position in a list rather than by a unique identifier. For example:

```
xml
<prescriptions
    patient="/0/@customers.1"
    medication="/0/@medications.0"/>
```

This means that to retrieve the details of a prescription, one must first traverse the list of prescriptions and find the one corresponding to the specified index. This method can become inefficient as the number of prescriptions increases, as it requires a linear search through the list, which can lead to poor performance, especially with large datasets.

At this stage, we proceed from simple architectures to aspect-oriented architectures and use fuzzy logic to model the uncertainty and variability in these architectural criteria.

Using the principle of fuzzy sets, a fuzzy logic system converts crisp inputs into crisp outputs. An inference engine in a fuzzy logic system, Fig. 2 employs fuzzy rules. The engine takes in inputs that may be fuzzy and produces outputs that may not be fuzzy. The inference engine's fuzzy core is flanked by one phase that converts crisp input to fuzzy data and another that does the opposite.

Returning to our example, in which we want to use fuzzy subsets to represent system performance and security and applying the 4 steps of Mamdani fuzzy inference system:

1. Fuzzification (translates the values of the non-functional properties into fuzzy values).
2. For each rule, the fuzzy sub-system computes the score based on its own fuzzy rules
3. Aggregation.
4. Defuzzification of the values of each rule.

The core of the approach is the execution of the transformation rules, which will unfold according to these steps:

4.1 Input Variables Fuzzification

The purpose of the fuzzification step is to transform numerical data into a linguistic variable. For this purpose, a membership function must be created. This function allows defining the degree of membership of the numerical data to a linguistic variable. In our example, digital data (security/performance) are transformed into linguistic variables. Several linguistic variables can be chosen to characterize a data type. Here, for the variable "performance", we can create a fuzzy subset with the following linguistic terms: {Very low, Low, Medium, High}.

Each term will be associated with a fuzzy membership degree, for example: Very low: 0, Low: 0.2, Medium: 0.5, High: 0.8, Very high: 1.

For the variable "security", we can create a fuzzy subset with the following linguistic terms {Very Low, Low, Medium, High}. Each term will also be associated with a fuzzy degree of membership, for example: Very Low: 0, Low: 0.3, Medium: 0.6, High: 1.

We can now combine the fuzzy subsets to create a fuzzy rule that will make a decision based on the state of the system. The following set of rules could be used to automatically adapt system quality as a function of performance and security:

- If performance is very low OR security is very low, THEN system quality is poor.
- If performance is low AND security is low, THEN system quality is poor.
- If performance is medium AND security is medium, THEN quality is medium.
- If performance is medium AND security is high, THEN system quality is medium.
- If performance is high AND security is high, THEN system quality is high.

Membership functions are defined for each fuzzy subset using triangular functions. A triangular function is a function in the shape of a triangle. It is defined by three parameters: a, b and c; a is the point at which the function begins to rise, b is the point at which the function reaches its maximum, and c is the point at which the function begins to fall.

To define a triangular function for a fuzzy subset, we choose the values of a, b and c according to the meaning of the fuzzy subset. For example, for the very weak fuzzy subset of the variable performance, we can define the triangular function f(x) as follows:

- a = 0 (because performance cannot be negative).
- b = 0 (because performance is very low).
- c = 0.2 (because performance starts to be low at 0.2).

Thus, the triangular function for the very weak fuzzy subset of the variable performance is:

$$\mathrm{F}(x) = \textit{triangular}(0.0,\ 0.0,\ 0.2) \tag{1}$$

Similarly, we defined triangular functions for other fuzzy subsets by choosing appropriate values for a, b and c.Our membership function for each fuzzy subset using triangular function are listed in Table 1.

Table 1. Linguistic variables associated to triangular values.

	Linguistic variable	Triangular value
Performance	*Very low*	(0.0, 0.0, 0.2)
	low	(0.0, 0.2, 0.5)
	medium	(0.2, 0.5, 1.0)
	high	(0.5, 1.0, 1.0)
Security	*Very low*	(0.0, 0.0, 0.3)
	low	(0.0, 0.3, 0.6)
	medium	(0.3, 0.6, 1.0)
	high	(0.6, 1.0, 1.0)

In our case study, security is measured by evaluating the system's ability to protect against unauthorized access by calculating the ratio of successful unauthorized access attempts to total access attempts and performance is evaluated by measuring the system's response time. After extracting and calculating the non-functional criteria security and performance, the values obtained are security S = 75% and performance P = 68%.

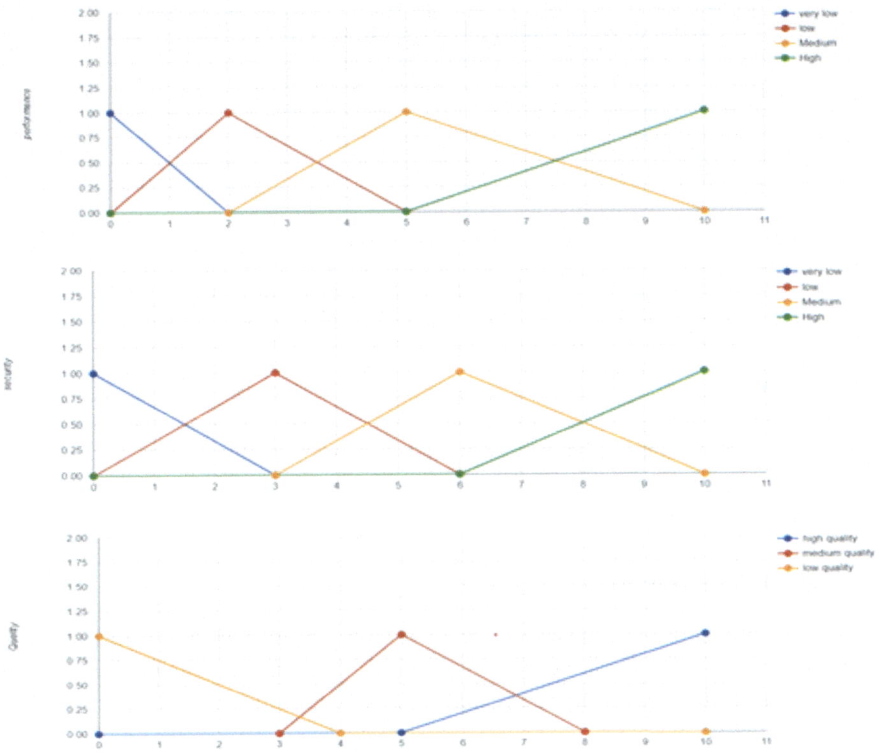

Fig. 6. Linguistic variables, performance, security and quality.

From Fig. 6, the value Security = 75% corresponds to:

- A membership of 60% to the set "medium", noted μm (S) = 0.60.
- A membership of 40% to the set "high", noted μh (S) = 0.40.
- A membership of 00% to the set "low", noted μl (S) = 0.00.
- A membership of 00% to the set "Very low", noted μV (S) = 0.00.

The value P = 68% corresponds to:

- A membership of 48% to the set "high", noted μh (P) = 0.48.
- A membership of 52% to the set "medium", noted μm (P) = 0.52.
- A membership of 00% to the set "low", noted μl (P) = 0.00.
- A membership of 00% to the set "Very low", noted μv (P) = 0.00.

4.2 Evaluating Rules Application of Fuzzy Operators

The result of evaluating of fuzzy rules according to the fuzzy inputs obtained are.

µh (Q) = 40%, µm (Q) = 52% and µl (Q) = 00%.

4.3 Aggregation

Using the cut-off method, we obtained the polygon in Fig. 7:

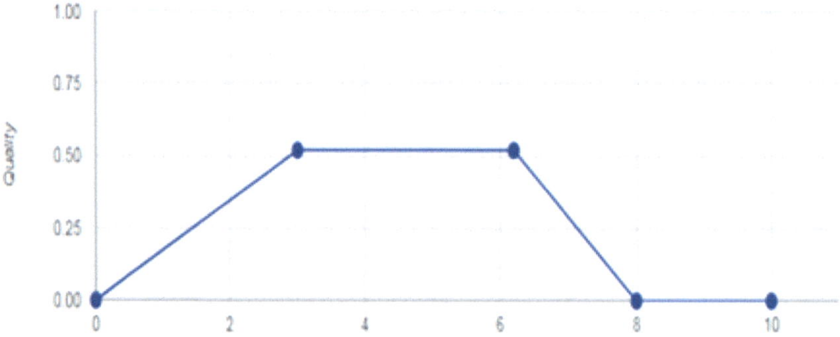

Fig. 7. Output variable and aggregation output.

4.4 Output Defuzzification

The goal of defuzzification is to transform the resulting parameters into numerical data. The primary purpose is to convert the aggregation's fuzzy set into a numerical number that the designer may utilize. By the center-of-gravity method as in (2):

$$GC = \sum_{x}^{b} a\mu A(\mathrm{x}).\mathrm{x} \Big/ \sum_{x}^{b} a\mu A(\mathrm{x}) \quad (2)$$

The ATL transformation script for fuzzy inference in our case study is described as:

```
helper def : getQuality(performance: Real, security: Real) : Real
= (
  -- Fuzzification
  let fuzzy_performance = fuzzifyPerformance(performance);
  let fuzzy_security = fuzzifySecurity(security);

  -- Rule evaluation
  let quality_rule1 = Math.min(fuzzy_performance, fuzzy_security);
  let quality_rule2 = Math.min(fuzzy_performance, fuzzy_security);
  let quality_rule3 = Math.min(fuzzy_performance, fuzzy_security);
  let quality_rule4 = Math.min(fuzzy_performance, fuzzy_security);
  let quality_rule5 = Math.min(fuzzy_performance, fuzzy_security);

  -- Aggregation
  let aggregated_quality = Math.max(Math.max(Math.max(qua-
lity_rule1, quality_rule2), quality_rule3),
   Math.max(quality_rule4, quality_rule5));

  -- Defuzzification using the centroid method
  let defuzzified_quality = defuzzify(aggregated_quality);

  return defuzzified_quality;
);
```

After transforming the model using an aspect-oriented approach, a target model is generated (Fig. 8) and significant improvements in both security and performance were achieved. Security enhancements were implemented by integrating aspect-oriented security measures into the target model, including the encryption of sensitive information and the enforcement of strict access controls. For instance, user information such as usernames and passwords are now securely stored in the following manner:

```
xml
<User name="MyPharmacy" username="encrypted_username" password="en-
crypted_password"/>
```

This approach minimizes exposure risks and enhances overall system security. Additionally, performance was improved by optimizing the target model to use unique identifiers for prescriptions instead of relying on positional references within lists. For example:

```
xml
<prescriptions id="prescription_123" patient="customer_1" medica-
tion="medication_0"/>
```

This adjustment contributes to a more efficient system, reducing the complexity and improving the processing speed of prescription data. This method allows for direct and fast access to prescription details, eliminating linear search and improving performance, even with large amounts of data.

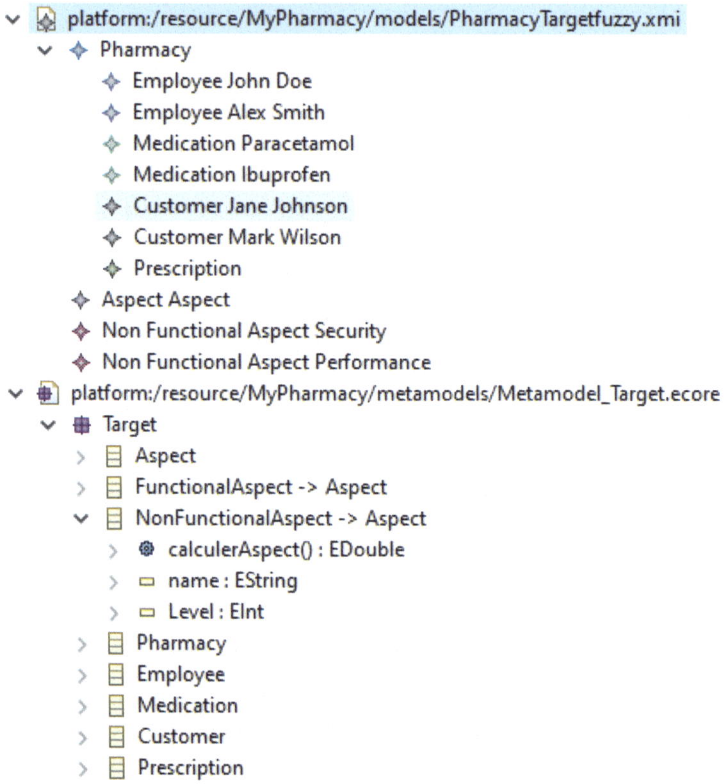

Fig. 8. Target model.

This case study highlights the significant improvements achieved through transformation with an aspect-oriented approach. The integration of fuzzy logic to model uncertainty and variability in architectural criteria addresses the security and performance issues observed with a simple transformation. By applying aspect-oriented security and performance aspects, the modernized system becomes more secure and efficient, providing better management of cross-cutting concerns and optimization of resources.

5 Conclusion

In this research, we have explored a comprehensive approach to modernizing legacy systems through model transformation, focusing on the application of aspect-oriented methodologies and fuzzy logic. Our study specifically addressed the transformation of a pharmacy management system, highlighting the benefits and challenges associated with this process.

We demonstrated how the integration of aspect-oriented principles into the Knowledge Discovery Metamodel (KDM) can significantly enhance the handling of crosscutting concerns, such as security and performance, by providing a more modular and adaptable framework. The use of fuzzy logic for evaluating non-functional criteria allowed us to manage uncertainties and variability effectively, offering a nuanced approach to assessing system quality.

Through a detailed case study, we illustrated the potential pitfalls of traditional transformation approaches, including security vulnerabilities and performance inefficiencies. Our proposed solution ensured that the transformed model met both functional and non-functional requirements.

However, the complexity introduced by integrating fuzzy logic and aspect-oriented principles into the transformation process could make the transformed model difficult to maintain. To address this, we have incorporated a continuous evaluation process for quality attributes, proactively managing the complexity and ensuring that the system remains adaptable and maintainable over time.

The insights gained from this research contribute to the advancement of model-driven engineering practices and offer practical solutions for modernizing complex systems. By addressing the critical challenges of transformation and validation, we have laid the groundwork for future work in this field and provided valuable guidelines for researchers and practitioners aiming to enhance the efficiency and reliability of their modernization efforts.

Our findings underscore the importance of integrating advanced methodologies into the transformation process and offer a compelling case for the continued exploration and development of aspect-oriented and fuzzy logic approaches in model-driven engineering. As we advance, these methodologies promise to play a pivotal role in shaping the future of software modernization and ensuring the continued relevance and effectiveness of legacy systems.

References

1. Levy, N., Losavio, F., Pollet, Y.: Architecture and quality of software systems. In: Software Architectures: Principles, Techniques and Tools. Architecture et qualité de systèmes logiciels. Architectures logicielles: Principes, techniques et outils (2014)
2. ADM Domain Task Force: Architecture Driven Modernization (ADM) Task Force (2003)
3. Kadri, S., Aouag, S., Hedjazi, D.: MS-QuAAF: a generic evaluation framework for monitoring software architecture quality (2021)
4. Trias, F., De Castro, V., Lopez-Sanz, M.: Migrating traditional web applications to CMS-based web applications. Electron. Notes Theor. Comput. Sci. **314**, 23–44 (2015)
5. Vasilecas, O., Normantas, K.: Deriving business rules from the models of existing information systems. In: Proceedings of the 12th International Conference on Computer Systems and Technologies. ACM, pp. 95–100 (2011)
6. Jiménez, Á., Vara, J.M., Bollati, V.A., et al.: MeTAGeM-trace: improving trace generation in model transformation by leveraging the role of transformation models. Sci. Comput. Program. **98**, 3–27 (2015)

7. Whittle, J., Jayaraman, P., Elkhodary, A., Moreira, A., Araújo, J.: MATA: a unified approach for composing UML aspect models based on graph transformation. In: Transactions on Aspect-Oriented Software Development VI: Special Issue on Aspects and Model-Driven Engineering, pp. 191–237 (2009)
8. Khan, M., Ali, I., Mehmood, W., et al.: CMMI compliant modernization framework to transform legacy systems. Intell. Autom. Soft Comput. **27**(4) (2021)
9. Pérez-Castillo, R., Delgado, A., Ruiz, F.: A method for transforming knowledge discovery metamodel to ArchiMate models. Softw. Syst. Model. **20**, 1–26 (2021)
10. Landi, A.S., San Martín, D., Santos, B.M., Cunha, W.S., Durelli, R.S., Camargo, V.V.: Architectural conformance checking for KDM-represented systems. J. Syst. Softw. **183**, 111116 (2022)
11. Pepin, J., Andre, P., Attiogbé, C., Breton, E.: An improved model facet method to support EA alignment. Complex Syst. Inf. Model. Q. **9**, 1–27 (2016)
12. Zadeh, L.A.: Fuzzy logic. Computer **21**(4), 83–93 (1988)
13. OMG: Knowledge Discovery Metamodel (KDM) Specification, Version 1.3, Object Management Group (OMG) (2016). https://www.omg.org/spec/KDM/1.3
14. Bruneliere, H., Cabot, J., Dupé, G., Madiot, F.: MoDisco: a model driven reverse engineering framework. Inf. Softw. Technol. **56**(8), 1012–1032 (2014)
15. Kiczales, G., des Rivieres, J., Bobrow, D.G.: The Art of the Metaobject Protocol. MIT Press, Cambridge, MA (1997)
16. Cheng, M.-Y., Tsai, H.-C., Ko, C.-H., et al.: Evolutionary fuzzy neural inference system for decision making in geotechnical engineering. J. Comput. Civ. Eng. **22**(4), 272–280 (2008)
17. Ross, T.J.: Fuzzy Logic with Engineering Applications. University of New Mexico, McGraw Hill Inc (1995)
18. Suman, S.K.: Basic concepts of fuzzy logic, computational science with Suman. https://computationalsciencewithsuman.blogspot.com/p/basic-concepts-of-fuzzy-logic.html. Accessed 2 Aug 2024

Constrained Mixed-ARDE for Optimizing LSTM Hyperparameters in Traffic Flow Forecasting

Meriem Talai[1,2(✉)], Nora Taleb[3], Hafed Zarzour[4], Iyad Kouloughli[1,2,3,4], and Mohand Kouloughli[1,2,3,4]

[1] University of Souk Ahras, Souk Ahras, Algeria
m.talai@univ-soukahras.dz, m.talai@centre-univ-mila.dz
[2] LISI Laboratory, University Center of Mila, Mila, Algeria
[3] LISCO Laboratory, Badji Mokhtar University, Annaba, Algeria
[4] LIM Research, University of Souk Ahras, Souk Ahras, Algeria

Abstract. This work contributes to the significant potential of Long Short-Term Memory (LSTM) for traffic flow prediction, which is highly dependent on complex; nonlinear and stochastic data. We propose an LSTM hyperparameters optimization approach using a variant of the Adaptive Reinitialized Differential Evolution (ARDE) algorithm; handling mixed types of genes and boundary constraints. Experiments of the proposed mixed-ARDE-based LSTM algorithm on three analyzed datasets demonstrate a considerable accuracy compared to baseline models, namely DE-LSTM, GA-LSTM and keras-tuner-LSTM.

Keywords: Differential Evolution optimization · LSTM · time series analysis · traffic flow forecasting

1 Introduction

The contribution of Intelligent Transportation System (ITS) in terms of road safety, flow fluidity, and pollution reduction is incontestable in real-world deployment [1]. To be efficient, ITS applications such as traffic management systems, adaptive signal control, and incident management systems require future perception of the traffic road conditions. To provide significant future findings about traffic patterns and traffic road states, predictions are made by analyzing specific time series data such as traffic flow or speed collected from IoT devices such as sensors, cameras, or GPS trackers.

However, road traffic time series used to extract main knowledge for prediction purposes reveals challenging characteristics including nonlinearity, stochasticity, unpredictable fluctuations known as residuals, and seasonality such as daily or weekly which make it a difficult task to predict accurately. Deep learning (DL) models and techniques based on recurrent neural networks (RNNs), like Long Short-Term Memory (LSTM) networks, have shown great promise in short-term traffic flow prediction [2]. These models can capture intricate patterns and dependencies in the data, making them suitable for

T. Marir et al. (Eds.): ICCSA 2024, ISEM 45, pp. 279–293, 2025.
https://doi.org/10.1007/978-3-031-90758-6_21

handling the temporal nonlinear dynamics of traffic flow. Also, these models can learn from vast amounts of historical data, enabling them to make accurate predictions even in the presence of nested seasonal patterns.

In this context, applying LSTM to short-term traffic forecasting involves further investigations to deepen its performances and provide more comprehensive models. Major investigations concern: first, the conjunction of multiple hyperparameters, such as the number of hidden units, learning rate, and batch size. An improper setting of even one of these parameters can reduce the model's efficiency or completely degrade its performance, leading to unpropitious predictions. Also, the computational cost of exploring the entire hyperparameter space, which size is mathematically approximated by the product of all possible values for all hyperparameters. Evaluating numerous combinations to find the optimal settings can be resource-intensive and time-consuming, adding to the complexity of applying LSTM models. In addition, the complex characteristics of road flow time series data still challenge the learning and generalization capabilities of forecasting models.

Previous research agrees that optimization algorithms belonging to evolutionary computation, namely differential evolution, are proficient in the task of the Deep Learning model's configuration to extract complex patterns [3]. Additionally, from the perspective of enforcing the LSTM performances in short-term traffic flow prediction, exploring the effect of the statistical characteristic and time resolution of the traffic flow time series dataset on the training and quality of LSTM models are important issues.

In this paper, we present a framework for road traffic flow prediction, based on an optimized LSTM model, using the ARDE algorithm. The main contributions are:

1 Associate the ARDE variant to the LSTM network as an efficient approach to realize an optimized deep learning model configuration.
2 Integrates approved mechanisms to handle mixed data types and data boundary constraints.
3 To the best of our knowledge, no previous research has explored these both aspects of LSTM hyperparameter configuration.
4 Presents a detailed methodology to progress toward reliable predictions.
5 Experimentally analyses the performances of Mixed-ARDE-based LSTM using different three datasets and baseline models.

The remainder of this paper is composed of the following sections: Sect. 2 delivers a literature review of related work and concepts. Then, details of the proposed ARDE-LSTM for mixed data type approach are presented in Sect. 3, including the dataset manipulations and the underlying mathematical model. Section 4 reports experimental results and discussion. Section 5 concludes this study and presents related perspectives.

2 Literature Review

2.1 LSTM Neural Network

An LSTM network is a Recurrent Neural Network (RNN) subsequent where each unit's output is influenced not only by the current input but also by its own previous outputs. However, RNNs are limited by the problems of vanishing and exploding gradient that occur when the feedback signal is backpropagated through a prolonged series of operations; the gradient becomes too small or too large. LSTM units include memory cells, gates, and recurrent connections to control the conservation of crucial information, even over extensive sequences. These mechanisms make them suitable to learn nonlinear temporal dependencies in time series data.

In more detail, for each sole data observation (x_t) received per timestep (which can be a single value or a vector if several features are observed each timestep), an LSTM unit updates and outputs two states: the hidden state (h_t) (as a conventional RNN unit) and the advanced memory cell state (c_t). . This cell state is in interaction with three internal gating mechanisms of the LSTM unit: the Input Gate (I_t), the Forget Gate (F_t), and the Output Gate (O_t) which respectively controls: information to be added to the cell state, information to be discarded from the cell state, and information from the cell state to be outputted. Notice that these gates are sigmoid activation function (σ) of weighted input (x_t) and previous hidden state (h_{t-1}), which is well appropriate to make binary decisions. Also, the tangent hyperbolic nonlinear activation function (tanh) is employed to normalize the weighted inputs (x_t) and the previous hidden state (h_{t-1}). Its result is filtered by the input gate and summed up to the previous cell state (c_{t-1}) filtered by the forget gate (F_t) to produce the final output of LSTM cell state (c_t). . Finally, this updated cell state is normalized using the (tanh) activation function and filtered by the output gate to produce the updated hidden state h_t.

$$\sigma(x) = \frac{1}{1 + e^{-x}} \tag{1}$$

$$\tanh(x) = \frac{\left(e^x - e^{-x}\right)}{\left(e^x + e^{-x}\right)} \tag{2}$$

The learning process of LSTM is effectively involved by adjusting the weights $\left(W_I, W_f, W_o, W_c\right)$ within sigmoid and tangent hyperbolic activation functions. These weights, also called parameters, are responsible of the amount of information passing through the LSTM gates. Within the same layer, LSTM units operate independently but share their parameters. Between different layers, LSTM units build hierarchical representations of the input data, and each layer contributes to learning and abstracting different features or patterns.

2.2 Adaptive Reinitialization Differential Evolution Algorithm

Differential Evolution (DE) algorithm belongs to Evolutionary Computation (EC) widely used as a numerical global optimizer. It was first proposed by [4] to work with continuous variables. Since, valuable research emerged on aspects such as the strategies to control its adaptation using its few parameters, strategies to choose specific improvements of its operators according to problem on hand, and hybridizing methods [5].

In this section, the optimization process of ARDE proposed in [6] is described. It focuses on adding a reinitialization phase, using an adaptive mutation factor in the mutation phase, and integrating an early stopping mechanism. In the following, the operator rand [0, 1] returns a random value between 0 and 1 with a uniform probability distribution.

Population Initialization. In this phase, P individuals or chromosomes are created to produce the first generation $G = 1$. Each chromosome $x_i^1 (i = 1, 2, \cdots, P)$ is modelized by a vector of D elements (genes) randomly initialized according to Eq. (3)

$$x_{ij}^1 = Xmin_j + \text{rand}\ [0, 1] \times \left(Xmax_j - Xmin_j\right) \tag{3}$$

the jth gene $(j = 1, \cdots, D)$ is constrained in the interval $[Xmin_j, Xmax_j]$. The population size is highly increased relatively to the original DE, to prepare the next phase.

Reinitialization Phase. This added phase aims to avoid unfavorable individuals produced in the stochastic initialization phase. ARDE keeps only NP best individuals among P individuals. They are selected by comparison of their fitness function evaluation with a fitness threshold ϵ;

$$\mathrm{f}\left(\mathrm{x}_\mathrm{i}^1\right) < \in \ \text{ where i} = 1, 2, \cdots, \mathrm{NP} \tag{4}$$

Mutation. The mutation phase applies a disturbance to a base vector v_m: a differential vector resulting from a scaled difference between two vectors v_k and v_r is added to the base vector. The result is the mutation vector v_i. Notice that the base vector and the two others are all different $(1 \leq i \neq k \neq r \neq m \leq NP)$ and randomly selected from the actual generation G,

$$v_i^G = v_m^G + F_i^G.\left(v_k^G - v_r^G\right) \tag{5}$$

F is the scale factor or the mutation factor, an important parameter since it controls the amplification of differential variation and thus the search for new individuals. Instead of keeping a constant mutation factor which becomes relatively wide when the research process tends to bring the studied population closer, ARDE algorithm heuristically modifies this factor, for each chromosome, in the interval $\left[F_{min}^G, F_{max}^G\right]$ which bounds decreases over generations as follow:

$$\left[F_{min}^G, F_{max}^G\right] \leq \left[F_{min}^{G-1}, F_{max}^{G-1}\right] \tag{6}$$

Crossover. This phase considers a binomial crossover where the D components of the trial vector u_i^G are produced depending on the constant probability given by the crossover factor Cr, where $Cr \in rand[0, 1]$. The combination of the mutation vector's components v_{ij}^G with those of the considered target vector x_{ij}^G is controlled by D random values $rand[0, 1]$, as expressed:

$$u_{ij}^G = \begin{cases} v_{ij}^G \text{ if } Cr \leq rand[0, 1] \\ x_{ij}^G \text{ otherwise} \end{cases} \tag{7}$$

Selection. The selection operator creates the offspring; it compares between the trial vector u_i^G and the target vector x_i^G without consideration of any other individual in the current population. It evaluates the fitness function f of both vectors to decide. It determines which of u_i^G or x_i^G subsists in the next generation, described in Eq. (8)

$$x_i^{G+1} = \begin{cases} u_i^G \text{ if} f\left(u_i^G\right) \leq f\left(x_i^G\right) \\ x_i^G \text{ otherwise} \end{cases} \tag{8}$$

Early Stopping. The optimization process is achieved if the best value of the fitness function f_G^* remains unchanged for k successive generations of the process where k is a fixed patience parameter.

$$f_{G+1}^* = f_G^* = \ldots = f_{G-k+2}^* \tag{9}$$

The global optimal vector x^* is determined according to:

$$x^* = arg\ \min_{x \in \mathbb{R}^D}\left(\left\{f\left(x_i^G\right)_{i=1,2,\ldots,NP}\right\}\right) \tag{10}$$

2.3 Related Work

In the following, five recent works, all based on LSTM for time-series forecasting with relatively simple extensions involving important prediction accuracy. Also, Table 1 specifies their technical concepts related to combining LSTM with several methods to determine its hyperparameters and identifies for each work the strengths and weaknesses of the whole proposed framework. Thus, indicating the potential applicability of LSTM since it is adapted to real-world data in several domains.

Authors of [7] explore the forecasting capabilities of an improved LSTM for multivariate, multi-domain and multi-scenario datasets. LSTM is optimally configured using Particle Swarm Optimization (PSO) algorithm. Also, its predictive performances are accurate using the bifold-Attention mechanism which selects the relevant parts of the input sequence for LSTM. Extensive experiments are conducted to explore the applicability of the solution in different scenarios. It works much better for short-term forecasting.

A three-phase framework, purely based on DL models, for road traffic prediction is proposed in [8]. The study determines the influential road segments using LSTM. It augments these segments' data with a deep generative model, and exploits the resulting data using Neural Basis Expansion Analysis for interpretable Time Series (N-BEATS) model. LSTM plays a crucial role since it is used in the first phase to keep the influential road segments while reducing data dimensionality. It is fitted to sequences of cross-correlation matrices that represent the time series dependency between m segments of the studied road. Efficient predictions for data roads with more than one segment and when using the Pearson coefficient.

In [9], an ensemble learning for wind speed forecasting. It employs a new Genetic Algorithm (GA) variant to configure at once, the architecture and hyper-parameters of the LSTM model. Each GA individual considers encoding a different number of LSTM hidden layers. Five different EvLSTM models are created and assembled in the EnEvLSTM model with weight coefficients determined by the DE algorithm and following the No Negative Constraint Theory (NNCT). For validation, two case studies, i.e. forecasting for ten minutes ahead and for one hour ahead. It shows that both proposed models outperform baseline models.

Grizzly model, proposed in [10], is based on a Bi-directional LSTM neural network approach that attaches both forward-pass and backward-pass of LSTM to the same output layer. In order to handle spatiotemporal large-scale datasets, a low-dimensional

Table 1. Details of the presented LSTM-based time series forecasting approaches

REF	Characteristics	
[7]	dataset	Six time-series datasets (Beijing)
	LSTM config	PSO particle (number of hidden layers, activation function, optimizer, batch size, epoch), min (MAE), min(MSE)
	Hyperparams	2-3layers, 24–93 neurons, Adam/RMSprop Batch = 32–64, Epoch = 17–96
	Evaluation	Baselines: ARIMA, MLP, LSTM, PSO-LSTM, Attention-LSTM Metrics: MAPE, RMSE
	Prons	• Consistent performance evaluation • Attention mechanism
	Cons	• No statistical study of datasets • High computational cost
[8]	Dataset	Baidu dataset (Beijing)
	LSTM config	Experiments
	Hyperparams	2 layers: 28, 64 neurons, Adam, RMSE, Epoch = 100, Batch = 72
	Evaluation	Baselines: ARIMA, SVM, CNN, SAE. Metrics: RMSE, MAE, MAPE
	Prons	• A multivariate prediction • LSTM's important role
	Cons	• No LSTM hyperparameters optimization • A single dataset
[9]	Dataset	Wind farms (China, Spain)
	LSTM config	GA individual (number of units per layer, optimizer, learning rate), min(MSE)
	Hyperparams	3 layers: 108, 133, 43 neurons, RMSprop, LR = 0.35, 60/20/20 split
	Evaluation	Baselines: ANN, WNN, ELM, DBN, LSTM. Metrics: MAE, RMSE, MAPE
	Prons	• Optimizes LSTM via GA. Ensemble learning improves accuracy
	Cons	• Limited statistical analysis. No complexity handling
[10]	dataset	PEMS (California), METR-LA (Los Angeles)
	LSTM config	Experiments
	Hyperparams	1–3 layers,128,64,32 units, ELU/sigmoid, Adam, MAE, Epoch = 60, 70/20/10
	Evaluation	Baselines: ARIMA, various GNNs. Metrics: MAE, RMSE
	Prons	• Improves performance • Hybrid spatiotemporal model

(*continued*)

Table 1. (*continued*)

REF	Characteristics	
	Cons	• Increases complexity • No data characteristics analysis
[11]	dataset	Gateway traffic time series
	LSTM config	Keras-Tuner
	Hyperparams	2 layers: 250, 70 neurons, Adam, Batch = 16, MSE
	Evaluation	Baselines: Previous Value, Historical Median/Mean, Linear Regression, ARIMA. Metrics: NRMSE, NMAE
	Prons	• Data preprocessing for a stable dataset • Simple anomaly det[ection
	Cons	• weak LSTM performance with Keras-Tuner • Simplistic model for irregular traffic

continuous input vector representation is generated by normalizing continuous values (speed) and by learning the embeddings of categorical features (the day of the week, the hour of the day and the sensor ID). It forecasts the same input size sequences of speeds. Grizzly surpasses all the baselines.

Authors of [11] study the use of specific real network data for traffic prediction, with LSTM Neural Networks. Also, the resulting predictions of the proposed model are used for anomaly detection in networks. The data, collected from an office environment, is analyzed and compared to baseline methods with lower training costs. The results show that LSTM-based predictions depend on the data composition while ARIMA method accomplishes the best performance for non-aggregated traffic.

The utilization of LSTM hyperparameter configuration is indispensable for two primary reasons: First, the results of deep learning (DL) models are highly variable depending on the dataset, necessitating the definition of a specific model for each dataset. However, identifying the appropriate model with a grid search approach is both time-consuming and costly, thereby justifying the use of optimization techniques. Second, hybridizing LSTM with hyperparameters optimization algorithms produces highly accurate results due to advanced search capabilities. This approach motivates further exploration of different algorithms, despite the eventual associated computational costs, as the benefits in terms of performance and accuracy are substantial.

3 Methodology

The following section details the Mixed-ARDE-based LSTM approach for traffic flow forecasting. First, the collected datasets are analyzed for validity and best-fit purposes. After, the ARDE optimization of LSTM hyperparameters is presented as a solution for an accurate traffic flow forecasting problem. It is detailed via a mathematical formalization which emphasizes the mixed aspects of the optimized variables and their boundary constraints. The results are compared to several LSTM baseline models which differ by

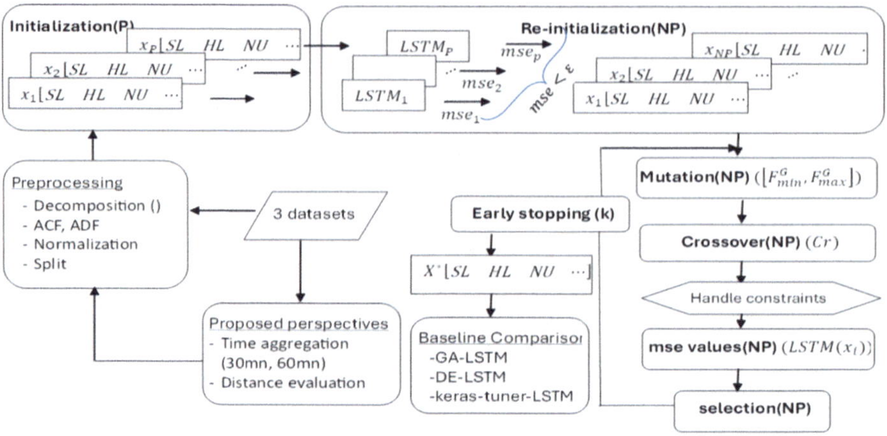

Fig. 1. Illustration of the proposed Mixed-ARDE-based LSTM approach.

their hyperparameters configuration. Figure 1 illustrates the Mixed-ARDE-based LSTM approach proposed within this paper.

3.1 Elementary Data Analysis

The three real-world datasets used to evaluate the Mixed-ARDE-based LSTM model are presented in Table 2.

Table 2. Traffic flow datasets characteristics.

Dataset	M50–11-12 -N	M1–02-03-S	1280-S
Database	Transport Infrastructure Ireland(TII) https://trafficdata.tii.ie	Transport Infrastructure Ireland(TII) https://trafficdata.tii.ie	Caltrans Performance Measurement System (PeMS) http://pems.dot.ca.gov
Road/Segment/Direction	Motorway (Dublin M50) Tallaght to Firhouse, North	Motorway (Dublin M01) Airport to Swords, South	District 3 (San Francisco) Bernal Heights to Ingleside, South
Sensors	Inductive loops		
Traffic period	1 April 2018 to 31 March 2019		
Time Resolution	15mn interval		
Missing (%)	0.03% (11 values)	0.02% (9 values)	none

To answer data integrity from the few missing values that are reported as None or NaN, the removal technique is forward-filling. It involves replacing NaN or None values with the most recent preceding non-missing value.

Before any manipulation of these datasets, the plausibility of their use for a DL forecasting model is verified. No forecasting technique, whether using statistical methods or deep learning techniques, can reasonably model a random walk process. To analyze data behavior, the Autocorrelation Function (ACF) measures the similarity of a time series to itself at different past steps (lags). Its values range from − 1 to 1. For data following a random walk process, the autocorrelation function exhibits a slow decay of significant autocorrelation coefficients (ρ_k) for a large number of lags, where the first-order coefficient ρ_1 is close to one.

[2]The ACF plot in Fig. 2 shows that the values of the time series are highly correlated even at distant lags. The peaks and troughs at regular intervals imply the presence of seasonality or cyclic behavior in the time series.

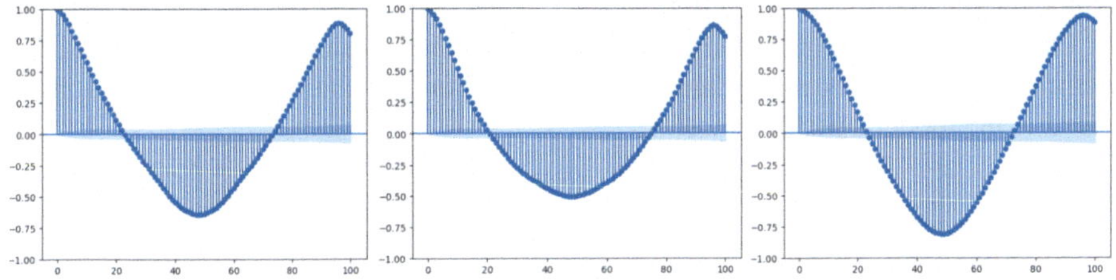

Fig. 2. ACF plots of M50–11-12 -N, M1–02-03-S and1280-S datasets, from the left to the right

Also, to confirm the dataset stationarity, a statistical mean is the Augmented Dickey-Fuller (ADF) test [12]. Results in Table 3, confirm the dataset stationarity since they all ensure the critical value (at 5% significance) exceeds the test statistic value and that the p-value is less than 0.05.

Table 3. ADF test results for the studied datasets

Dataset	Test Statistic	Critical value (5%)	p-value
M50–11-12 -N	− 41.745072	− 2.861622	0.000000
M1–02-03-S	− 43.920478	− 2.861622	0.000000
1280-S	− 38.770697	− 2.861622	0.000000

Another important preprocessing step, is to decompose the time series into trend, seasonality, and residual components.

Figure 3 depicts one month traffic flow from 28 March to 26 April 2018, for the three datasets, equivalent to approximately 2820 raw time series. As it can be seen, there is a repeating pattern in this illustrative data, corresponding to daily traffic cycles. The seasonal component depicts strongly regular peaks and troughs, especially in the

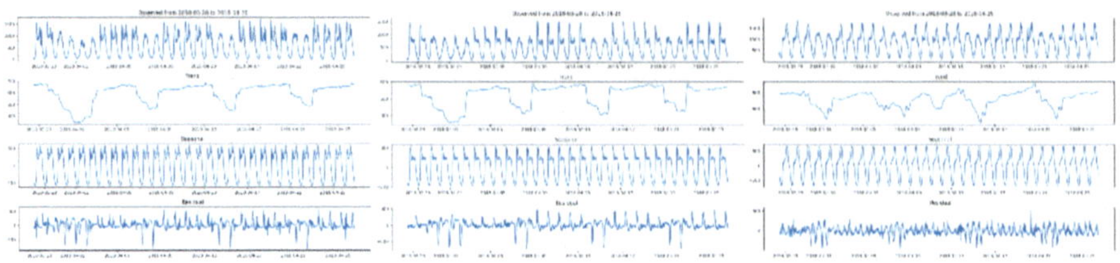

Fig. 3. Illustrative datasets decomposition (trend, seasonal and residual components) of M50–11-12 -N, M1–02-03-S, 1280-S datasets from the left to the right

M50-11-12-N dataset, which corresponds to traffic flow during times of days. The plot of the trend component, of the two first datasets, shows successive periods of noticeable downward trend at the beginning, followed by a stabilized trend that gradually increases. The third dataset trend is more fluctuated indicating a high level of anomalies. Remaining variabilities are depicted in the residual component which indicates more frequent short-term anomalies for the third dataset.

Finally, normalization of the data using the MinMaxScaler Eq. (9) to improve the performance of LSTM training and help handle outliers preventing extreme values from affecting the model's performance,

$$X' = \frac{X - X_{min}}{X_{max} - X_{min}} \tag{11}$$

where X' is the normalized value, X is the original value, X_{min} and X_{max} are respectively the minimum and maximum flow values.

3.2 Mathematical Formalization

Considering the powerful capacity of LSTM in time series data forecasting and the ARDE results, we combine the ARDE evolutionary algorithm with LSTM for traffic flow prediction. In fact, to capture temporal dependencies and patterns of time series data, LSTM requires careful selection of its hyperparameters set.

During its iterations, ARDE is used to search the optimal LSTM hyperparameters set, among eventual solutions in the search space, all encoded in chromosomes $x_i, i = 1, \cdots NP$ of six genes each determines an LSTM hyperparameter value for the learning process. The genes of an ARDE chromosome are the number of hidden layers (HL), number of units per layer (NU), learning rate (LR), sequence length (SL), optimizer (Op), and batch size (BS) which have a significant influence on the performance of LSTM. Considering the type of these hyperparameters, an ARDE chromosome is composed by the concatenation of integer $[x]^z$, discrete $[x]^d$, and continuous $[x]^c$ sub-vectors.

$$x_i = \left[\ [x_i]^z \ [x_i]^d \ [x_i]^c \ \right], \ i = 1, \cdots NP \tag{12}$$

ARDE is directed by the minimization of the loss function $f(x_i)$, typically measured by the Mean Squared Error (MSE) which is ideal to punish high error predictions.

MSE is calculated between LSTM predictions (for each set of hyperparameters) and the corresponding ground truth y, based on the validation dataset of size N.

$$\mathrm{MSE}(x_i) = \frac{1}{N}\sum\nolimits_{k=1}^{N} (LSTM_k(x_i) - (y_k))^2 \tag{13}$$

Thus, the problem considered by Mixed-ARDE-based LSTM is a mixed integer-discrete-continuous boundary-constrained non-linear programming problem, expressed:

Find an optimized solution;

$$x^* \in \left\{x_i / x_i = \left[\ [x_i]^z\ [x_i]^d\ [x_i]^c \right]\right\},\ i = 1, \cdots NP$$

that minimizes the loss function of LSTM model;

$$\min(f(x_i) = \mathrm{MSE}(\mathrm{LSTM}(x_i)))$$

subject to the constraints;

$$xmin_j \le x_{\cdot j} \le xmax_j \ \ j = 1, \cdots, D$$

where:

D is the dimension of the search space ($x_i = [x_1, x_2, \ldots x_D]_i$).

z, d, c respectively for integer, discrete and continuous types.

$xmin_j,\ xmax_j$ are the lower and upper bounds of the jth component of x_i.

In their basic form, DE algorithms are limited to continuous variables. In [13], a DE algorithm is extended to mixed continuous, integer, and discrete variables with boundary constraints handling. First, discrete variables are represented using the index of the values instead of the values themselves. Thus, the problem is reduced to a mixture of integer and continuous values. Second, three important complementary modifications are introduced:

1. at the population initialization, random values of genes are depicted in the interval $\left[Xmin_j,\ Xmax_j + 1\right]$ as follows:

$$x_{ij}^1 = Xmin_j + \mathrm{rand}[0, 1] \times \left(Xmax_j + 1 - Xmin_j\right) \tag{14}$$

2. the continuous representation of the genes is maintained except during the evaluation step of the function, where a truncation operation ensures the use of the integer part for the defined integer values.
3. to handle mutation operations that extend the search outside the initialized range of the search space, and to ensure that hyperparameter values remain within their allowed ranges, any parameter value that violates boundary constraints is replaced with a random value generated within the feasible range.

4 Experiments and Analysis

4.1 Computational Environment

This study utilized Python software and Google Collab for all forecasting models. Data analysis is processed on a Windows 10x64 OS with an Intel Core i5-8265U CPU (1.60 GHz, 4 cores, 8 logical processors), and integrated Intel UHD Graphics 620 alongside Radeon 520 GPU. The ADF test is performed using statsmodels.tsa.stattools library, on all three datasets. Also, statsmodels.graphics.tsaplots is used for ACF plots. Missing values are processed with Pandas library.

4.2 ARDE Parameters

As shown in Fig. 1, the population is first composed of $P = 100$ random individuals. After the reinitialization phase, only $NP = 10$ individuals survive to start the evolutionary mechanism depending on the threshold fixed to $\varepsilon = 0,\ 1$. The mutation factor bounds are first set to $[0.8,\ 1]$ and decrease by 0.2 at each generation and answer a 0.2 minimal interval size at final generations. The random crossover factor $Cr = 0.7$ and ARDE iterates at most for $G_{max} = 10$ generations, with $k = 3$ to control the early stopping mechanism for all three data sets.

4.3 LSTM Configuration

The upper and lower bounds of the proposed gene encoding are set, given of the related work and many other experimental results, as follows:

- Hidden Layers (HL): An optimized network can contain from only one hidden LSTM layer to three layers to reduce the search space without degradation in network capabilities.
- Number of units per layer (NU): starting with eventual one unit and is limited to 75 units per layer
- Optimizer (Op): three optimizers are suggested; Adam, SGD, and RMSprop coded respectively from 1 to 3.
- Batch size (BS): is a discrete value in the set $\{32, 64, 128, 256\}$ encoded from 1 to 4
- Learning rate (LR): is comprised in the interval $[0.0001, 0.1]$
- The sequence length (SL): is the number of past timesteps learned with LSTM, the upper value is set according to the ACF plot regarding the first significant lagged values of each plot. It is set to 20, 18 and 20 for M50–11-12 -N, M1–02-03-S, 1280-S datasets respectively.

Each dataset is split into train, validation, and test sets, allocating 80% for training, 10% for validation, and 10% for testing according to chronological order. This ensures that the model is trained on a sufficient amount of data while allowing for evaluation on unseen data.

4.4 Results and Discussion

Mixed-ARDE-based LSTM model's performance is compared DE-LSTM, GA-LSTM [9], and Keras-tuner-LSTM [11] for the three datasets and evaluated with several metrics: Mean Squared Error (MSE), Mean Absolute Error (MAE), Mean Absolute Percentage Error (MAPE), and Root Mean Squared Error (RMSE) depicted in Table 4. DE-LSTM, GA-LSTM, and Keras-tuner-LSTM are made by using, respectively; Differential Evolution (DE), Genetic Algorithm (GA), and Keras-tuner to optimize the hyperparameters of LSTM.

Table 4. Mixed-ARDE-based LSTM results compared to baseline models for three datasets

Datasets	Mixed-ARDE-based LSTM				DE-LSTM				GA-LSTM				Keras-tuner-LSTM			
	MSE	MAPE	MAE	RMSE	MSE	MAPE	MAE	RMSE	MSE	MAPE	MAE	RMSE	MSE	MAPE	MAE	RMSE
M50–11-12	**3013.82**	11.50	**36.11**	**54.89**	3271.98	**10.50**	37.64	57.20	3035.54	14.15	36.81	55.09	4984.53	14.83	47.08	70.60
M1–02-03	**1745.31**	**8.42**	**29.26**	**41.77**	2096.12	10.10	32.73	45.78	1925.28	8.71	31.13	43.87	3351.50	11.64	40.51	57.89
1280-S	**3872.13**	**6.36**	**43.19**	**62.22**	3994.99	7.03	44.90	63.20	3879.59	6.49	43.34	62.28	4919.88	8.04	50.72	70.14

Overall, Mixed-ARDE-based LSTM results defeat baselines for 91.66% of the evaluation metrics (4 metrics x 3 datasets). The second-best approach considering best metrics is DE-LSTM with best MAPE value for the M50–11-12 dataset. However, considering distance metrics to Mixed-ARDE-based LSTM, GA-LSTM is the second-best model.

High values of MSE for all datasets are a direct consequence of its responsiveness to errors: In fact, its mathematical expression squares errors, especially when considering data outliers' rates for the 1280-S dataset and the effect of seasonality component for the M50–11-12 dataset.

Considering their execution time, Mixed-ARDE-based LSTM, has a significantly higher computational cost than GA-LSTM and keras-tuner-LSTM models, as depicted in Fig. 4. However, it spends significantly less time than the original DE algorithm due to its acceleration mechanism in the reinitialization phase and the early stopping mechanism.

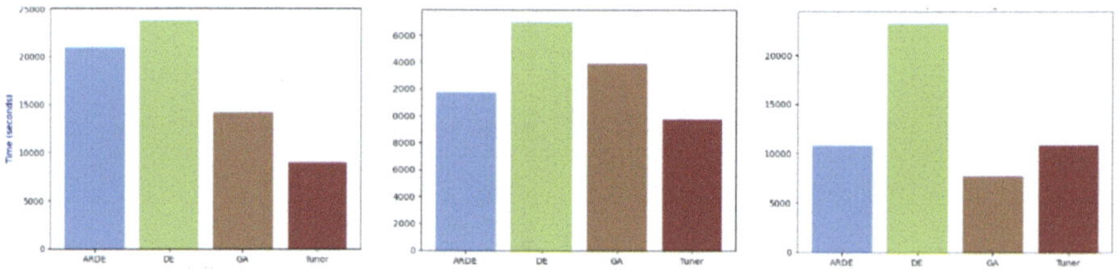

Fig. 4. Comparative of time-consuming results from the three datasets: M50–11-12 -N, M1–02-03-S and 1280-S; from the left to the right.

5 Conclusion

In this study, we explored the results of a novel approach for traffic flow forecasting by integrating the Adaptive Reinitialized Differential Evolution (ARDE) algorithm with Long Short-Term Memory (LSTM) networks. This variant, termed Mixed-ARDE-based LSTM, strengthens the process of LSTM hyperparameters search. It improves the original DE algorithm by an adaptive mutation factor to enhance the exploitation process of optimization. Also, it details mechanisms to handle various type of genes within a DE individual even though it is designed for continuous variables. In addition, boundary constraints management is specified. Three datasets for traffic flow forecasting are first analyzed and processed before experiments. The empirical results demonstrated that the forecasting accuracy of Mixed-ARDE-based LSTM significantly outperforms standard DE by an average of 9.5% and keras-tuner optimizers by an average of 23.7% for LSTM hyperparameters configuration. The Genetic Algorithm indicates a nearby lower performance by an average of 3.4%. However, the Mixed-ARDE-based LSTM fails behind both GA and Keras-tuner, in terms of computational time. Future works can consider improvement of the worst aspects of Mixed-ARDE based LSTM by using architectures and approaches that are more favorable to parallel computation and real-time performance systems as [14]. Exploring results with different time resolutions and other time series prediction problems is another consideration for future research.

References

1. Kołodziej, J., Hopmann, C., Coppa, G., Grzonka, D., Widłak, A.: Intelligent transportation systems – models, challenges, security aspects. In: Kołodziej, J., Repetto, M., and Duzha, A. (eds.) Cybersecurity of Digital Service Chains, pp. 56–82. Springer, Cham (2022). https://doi.org/10.1007/978-3-031-04036-8_3
2. Doğan, E.: LSTM training set analysis and clustering model development for short-term traffic flow prediction. Neural Comput. Appl. **33**, 11175–11188 (2021). https://doi.org/10.1007/s00521-020-05564-5
3. Roy, S., Mehera, R., Pal, R.K., Bandyopadhyay, S.K.: Hyperparameter optimization for deep neural network models: a comprehensive study on methods and techniques. Innov. Syst. Softw. Eng. (2023). https://doi.org/10.1007/s11334-023-00540-3
4. Storn, R., Price, K.: Differential evolution – a simple and efficient heuristic for global optimization over continuous spaces. J. Glob. Optim. **11**, 341–359 (1997). https://doi.org/10.1023/A:1008202821328
5. Zhang, Y., Chen, G., Cheng, L., Wang, Q., Li, Q.: Methods to balance the exploration and exploitation in differential evolution from different scales: a survey. Neurocomputing **561**, 126899 (2023). https://doi.org/10.1016/j.neucom.2023.126899
6. Zhang, X., Zhao, Z., Li, J.: ARDE-N-BEATS: an evolutionary deep learning framework for urban traffic flow prediction. IEEE Internet Things J. **10**, 2391–2403 (2023). https://doi.org/10.1109/JIOT.2022.3212056
7. Pranolo, A., Mao, Y., Wibawa, A.P., Utama, A.B.P., Dwiyanto, F.A.: Robust LSTM with tuned-PSO and Bifold-attention mechanism for analyzing multivariate time-series. IEEE Access **10**, 78423–78434 (2022). https://doi.org/10.1109/ACCESS.2022.3193643
8. Benabdallah Benarmas, R., Beghdad Bey, K.: A deep learning-based framework for road traffic prediction. J. Supercomput. **80**, 6891–6916 (2024). https://doi.org/10.1007/s11227-023-05718-x
9. Huang, C., Karimi, H.R., Mei, P., Yang, D., Shi, Q.: Evolving long short-term memory neural network for wind speed forecasting. Inf. Sci. **632**, 390–410 (2023). https://doi.org/10.1016/j.ins.2023.03.031
10. Ounoughi, C., Ben Yahia, S.: Sequence to sequence hybrid Bi-LSTM model for traffic speed prediction. Expert Syst. Appl. **236**, 121325 (2024). https://doi.org/10.1016/j.eswa.2023.121325
11. Bahe, M., et al.: Traffic prediction of real network traffic data with LSTM neural networks. In: Arai, K. (ed.) Advances in Information and Communication, pp. 311–326. Springer, Cham (2024). https://doi.org/10.1007/978-3-031-53960-2_20
12. Dickey, D.A., Fuller, W.A.: Distribution of the estimators for autoregressive time series with a unit root. J. Am. Stat. Assoc. **74**, 427–431 (1979). https://doi.org/10.1080/01621459.1979.10482531
13. Jouni, L., Ivan Zelinka, I.Z.: Mixed integer-discrete-continuous optimization by differential evolution. In: Proceedings of the 5th International Conference on Soft Computing, pp. 71–76 (1999)
14. Özdemir, R., Taşyürek, M., Aslantaş, V.: Improved marine predators algorithm and extreme gradient boosting (XGBoost) for shipment status time prediction. Knowl.-Based Syst. **294**, 111775 (2024). https://doi.org/10.1016/j.knosys.2024.111775

Disaster Information Extraction: Evaluation of NLP Techniques Using JAPE Rules, Ontologies and Machine Learning Approaches

Atmane Hadji[1,2](✉) and Mohamed Khireddine Kholladi[3,4]

[1] Faculty of Exact Sciences, Department of Computer Science, University of Bejaia, 06000 Bejaia, Algeria
a.hadji@centre-univ-mila.dz
[2] LISI Laboratory of University Center A. Boussouf Mila, 43000 Mila, Algeria
[3] Faculty of Exact Sciences, Department of Mathematics and Computer Science, University of El-Oued, HAMMA Lakhdar, El-Oued, Algeria
[4] MISC Laboratory of Abdelhamid, Mehri University of Constantine 2, Constantine, Algeria

Abstract. In this paper, we present a comparative study of different approaches to extract disaster information based on news and texts provided by the Internet, which rely on Natural Language Processing (NLP) techniques. We evaluated these distinct approaches: the first approach uses JAPE rules, the second approach relies on ontologies due to their advantages in knowledge sharing and reuse, and the last approach uses a number of machine learning algorithms. JAPE rules are easy to create and are able to identify and extract named entities and specific linguistic patterns in simple systems. These rules are designed to provide broad identification and targeted application of disaster-related texts, but we find poor responsiveness at the level of complex systems. Ontologies were used to organize, structure and index knowledge, to improve the contextual understanding of information, and to increase accuracy and recall. Machine learning was explored using four algorithms to evaluate their ability to efficiently identify, predict and extract disaster information (location, type and damage). The performance evaluation of each approach showed that the ontology-based approach was the best and most effective overall, combining speed and accuracy in extracting relevant information. JAPE rules, although easy to construct and cover a wide range of domains, are limited in complex systems. Machine learning algorithms show good and convergent results, depending on the algorithm used: (SVM, LR, RFC and DTC), which indicates flexibility, adaptability, and effective integration and can provide better performance. In summary, this study shows that ontology-based approaches are the most effective for extracting disaster-related information. The integration of these techniques, especially by combining the strengths of different approaches and creating hybrid approaches, would provide more robust solutions for improving information extraction systems in disaster contexts.

Keywords: Information Extraction · Disaster Information · JAPE Rules · Ontology · Machine Learning

T. Marir et al. (Eds.): ICCSA 2024, ISEM 45, pp. 294–313, 2025.
https://doi.org/10.1007/978-3-031-90758-6_22

1 Introduction

With the development of the Internet and mobile media technology, social media networks such as Facebook and Twitter have become powerful and indispensable communication tools in today's era and important sources of information in disaster situations [1]. When disasters happen, regardless of whether they are natural disasters such as earthquakes, floods, hurricanes, or forest fires, their social impact as well as that on both the economy and environment is immense being global in scale. Therefore, it is a matter of urgency to direct the texts of such events as succinct as possible and with total data accuracy towards disaster mitigation, coordinating relief and research in other fields [2]. As this is crucial for analyzing the impact of natural disasters, for drawing connections to insights gained about their occurrence or impacts and whenever possible, then realizing new preventive measures that may be taken [3].

Automatic extraction of information, including such critical details as place names, types of events, times and other relevant information, plays an indispensable role in understanding the context and scale of natural disasters by means of a precise social index. However, when it comes to distilling such information from massive textual databases, this is extremely time-consuming and expensive. Recent years have witnessed prominent progress in natural language processing (NLP) and linguistic analysis, leading to the integration of automatic methodologies for detecting named entities from large-scale text data [4]. Use of gazetteer names--those are carefully compiled lists of words-of-interest and rules based on Java Annotation Patterns Engine (JAPE) to make more efficient and faster matching between patterns have been promising in improving the precision of named entity extraction. Not relying on traditional information extraction techniques, Ontology-based Information Extraction Systems (OBIE) [5] classify the type of entities recognized and chain them together with their semantic descriptions from the ontology.

Making full use of the above-mentioned social media developments, this study has aimed to present three automatic extraction approaches designed specifically for natural catastrophe information. By leveraging the synergistic potential of gazetteers and JAPE rules, alongside another approach based on ontologies and a third based on machine learning, this study seeks to streamline the process of extracting critical information from disaster-related texts. Ultimately, this contribution seeks to increase the strategies for disaster response and management by getting better information out of disaster-influenced texts.

2 Related Works

This section presents the state of the art in information extraction from rule-based systems, ontologies and machine learning. Rule-based methods have proven to be effective in various areas of information extraction, in particular due to their ability to capture specific relationships by applying defined syntactic and semantic rules. Operating on the principles of logic and conditional statements, these systems encode decision-making processes in an automatic and structured way. These systems depend on the use of specific rules to direct the behavior of the system itself, and they are particularly useful in cases where clear instructions and conditions dictate the operations of lower-level processes. Rule-based systems are accountable and, hence, provide more transparency in tracking which makes them suitable for certain applications. The entities are defined and the relationships between them are specified, thereby enabling the development of a more precise understanding of the information [6]. One of the main advantages of this approach is capturing and understanding the meaning of the domain which allows intelligent searching and exchange of information between different systems. However, ontology-based approaches were inspired by web semantics techniques to visualize knowledge of hierarchical and interconnected methods, and ontologies are fundamental in several fields such as information extraction, natural language processing and web semantics. Machine Learning (ML) based information extraction has been introduced in different domains and applications, showing the very high flexibility, efficiency, and effectiveness in processing complex and unstructured data. There are several algorithms (Support Vector Machine (SVM), K-Nearest Neighbors (KNN), K-Means, Random Forest, Logistic Regression,...).

This work reveals the flexibility and power of these methods in the different contexts of various domains. In what follows, we will look at some related work:

The work proposed by [7] presents a hybrid rule-based approach for extracting technical entities from software engineering documents. The study highlights the effectiveness of this approach in accurately identifying and extracting relevant information. The authors in [8] presented the VALET (Very Agile Language Extraction Toolkit) framework, a rule-based information extraction system that combines lexical, orthographic, syntactic and corpus-analytic information in a flexible syntax. The work in [9] addresses the challenge of automatically extracting information structure from PDF books, which are widely used due to their ease of access on the Internet. The extraction process is essential for efficient knowledge representation, indexing, archiving and retrieval in digital libraries on the semantic web. This study proposes an intelligent rule-based approach to accurately extract logical metadata from PDF book. The study realized by [10] applies rule-based information extraction process using selected data from a bibliographic database. The objective is to establish a database of pertinent concepts, refine retrieved data, and automate the local retrieval process. In this study [11], the author examines research since 2017 that uses machine learning (ML) and deep learning (DL) for disaster management. It focuses on disaster prediction, risk assessment, early warning systems, and post-disaster response. The analysis includes recent ML and DL applications, as well as results and future research directions. The article in [12] introduces a new rule-based Named Entity Recognition (NER) method called "drNER," marking the first attempt to extract dietary concepts. Evaluated on diverse text corpora, including scientifically validated websites and scientific publications, drNER demonstrates promising results for extracting evidence-based dietary recommendations.

The author in [13] proposed an approach that starts with a domain-specific ontology created by human experts. It extracts appropriate information from unstructured text, using techniques to identify and apply heuristics to extract that information and it is added as structured information to the selected ontology. The proposed system [14] integrates AI and ontology creation to facilitate clinical information extraction and visualization. The study approach [15] presents the extraction of relevant tables from technical documents in PDF or HTML format through an ontology-based method.

OntoHuman [16] introduces an automated method for extracting key-value pairs from documents based on ontologies. It also allows users to provide feedback to enhance the ontologies. This system was applied in the field of spatial engineering. OBIESOF [17] is an ontology-driven information retrieval system designed for organic agriculture. It utilizes an ontology along with a conceptual schema for comprehensive formalization of any domain. The goal is to gather, store, and share information about organic farming for current or future software developers aiming to create applications for farmers. The system employs information extraction (IE) techniques and ontology methods. The ontology's ability to respond to user queries was evaluated through skill-based questions using the DL query interface. The results showed promising answers, indicating the ontology's potential as a knowledge repository. Another study [18] proposed an ontology-based information retrieval system focused on extracting features and values relevant to land use analysis from legal and regulatory documents pertaining to geographical regions. The system generates a generalized ontology comprising instances, criteria, and property values. This resulting ontology was integrated into a multimodal automated decision model applied to various maps representing different land use types.

In the following, we will present some related work based on machine learning:

This research [19], focused on combining machine learning and natural language processing (NLP) techniques to qualify Weber fractures from radiology data, and the study showed great effectiveness in improving clinical data management, making a practical application of machine learning techniques for medical data processing. Furthermore, a context schema-based information extraction system was developed by [20] to extract facts from unstructured radiology reports. The creation of a model based on machine learning techniques to structure contextual information demonstrated the feasibility of this system for clinical applications such as abnormality tracking and report writing. In contrast, [21] focused on automating invoice processing by using machine learning to extract information from unstructured invoice formats. The results showed that the LayoutLM model offered superior predictive performance to Chargrid and Random Forest, dealing with layout bias in datasets. [22] presented a new approach to detecting COVID-19-related "fake news" by combining biomedical information extraction with ML models. By extracting medical features from over a thousand press articles, the study showed that integrating this information improves the efficiency of detection models, with a notable performance of the Random Forest model. In addition, [23] proposed deep learning models for geospatial relationship extraction, demonstrating that the pipeline approach, separating entities and relationships, outperformed joint approaches in terms of performance.

In conclusion, new and evolving models, namely transformers, have been highlighted and used for information extraction [24, 25] in recent studies. [24] demonstrated the effectiveness of transformer models for processing handwritten digital documents. Meanwhile, [25] demonstrated a model for extracting important information from Chinese-language resumes. This research demonstrates the ability of pre-trained models to extract complex knowledge and enrich specialized databases. These studies offer significant advances in information extraction, demonstrating the power of machine learning algorithms to transform unstructured data into consistent information that can be used in a variety of contexts.

3 Proposed Methods

Risks and disasters are becoming less and less tolerated by society. These phenomena are now taken into account in various fields and at different scales: from individual to population, from territory to the entire planet. Regardless of the scale of analysis, significant efforts are made to mitigate risks. Specific professions have been created to limit them, both in the fields of environment, industry, agriculture, and services. Therefore, when an accident or disaster occurs, the event has a significant impact, as societies are increasingly less inclined to accept them.

Natural disasters refer to abrupt and catastrophic events that result in extensive damage to both the environment and human populations. These events encompass hurricanes, volcanoes, earthquakes, forest fires, avalanches, floods, and tsunamis. We have created Gazetteers in the field of natural disasters to extract important information from files on the web. In this work, we studied information and news published on the internet to extract information about natural disasters, their characteristics, spatial indicators, and the damages to victims of natural disasters.

3.1 Spatial Indicators

These are the pieces of information that help us identify and locate a place in the publication. The locations in the text published in electronic newspapers can be symbols, dialects, or abbreviations. Therefore, the task of extracting spatial information is generally difficult because it is ambiguous, and aliases and acronyms are used instead of the original name.

3.2 Natural Disaster Damages

Extraction of information contained in the published text regarding the damages caused by natural disasters in terms of human and material losses, collapses, and other losses.

3.3 Characteristics of Natural Disasters

These are the pieces of information contained in the text regarding the specific features of natural disasters, i.e., the characteristic that distinguishes each natural disaster. For example, the intensity of an earthquake, high temperature, quantity of precipitation, and other characteristics.

4 Architecture Approaches

The following architecture (Fig. 1) depicts the detailed design of first approach based JAPE rules and second based ontologies. We studied both approaches in parallel, each based on natural language processing (NLP) techniques. Both approaches were also experimented with on the same dataset to ensure consistency in results and comparisons. Additionally, they were implemented on the GATE platform, which allowed for uniform integration and evaluation of the performance of the different methods within a robust text processing environment. For the third approach, we independently studied another dataset and a different Python development environment, which we will explain further below. The proposed system consists of several stages:

4.1 Data Collection

The data collection was carried out manually and semi-automated techniques through several important information sources: News websites, social networks (Facebook, Twitter,...) as well as some digital newspapers. The final corpus covers different fields such as humanitarian disasters (accidents, fires,...) and natural disasters (floods, earthquakes,...). We have completed the first set of this work, which is the dataset of a series of Facebook comments posted on the Mila earthquake, fires in Tizi Ouzou, among others. Additionally, we leverage the GATE[1] platform (General Architecture for Text Engineering) to proficiently extract relevant comments from popular social media platforms such as Facebook and Twitter. The filtering techniques applied to the corpus include more than one baseband. We filter the data by bypassing extra spaces and formatting elements to obtain plain text. Consequently, typos are corrected using automated and manual tools, and text normalization is followed, including the removal of special characters, spaces and punctuation. We therefore have to carry out manual coding, where texts are manually broken down into functional units (sentences or paragraphs) to guarantee the accuracy and precision of the final analysis. Filtering and normalization steps are indispensable. After this phase, we analyzed the text within the framework of automatic natural language functionalities, following the order of the bars: (tokenization, sentence splitter, part-of-speech tagging...) [26]. Our data acquisition process involves meticulous information gathering from various online sources. We employ events leading to casualties caused by natural disasters. Examples include the Mila earthquake.

4.2 Pretreatment

In this step, noise is reduced, and all non-alphabetic characters, symbols, links, etc. are removed.

4.3 Processing and Text Analysis (Rule-based and Ontology-based Methods)

We will then proceed with the TALN automatic natural language processing technique in the following way:

[1] https://gate.ac.uk/.

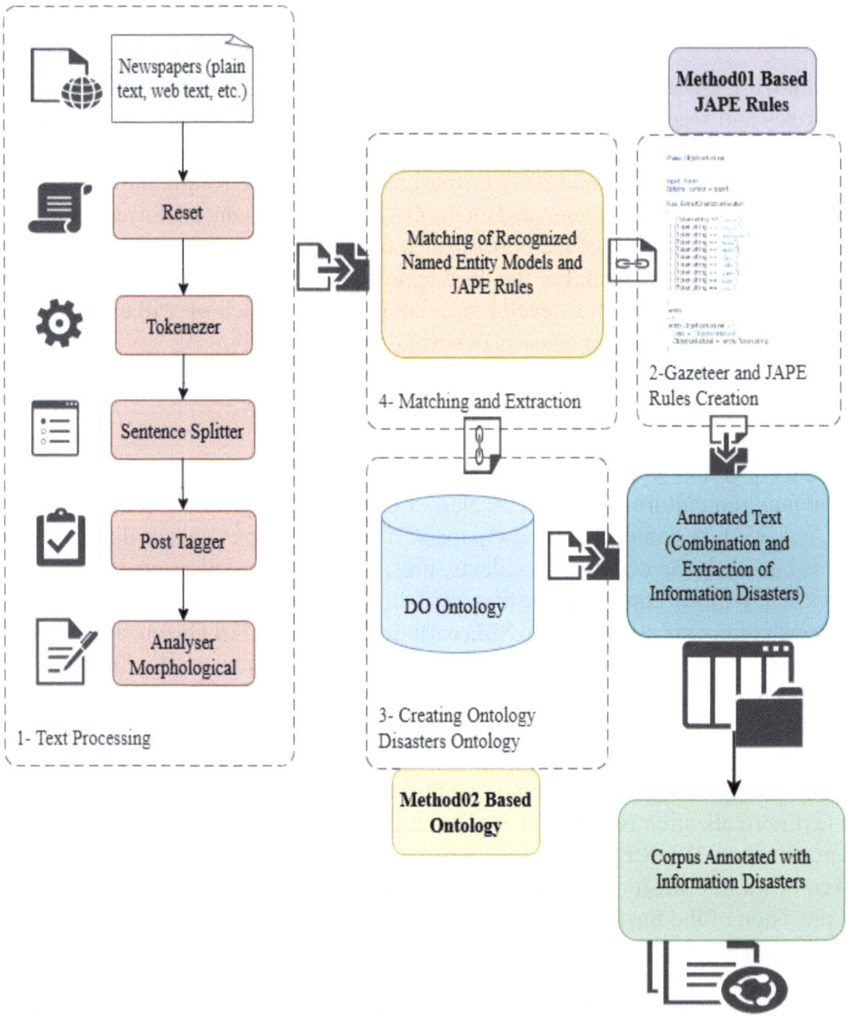

Fig. 1. General architecture of the evaluated methods: method one based JAPE Rules and second method based ontology.

Tokenization: The text is divided into single words such as names, symbols, punctuation marks and many other types. Each word is annotated with a token that cannot be modified by applications or text types.

Sentence Splitter: This component segments text into phrases using a cascade of finite-state transducers. This is a prerequisite for using the tagger. A list of gazetteer abbreviations is used to differentiate phrase tags from other tag types.

Part of Speech Tagger: The tagger used in GATE is the Brill tagger, which assigns a speech tag to each word or symbol in the text. It is based on a Lexicon and a predefined set of rules derived from a large corpus. Two additional lexicons are also available: one for uppercase text, the other for lowercase text.

Semantic Tagger: The ANNIE's semantic tagging program (a semi-new information extraction system) runs on JAPE (Java Annotation Pattern Engine). It uses rules that operate on annotations, properties, and values assigned in previous sentences. Additionally, it includes rules designed for annotations, features, and values that require manual mapping [27]. In our study, the first method is based on JAPE rules and lexicons, independently of the second method, which is based on ontologies, and the third method tests four machine learning algorithms as follows:

4.4 First Method Based Rules

JAPE Rules: A JAPE grammar [28] consists of a set of phases, each of which is made up of a set of model/action rules. Phases run sequentially and form a cascade of finite-state transducers on annotations. The left-hand side (LHS) of the rules consists of a description of the annotation pattern. The right-hand side (RHS) consists of annotation manipulation instructions. The corresponding annotations on the LHS of a rule can be referenced on the RHS by means of labels attached to the pattern elements. Below is an example of a JAPE rule (Fig. 2).

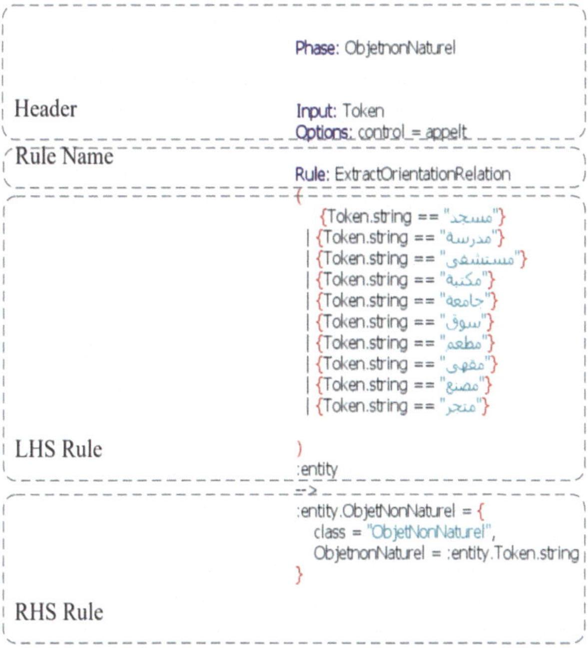

Fig. 2. JAPE Rules Example.

Gazetteer Creation

The gazetteer is an essential tool for identifying, indexing and classifying named entities in text, based on predefined lists. It consists of various lists, such as names of people, cities, names of organizations, dates and days of the week, etc. These lists (see Table 1) are plain text files, with each entry on a separate line. The Gazetteer contains not only specific entities, but also relevant keywords such as company abbreviations, designations, titles (Dr., Pr.,..."), etc. The lists are then compiled into automaton structures that associate text tokens with named entities.

Table 1. Gazetteer List Example.

Catastrophes	Informations
Damages	Material, Economic, Physical, Environmental
Collapse	Structural, Building, Catastrophe
Financial losses	Monetary, Economic, Costs;.
Human Damages	Injuries, Human losses, Victims
Of climatic origin	Meteorological, phenomenon, Climate
Cyclone	Tropical storm, Wind, Hurricane
Fires	Fire, Burn, Wildfire
Of terrestrial origin	Overflow, Flood, Natural disaster
Earthquake	Seismic, Geological, Telluric
Volcano	Seismic, Tremor, Epicente
Landslide	Eruption, Lava, Crater

4.5 Second Method Based Ontology

Ontology Development

As previously outlined in the ontology construction process, manual ontology development involves the following steps [29]:

Step 1: Determining the Domain and Scope of the Ontology

The initial step in ontology development is to define the domain and scope of the ontology. To ascertain this, it is imperative to understand the domain the ontology will cover, its purpose, the types of questions it aims to address, and who will be utilizing the ontology. By addressing these questions, we establish that the ontology's domain will encompass areas of current events, which is pivotal to our annotation approach.

Step 2: Considering the Reuse of Existing Ontologies

This step involves checking if there exists a previously developed ontology in the same domain. If such an ontology exists, it is often more expedient to adapt an existing ontology to suit one's needs rather than creating a new one from scratch. Since we did not find a previously created ontology specifically tailored to the domain of Natural Disasters (Disaster Ontology or DO), we bypassed this step.

Step 3: Enumerating Key Terms in the Ontology

This step can be likened to a brainstorming activity, where we compile a list of words we intend to use to represent ontology terms and their associated properties.

Step 4: Defining Classes and Class Hierarchy

In this stage, we establish the classes (concepts) employed within our ontological domain. We define classes and sub-classes relevant to our domain. Figure 3 provides a description of the class ontology.

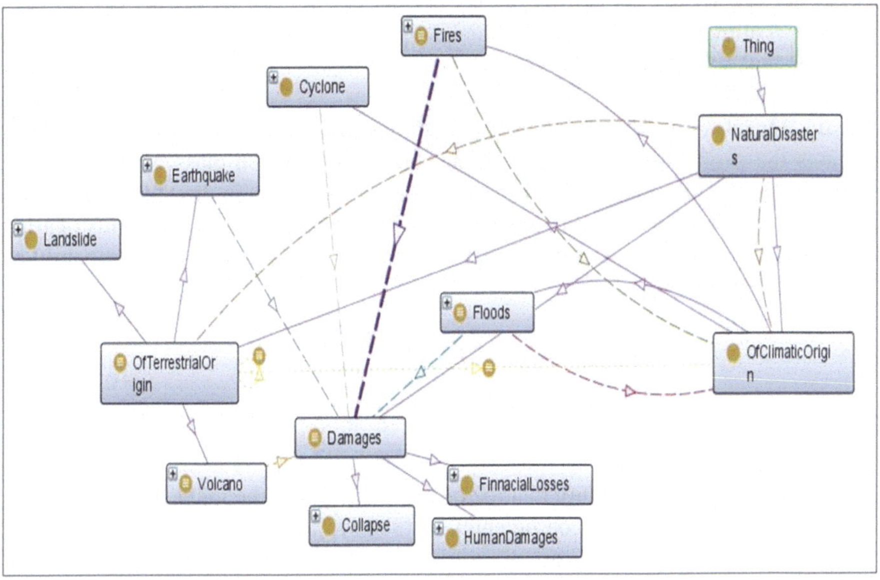

Fig. 3. High-level ontology classes of DO (Disasters Ontology)

4.6 Third Approach Based Machine Learning

Extracting relevant information from texts related to natural disasters is crucial for enhancing crisis management and providing timely and effective assistance. Among the various approaches used for this task, machine learning has emerged as a powerful and adaptable method capable of processing large volumes of textual data with increased accuracy. Unlike rule-based or predefined models, machine learning utilizes algorithms to learn patterns from the data itself, allowing for continuous adaptation and improvement of performance.

In this study, we explore the application of machine learning for disaster information extraction using four distinct algorithms: Support Vector Machines (SVM), Logistic Regression, Random Forest Classifier, and Decision Tree Classifier. Each algorithm presents unique characteristics and specific advantages that influence their effectiveness in extracting relevant information:

Support Vector Machines (SVM): The SVM algorithm is known for its ability to classify data by finding the optimal hyperplane that separates classes with the maximum margin. In the context of natural disasters, it is used to distinguish different categories of information from complex textual data.

For linearly separable data, the separation hyper plane can be determined by:

$$W^T + b = 0 \tag{1}$$

- W is the hyperplane's weight vector (or normal).
- x is the feature vector of a data point.
- b is the bias (offset) of the hyperplane.

Logistic Regression: This model is often employed for binary classification tasks. It estimates the probability that a data point belongs to a given class, which is useful for predicting the presence or absence of specific information in disaster-related texts. Logistic Regression is used for binary classification by estimating the probability of a data point belonging to a class. It uses a logistic function to map predictions to probabilities.

Mathematical Expression:

$$p(y = 1|x) = \frac{1}{1 + e^{-(W^T x + b)}} \tag{2}$$

where:

- $p\,(y = 1 \mid x)$ is the probability that the output y is 1 given the input features x.
- w is the vector of weights (coefficients),
- x is the vector of input features,
- b is the bias term,
- *exp* is the base of the natural logarithm.

The goal is to find the weights w and bias b that maximize the likelihood of the observed data.

Random Forest Classifier: The Random Forest algorithm uses a combination of multiple decision trees to enhance classification accuracy. By aggregating the decisions of several trees, it offers improved generalization and robustness against overfitting, which is crucial for handling varied and noisy text data. Random Forest Classifier is an ensemble learning method that combines multiple decision trees to improve classification accuracy and control overfitting. Each tree is trained on a random subset of the data and features.

Mathematical Expression: There isn't a single mathematical expression for Random Forests due to their ensemble nature, but the prediction for an input x is given by:

$$\hat{\mathrm{Y}} = \mathrm{mode}(\{\mathrm{T}(\mathrm{x_i})|i = 1, 2, \ldots, N\}) \tag{3}$$

where:

- $p(y = \frac{1}{x}) = T_i(x)$ is the prediction of the i-th decision tree,

- N is the number of trees,
- The mode function returns the most common class label among all trees' predictions.

Decision Tree Classifier: Decision trees are interpretable models that make decisions based on successive binary questions. They are particularly effective for identifying simple rules within textual data and understanding the classification criteria. Decision Tree Classifier splits data into subsets based on feature values, creating a tree structure where each node represents a decision based on an attribute.

Mathematical Expression:

$$Gini(D) = 1 - \sum\nolimits_{i=1}^{K} (P_i)^2 \tag{4}$$

where:

- D is a dataset,
- k is the number of classes,
- pi is the proportion of instances belonging to class i.

The tree splits until it reaches a stopping criterion, such as a maximum depth or minimum number of samples per leaf.

5 Results and Evaluation

After running the corpus with the use of JAPE and Gazetteer rules (Fig. 4), the system is now able to detect the entities named spatial and Characteristics, Damages of Natural Disasters.

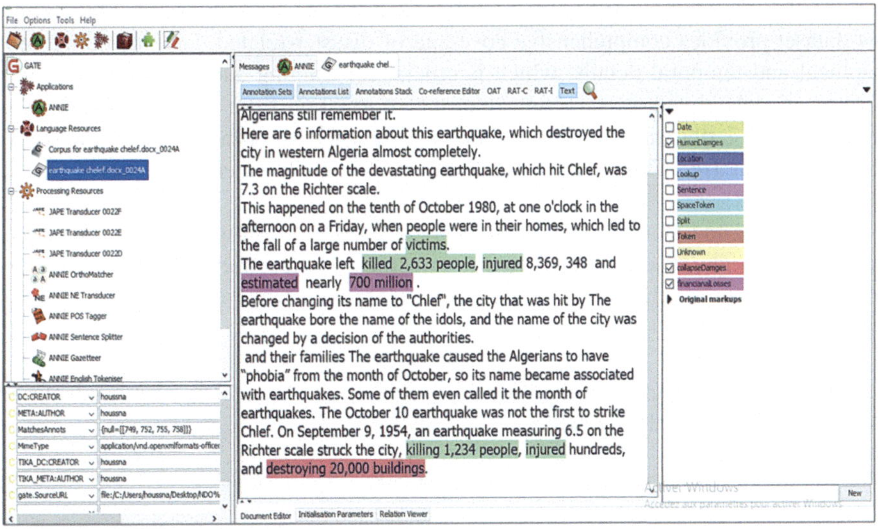

Fig. 4. Execution of Named Entities (Damages) based JAPE rules.

Following the application of the DO ontology to the corpus (Fig. 5), the system can now identify named entities related to spatial information, characteristics, and damages associated with natural disasters.

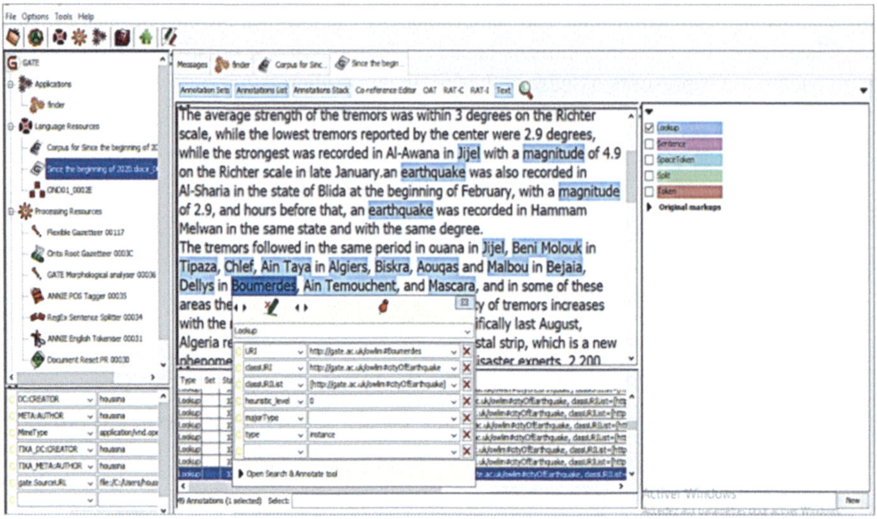

Fig. 5. Execution of Extraction Desasters Information based Ontology.

The dataset[2] used for our machine learning approach is annotated with a range of attributes to support effective information extraction. It includes fields such as id, country, iso3, gwno, year, geo_id, geolocation, level, adm1, adm2, adm3, location, historical, hist_country, disastertype, disasterno, latitude, and longitude. With over 30,000 entries, this dataset provides comprehensive coverage of disaster-related data, including geographical and temporal details, which is crucial for training and evaluating machine learning models aimed at extracting relevant information from disaster-related texts. Figures 6 and 7 illustrate the results obtained for each algorithm used in our study: SVM, Decision Tree Classifier (DTC), Random Forest Classifier (RFC), and Logistic Regression (LR).

To evaluate and compare the methods we studied, we will use metrics: Precision, Recall, and F-scale. Precision refers to the correctness of the retrieval, while recall refers to the completeness of the retrieval. The F-measure provides the harmonic mean between precision and recall [30].

According to [31]:

Precision is the fraction of the valid annotations over the total number of identified annotations. It is formally defined as:

$$\text{Precision} = (\text{Correct}) / (\text{Correct} + \text{Spurious}) \tag{5}$$

[2] https://ourworldindata.org/natural-disasters.

```
Decision Tree Classifier:
Accuracy: 0.92
Precision: 0.92
Recall: 0.92
F1 Score: 0.92
Classification Report:
              precision    recall  f1-score   support

  earthquake       0.80      1.00      0.89        12
       flood       0.96      0.96      0.96       185
       storm       0.58      0.47      0.52        15

    accuracy                           0.92       212
   macro avg       0.78      0.81      0.79       212
weighted avg       0.92      0.92      0.92       212


============================================================
Support Vector Machine (SVM):
Accuracy: 0.87
Precision: 0.89
Recall: 0.87
F1 Score: 0.81
Classification Report:
              precision    recall  f1-score   support

  earthquake       1.00      0.00      0.00        12
       flood       0.87      1.00      0.93       185
       storm       1.00      0.00      0.00        15

    accuracy                           0.87       212
   macro avg       0.96      0.33      0.31       212
weighted avg       0.89      0.87      0.81       212
```

Fig. 6. Results of Extraction Desasters Information Decision Tree Classifier (DTC) and Support Vector Machine (SVM).

Recall is the fraction of the valid annotations over the total amount of annotations. It is formally defined as:

$$\text{Recall} = \text{Correct} / (\text{Correct} + \text{Missing}) \tag{6}$$

F-measure is defined as the harmonic mean of two factors, precision and recall. It is formally as:

$$\text{F - score} = (2 * \text{Precision} * \text{Recall}) / ((\text{Precision} + \text{Recall})) \tag{7}$$

The following Tables present the evaluation results for the extraction of information related to natural disasters.

```
Logistic Regression:
Accuracy: 0.86
Precision: 0.83
Recall: 0.86
F1 Score: 0.81
Classification Report:
              precision    recall  f1-score   support

  earthquake       0.00      0.00      0.00        12
       flood       0.87      0.98      0.92       185
       storm       1.00      0.00      0.00        15

    accuracy                           0.86       212
   macro avg       0.62      0.33      0.31       212
weighted avg       0.83      0.86      0.81       212


==========================================================
Random Forest Classifier:
Accuracy: 0.93
Precision: 0.92
Recall: 0.93
F1 Score: 0.92
Classification Report:
              precision    recall  f1-score   support

  earthquake       1.00      0.75      0.86        12
       flood       0.94      0.98      0.96       185
       storm       0.67      0.40      0.50        15

    accuracy                           0.93       212
   macro avg       0.87      0.71      0.77       212
weighted avg       0.92      0.93      0.92       212
```

Fig. 7. Results of Extraction Desasters Information Logistic Regression (LR) and Random Forest Classifier (RFC).

6 Analysis and Discussion

6.1 Analysis and Discussion Machine-Learning

In this paper, we evaluated four different machine learning algorithms - support vector machines (SVM), decision tree classifier (DTC), random forest classifier (RFC) and logistic regression (LR) - for the disaster information extraction task. This evaluation was carried out using performance measures such as precision, recall and F-measure. The results are summarized in Table 2. The results are summarized in Table 2.

Support Vector Machines (SVM): The SVM algorithm's precision is 0.89 and recall 0.87, giving an F-measure evaluated at 0.81. SVM showed good precision and strong recall performance, but the F-measure indicates that there is room for performance improvement and a greater balance between precision and recall. The small decrease in F-measure compared to DTC and RFC suggests that SVM may not be at the right level to handle dataset variability.

Decision Tree Classifier (DTC): The decision tree classifier gave excellent results, with precision and recall values of 0.92, corresponding to an F-measure of 0.92. This

Table 2. RESULT and evaluation for m-learning algorithms

	Precision	Recall	F-score
SVM	0,89	0,87	0,81
D TC	0,92	0,92	0,92
R FC	0,92	0,93	0,92
LR	0,83	0,86	0,81

consistency in precision and recall shows that the decision tree classifier is much more effective at correctly identifying relevant information, with very few false positive extraction results. The high F-measure also testifies to high confidence and consistency in the extraction of disaster information.

Random Forest Classifier (RFC): The Random Forest Classifier achieved the highest recall (0.93) and precision as in DTC (0.92), which also gave an F-measure of 0.92. The performance of RFC is similar to that of DTC, but with a slight advantage in terms of recall. This suggests that RFC is quite adept at capturing relevant information, probably due to its ensemble nature, which reduces over-fitting and improves generalization.

Logistic Regression (LR): Logistic Regression has the lowest precision (0.83) and the lowest recall (0.86), resulting in an F-measure of 0.81. Although LR is effective for binary classification, its performance in this study is weaker than that of the other algorithms. This may be due to its linear characteristic, which may not capture complex information as effectively as non-linear models such as DTC and RFC.

6.2 Discussion ML

The SVM demonstrates a significant precision rate and also a remarkable recall rate, but a more reduced F-measure, solution is the need for integration or possible combination with other methods to improve performance. The evaluation shows that both the DTC and the RFC are very effective in extracting disaster information, as they both achieve high precision, recall and F-measure rates. The RFC ensemble approach offers a slight advantage in terms of recall, making it slightly more reliable in identifying all relevant information. LR, while useful, shows poorer overall performance, suggesting that it may not be the best stand-alone method for this particular task, but could be beneficial in a hybrid approach.

6.3 Analysis of Three Methods

In this study, we evaluated three distinct approaches for disaster information extraction: JAPE rules, ontologies, and machine learning algorithms. The performance of each method was measured in terms of precision, recall, and F-measure. The results are summarized in Table 3.

Table 3. Results and evaluation for thirds methods

	Precision	Recall	F-score
Method1 Rules	0.71	0.94	0,81
Method2 Ontology	0.85	0.96	0,90
Method3 M.Learning	0,89	0,89	0,86

Method 1 JAPE Rules: JAPE rules The JAPE rules-based approach achieved a precision of 0.71 and a recall of 0.94, resulting in an F-measure of 0.81. We conclude that this method has a very high ability to identify relevant information (high recall) but with a relatively low precision, meaning a relatively large number of false positives. The high F-measure shows that despite the low precision, the JAPE rule-based method is effective and has the ability to cover a wide range of relevant information, which is crucial in such disaster situations where every piece of information can be very vital and useful.

Method 2 Ontology: The ontology-based method achieved high results in precision with a precision of 0.85 and recall with a value of 0.96, resulting in an F-value of 0.90. This method shows that there is a balance in performance between precision and recall. High precision shows a significant reduction in false positives, while high recall shows that the method has the ability to extract and capture most of the relevant information. Thus, the ontology-based method is very robust, providing an excellent balance between precision and recall, and is particularly well suited for scenarios where accurate and comprehensive information extraction is critical.

Method 3 Machine Learning: The results of the machine learning-based approach achieved an equalization between the precision and recall indices with a value of 0.89, with a slightly lower F-measure of 0.86. This method shows a balanced and high performance for both metrics, indicating its ability to integrate, harmonize and maintain good precision while being able to capture the majority of relevant information. The slight decrease in the F-measure when compared to the ontology-based method indicates that although machine learning algorithms are very effective, there may be cases that require better performance and thus modifications and improvements can be made such as combining them with other methods to create a more powerful hybrid approach.

6.4 Discussion of Three Methods

In the evaluation of the three methods for extracting disaster-related information, we highlight the strengths and weaknesses of each of them:

JAPE Rules-Based Method: This method is highly effective in extracting and capturing the majority of relevant information due to the high level of recall but the precision was lower, resulting in a higher number of false positives.

Ontology-Based Method: It was the best method compared to the other methods and provided the best balance between precision and recall, giving a higher value of F. This method is particularly useful for performing tasks that require information accuracy and very comprehensive coverage.

Machine Learning-Based Method: The ML-based method showed high performance in terms of precision and recall, but the ontological approach performed slightly better than the ML-based method. Machine learning algorithms are flexible and easy to adapt and integrate with other methods to obtain new, more efficient approaches. The results indicate that each method has its advantages, but the ontology-based approach provides the most balanced and robust performance for disaster information extraction, with complete and accurate coverage. The JAPE rule-based and machine learning-based methods also demonstrate great effectiveness and may be favoured depending on the specific requirements and needs of the application.

7 Conclusion

In conclusion, we conducted a comprehensive evaluation of three disaster information extraction approaches and the results were ranked according to each method on the basis of precision, recall and F-measure. The rule-based approach performed worse than all methods. The ontology-based approach performed better, and the machine learning algorithms also performed well, but slightly worse than the ontology. From the results obtained, the ontology-based approach shows superiority over other methods in terms of performance and generalization because it has the best balance between precision and recall resulting in the highest F-score. This indicates its flexibility and efficiency in extracting disaster-related information accurately and comprehensively. The JAPE rule-based method showed a high extraction rate, giving this approach the potential to be powerful in identifying a wide range of relevant information, but it showed a low accuracy rate implying a high false positive rate. The machine learning-based approach gave robust and balanced results for all measures and proved to be a very effective method, although less effective than the ontology-based approach for the F-measure. These results highlight the importance of selecting the appropriate method according to the requirements and needs of the specific disaster information extraction task. The ontology-based approach is well suited to applications requiring precise and complete information. Conversely, for situations requiring large-scale coverage, the JAPE rule-based approach may be more appropriate. The ML-based approach is a flexible and robust alternative, capable of delivering high performance for different measures.

Future work could explore the integration of these approaches in order to capitalize on their respective strengths, which could lead to more efficient disaster information retrieval systems. Specifically, we propose to develop a hybrid 'learning' approach that combines the structured knowledge and semantic richness of ontologies with the predictive power and adaptability of machine learning algorithms. This hybrid approach aims to improve the accuracy and completeness of information extraction by using ontological frameworks to guide machine learning models, thereby improving their ability to understand and process complex disaster-related data. In addition, further research could focus on optimizing these methods and evaluating their performance in various disaster scenarios to improve their practicality and flexibility.

References

1. Shen, H., Ju, Y., Zhu, Z.: Extracting useful emergency information from social media: a method integrating machine learning and rule-based classification. Int. J. Environ. Res. Public Health **20**(3), 1862 (2023)
2. Tellier, I.: TAL et extraction d'information. Université Paris 3 - Sorbonne, France (2014)
3. Téllez Valero, A., Montes y Gómez, M., Villaseñor Pineda, L.: Using machine learning for extracting information from natural disaster news reports. Computación Sistemas **13**(1), 33–44 (2009)
4. Neuberger, J., Ackermann, L., Jablonski, S.: Beyond rule-based named entity recognition and relation extraction for process model generation from natural language text. arXiv preprint arXiv:2305.03960 (2023)
5. Konys, A.: Towards knowledge handling in ontology-based information extraction systems. Procedia Comput. Sci. **126**, 2208–2218 (2018)
6. Hadzic, M., Wongthongtham, P., Dillon, T., Chang, E.: Introduction to ontology. In: Ontology-Based Multi-Agent Systems, pp. 37–60. Springer, Heidelberg (2009)
7. Sharma, A., Saha, S., Mathew, A.T.: Technical entity extraction in software engineering documents: a rule-based approach. In: 40th International Conference on Software Engineering (ICSE), pp. 1033–1043. IEEE/ACM (2018)
8. Freitag, D., Cadigan, J., Sasseen, R., Kalmar, P.: VALET: rule-based information extraction for rapid deployment. In: Proceedings of the 13th Conference on Language Resources and Evaluation (LREC), pp. 524–533 (2022)
9. Alamoudi, A., Alomari, A., Alwarthan, S., Ur-Rahman, A.: A rule-based information extraction approach for extracting metadata from PDF Books. ICIC Express Lett. **12**, 121–132 (2021)
10. Makhija, V., Swapnil, A.: Rule based text extraction from a bibliographic database. J. Libr. Inf. Technol. **38**, 5–10 (2018)
11. Linardos, V., Drakaki, M., Tzionas, P., Karnavas, Y.L.: Machine learning in disaster management: recent developments in methods and applications. Mach. Learn. Knowl. Extr. **4**(2) (2022)
12. Eftimov, T., Koroušić Seljak, B., Korošec, P.: A rule-based named-entity recognition method for knowledge extraction of evidence-based dietary recommendations. PLoS ONE **12**(6), e0179488 (2017)
13. Anantharangachar, R., Ramani, S., Rajagopalan, S.: Ontology guided information extraction from unstructured text. arXiv preprint arXiv:1302.1335 (2013)
14. Jusoh, S., Awajan, A., Obeid, N.: The use of ontology in clinical information extraction. J. Phys. Conf. Ser. **1529**, 052083 (2020)
15. Rizvi, S.T.R., Mercier, D., Agne, S., Erkel, S., Dengel, A., Ahmed, S.: Ontology-based information extraction from technical documents. In: ICAART, vol. 2, pp. 493–500 (2018)
16. Opasjumruskit, K., Böning, S., Schindler, S., Peters, S.: OntoHuman: Ontology-based information extraction tools with human-in-the-loop interaction. In: International Conference on Cooperative Design, Visualization and Engineering (CDVE). Springer (2022)
17. Abayomi-Alli, A., Oluwasefunmi, Misra, S., Akala, M.O., Ikotun, A.M., Ojokoh, B.A.: An ontology-based information extraction system for organic farming. Int. J. Semantic Web Inf. Syst. **17**(7), 79–99 (2021)
18. Al-Ageili, M., Mouhoub, M.: An ontology-based information extraction system for residential land-use suitability analysis. Int. J. Softw. Eng. Knowl. Eng. **32**(07), 1019–1042 (2022)
19. Fiebeck, J., Laser, H., Winther, H.B., Gerbel, S.: Leaving no stone unturned: using machine learning based approaches for information extraction from full texts of a research data warehouse. In: International Conference on Data Integration in the Life Sciences, pp. 50–58. Springer, Cham (2018)

20. Steinkamp, J.M., Chambers, C., Lalevic, D., Zafar, H.M., Cook, T.S.: Toward complete structured information extraction from radiology reports using machine learning. J. Digital Imag. **32**, 554–564 (2019)
21. Krieger, F., Drews, P., Funk, B.: Automated invoice processing: machine learning-based information extraction for long tail suppliers. Intell. Syst. Appl. **20**, 200285 (2023)
22. Fifita, F., Smith, J., Hanzsek-Brill, M.B., Li, X., Zhou, M.: Machine learning-based identifications of COVID-19 fake news using biomedical information extraction. Big Data Cogn. Comput. **7**(1), 46 (2023)
23. Yang, Y., Wu, Z., Yang, Y., Lian, S., Guo, F., Wang, Z.: A survey of information extraction based on deep learning. Appl. Sci. **12**(19), 9691 (2022)
24. Dagdelen, J., et al.: Structured information extraction from scientific text with large language models. Nat. Commun. **15**(1), 1418 (2024)
25. Luo, S., Yu, J.: ESGNet: a multimodal network model incorporating entity semantic graphs for information extraction from Chinese resumes. Inf. Process. Manag. **61**(1), 103524 (2024)
26. Navigli, R., Velardi, P.: Structural semantic interconnections: a knowledge-based approach to word sense disambiguation. IEEE Trans. Pattern Anal. Mach. Intell. **27**(7), 1075–1086 (2005)
27. Cunningham, H., Maynard, D., Tablan, V.: JAPE: a java annotation patterns engine, Research Memorandum, CS–00–10 (1999)
28. Thakker, D., Osman, T., Lakin, P.: Gate Jape Grammar Tutorial, Nottingham Trent University, UK, Phil Lakin, UK, Version 1 (2009)
29. Hadji, A., Kholladi, M.K., Borisova, N.: Enhancing spatial information extraction from Arabic text: a hybrid approach with ontology and rule-based. Ingénierie des Systèmes d'Information **29**(4), 1261–1273 (2024)
30. Gutierrez, F., Dou, D., Fickas, S., Lanka, S., Zong, H.: Hybrid ontology-based information extraction system. J. Inf. Sci. **42**(6), 798–820 (2016)
31. Maynard, D., Peters, W., Li, Y.: Metrics for evaluation of ontology-based information extraction. In: Proceedings of the 15th International Conference on World Wide Web, Workshop on Evaluation of Ontologies for the Web (2006)

TIASM-Platform: A New AI Powered Platform For E-Health

Fadia Baissi, Elhadj Abdelkader Abdelbaki, Laid Kahloul(✉), Amira Mohammedi, and Asma Ammari

LINFI Laboratory, Mohamed Khider Biskra Univeristy, Biskra, Algeria
{l.kahloul,amira.mohammedi,asma.ammari}@univ-biskra.dz

Abstract. In recent years, digitization has transformed medical practices globally, yet Algeria still faces significant healthcare challenges, including unequal access to specialized services, a shortage of medical professionals, and the compounded effects of an aging population and complex diseases. For instance, a Poland study found that approximately 64.9% of participants preferred using the Internet to seek health information, while 63.4% used it to find doctors or medical services, and about 65.3% relied on it to locate medical facilities. In response to these challenges, digitization and artificial intelligence (AI) offer promising solutions to enhance healthcare delivery. That's where our paper comes in with proposing an e-health platform (TIASM-platform) to overcome existing barriers by providing a dedicated space for doctors and researchers. This platform features secure storage and access to a rich repository of healthcare-related academic papers, datasets, and models realized initially at LINFI laboratory and eventually at other laboratories. The findings of this research promise substantial improvements in healthcare resource management, ultimately contributing to better healthcare outcomes in Algeria and beyond.

Keywords: Digitization · E-Health · Artificial Intelligence · Machine Learning · Deep Learning

1 Introduction

In recent years, digitization has revolutionized numerous domains, and healthcare is not an exception. This digital transformation is reshaping how medical services are delivered and accessed globally. The COVID-19 pandemic has highlighted the urgent need for accessible healthcare solutions, bringing to the forefront the importance of digital health technologies. These advancements have enabled remote consultations, telemedicine, and the use of artificial intelligence

TIASM: **T**échnique de l'**IA** pour le **S**outien de la **M**édecine
Laboratoire de l'**INF**ormatique **I**ntelligente: https://lab.univ-biskra.dz/linfi/.

T. Marir et al. (Eds.): ICCSA 2024, ISEM 45, pp. 314–328, 2025.
https://doi.org/10.1007/978-3-031-90758-6_23

in diagnostics, significantly improving healthcare delivery. Additionally, the integration of big data and machine learning into healthcare systems has enhanced disease prediction, patient management, and treatment outcomes. However, these technological advancements have not been evenly distributed, leading to significant disparities in healthcare access, especially in developing countries like Algeria.

Despite the global progress in healthcare digitization, Algeria's healthcare system faces significant challenges. These include an uneven distribution of resources, particularly in underserved areas, which leads to limited access to specialized care. This issue is compounded by a shortage of medical professionals and inadequate infrastructure, making it difficult to manage and treat diseases that require advanced medical examinations. This disparity creates significant obstacles for patients in remote areas who need specialized care available only in major institutions. Moreover, a recent study found that nearly two-thirds of Internet users (64.9%) rely on the Internet as a primary source for health information, highlighting the growing reliance on digital platforms [12]. These issues underscore the critical need for innovative digital solutions to bridge the gap in healthcare delivery. By leveraging the power of digitization, we can ensure that all populations benefit from advancements in medical technology and data science, ultimately leading to a more robust and accessible healthcare system in Algeria, Africa, and worldwide.

On the other hand, Artificial Intelligence (AI) is driving a major revolution in healthcare, and this transformation has been underway for several decades. Indeed, one can find in literature several platforms have been proposed to support diagnosis. One of the pioneering platforms in this field, developed in the 1970 s was MYCIN, which demonstrated the potential of AI to provide decision support in medical contexts, laying the groundwork for future developments [23]. Today, there are lot of AI-powered E-Health platforms, which can be broadly classified into two major categories: general use platforms and specialized use platforms.

General-use platforms, such as Enlitic [15] for Managing medical imaging data, Merative [13] for clinical decision support and healthcare analytics, Regard [9] for Automating clinical tasks, Google Vertex AI [2] for information retrieval, Care.ai [11] for AI-assisted virtual care, and Kaggle, which is commonly used to collect datasets, models, and research on many topics. Conversely, specialized platforms focus on specific medical needs, such as Freenome [19] for early cancer detection, Iterative Health [10] for detection of GI (gastrointestinal) diseases, and VirtuSense [21] for predictive fall prevention.

However, none of these existing platforms or studies offer the specialized and shared gateway tailored to healthcare actors and researchers that our platform aims to provide. Our platform not only aims to support the testing of new artificial intelligence models but also to facilitate the sharing and collaboration on various healthcare projects.

The motivation for our project arises from the critical need to improve healthcare accessibility and diagnostic accuracy in Algeria and beyond. Also, numerous efforts are made each year in research projects, particularly in our LINFI laboratory, to improve healthcare. However, the results of these research efforts often remain underutilized and are lost among the multitude of publications. Through this project, we aim to benefit researchers and professionals by making these works accessible via our platform. By leveraging AI, there is an opportunity to develop a robust and scalable platform that can support: (i) healthcare professionals in diagnosing and managing diseases more effectively, (ii) researchers in sharing and adding new datasets, papers and models, and (iii) promoting and making research projects realized more visible. Consequently, the success of our work in this context will provide a strong foundation for adapting and enhancing these solutions to meet the specific needs of the Algerian healthcare landscape. In which our study will bring a key contributions to the field of AI-powered healthcare solution. At the current stage, our AI-Powered Healthcare Platform provides a specialized digital platform tailored for the Algerian context, granting access to a meticulously curated collection of healthcare-focused academic papers, datasets, and AI applications. Users will have the ability to store their applications, with the added functionality of managing multiple versions of AI-models within each application. Our platform is designed to host several applications, already realized by LINFI members, including bone age assessment [25], pneumonia detection [14], diabetic retinopathy detection ([5,18]), paludism diagnosis using deep learning [1], etc. Furthermore, our platform will contain over 100 research papers most of which are recently published (between 2022 and 2024). Moreover, our platform will facilitate comprehensive testing of each model, offering an unparalleled solution for research, and resource access within the healthcare sector. These contributions collectively aim to enhance healthcare delivery in Algeria, leveraging AI to provide accurate diagnostics and facilitate effective medical research and collaboration.

The rest of this paper is organized as follows: Sect. 2 presents the required background of this work, Sect. 3 presents some related work and positions the current work, Sect. 4 details the platform development process by discussing the requirement analysis, the platform design, and the realization stage, Sect. 5 presents a comparative study between our platform and existing ones, and finally Sect. 6 concludes this paper.

2 Background

The following subsections define digitization and explore the roles of Artificial Intelligence (AI), Machine Learning (ML), and Deep Learning (DL) in medical diagnostics.

2.1 Digitization

Digitization is a major trend: It's having a significant impact on society and businesses, similar to the industrial revolution. It's more than just converting things to digital: it refers to simply changing analog data (like paper documents) to a digital format (like PDFs) [22]. However, it's about using digital technologies to fundamentally change how things work. Also, Digitization and digital transformation are often used interchangeably [20]. They both refer to the broad application of digital technology across all aspects of society. In Addition to that Digitization allows us to turn existing products or services into digital versions, offering benefits over traditional physical products [6] [7]. Also, it increased use of technology From an organizational standpoint, digitization means adopting and using more digital technologies [3]. Overall, the passage emphasizes that digitization is a powerful force transforming how we live and work. It goes beyond simply converting things to digital, it's about leveraging technology to create entirely new ways of doing things which fuels the proliferation of data, which in turn fuels the advancement of artificial intelligence, creating a symbiotic relationship between digitization and AI.

2.2 Artificial Intelligence (AI)

AI entails the replication of human intelligence through machines or systems, with the goal of developing machines that can engage in reasoning, learning, planning, and prediction, reflecting human cognitive capacities and behaviors. The realm of AI methods encompasses a wide array of techniques designed to empower machines to assimilate knowledge from data, make decisions, and carry out tasks autonomously. Numerous scholars are committed to exploring artificial intelligence such as search algorithms, knowledge graphs, natural language processing, expert systems, evolutionary algorithms, machine learning (ML), and deep learning (DL) [17].

2.3 Machine Learning (ML)

In the era of the Fourth Industrial Revolution (4IR or Industry 4.0), the digital realm is abundant with diverse datasets, including Internet of Things (IoT) data, cybersecurity data, mobile data, and health data. Harnessing the power of artificial intelligence (AI), especially machine learning (ML), is essential for intelligently analyzing this data and developing smart and automated applications [24]. Machine learning algorithms are crucial in unlocking the insights hidden within these vast datasets, enabling organizations to make informed decisions, enhance efficiency, and drive innovation across various industries.

2.4 Deep Learning (DL)

Deep learning is a hot trend in machine learning, Artificial Intelligence, data science, and analytics. Its ability to learn from data makes it a powerful tool. Major companies like Google, Microsoft are heavily invested in DL research [8]. In Addition to that, deep learning models are powerful due to their layered architecture. However, training these models takes a significant amount of time because they have many parameters to adjust. The good news is that once trained, they can make predictions very quickly, unlike some other machine learning algorithms [24]. In the medical field, deep learning has shown remarkable promise in applications such as medical image analysis, disease diagnosis, and personalized medicine. Its ability to extract intricate patterns from large datasets has led to breakthroughs in cancer detection, MRI analysis, predicting patient outcomes for example Vital Sign Predictions Using Long-Short-Term Memory Networks [4], and the Detection of Fractures in Cervical Spine [16].

3 Related Works

In this section we will talk about several examples of applications in medicine field which in 1977, AI made its debut in the field of medicine with the introduction of MYCIN, a renowned computer-based consultation system. MYCIN was specifically developed to aid physicians in diagnosing and selecting therapies for patients with bacterial infections [23].

Today, AI applications in healthcare have involved to become even more efficient, offering a myriad of advanced capabilities and it can be broadly categorized into two major categories: general use and specialized use. General-use platforms such as Enlitic's range of AI technologies seamlessly integrates with current workflows, enabling healthcare providers to improve clinical processes, and extend their capabilities [15], Merative for clinical decision support and healthcare analytics which Francisco Partners has finalized its acquisition of Watson Health's data analytics tools [13], Regard for automating clinical tasks and Developing augmented reality (AR) functionalities for an Electronic Clinical Trial Management System (ECTMS) and assess their influence on user experience [9], Google Vertex AI provides two options: AutoML and custom training. AutoML eliminates the need for data science. On the other hand, custom training necessitates familiarity with TensorFlow or scikit-learn models [2], Care.ai for AI-assisted virtual care, Primary care physicians are ideally positioned to spearhead the healthcare AI revolution. They can play a pivotal role in advancing AI in healthcare by collaborating with technologists to ensure the relevance and human-centeredness of AI applications, employing quality improvement methodologies in AI deployments [11]. Finally, the most used software Kaggle that assist professionals and learners in achieving their data science objectives through its set of tools and resources. On the other hand, specialized platforms such as Freenome for early cancer detection [19]. Iterative Health for detecting gastrointestinal(GI) illness, offers a systematic guide for the iterative medical decision-making process, presenting a clear framework and detailing the key iterative stages of model-based decision-making [10], VirtuSense for predictive fall prevention [21]

However, None of the existing platforms possess the comprehensive capabilities of our revolutionary platform, which is specifically designed to cater to the unique needs of both doctors and researchers.

4 Platform Devlopment

4.1 Requirements Analysis

This section presents the requirements analysis for our AI-powered healthcare platform. This part will be showed as a set of use case diagrams depicting actors and uses case of the systems.

Potential Users and Their Roles: In designing our AI-powered healthcare platform, it is essential to consider the diverse range of users who will interact within it. It is important to clarify the subtle difference between "user" and "actor", as these terms are sometimes confused. A "user" refers to an individual interested in utilizing or benefiting from the platform, while an "actor" specifies a role played by an external entity interacting with the system. Users interact with the system by playing different roles (actors). A single user can interact with the system under different roles at different times, as a role can be played by multiple users depending on the context of system usage. This distinction helps ensure a clear understanding of the interactions and requirements within the platform.

Users: The following are the primary categories of users who will interact with our platform, each utilizing its functionalities to achieve their specific goals: ***Researchers:*** Conduct research activities using datasets and AI models. ***Students:*** Explore and experiment with AI models and medical data. ***Teachers:*** Use the platform for educational purposes and to demonstrate AI applications in medical fields. ***Doctors:*** Use AI models to assist in diagnosis and analysis of medical images and data. ***Biologists:*** Apply AI models in biological and medical research.

Actors: The following actors represent the diverse roles and responsibilities for users interacting with our platform: ***Visitor:*** Unregistered users who can access limited functionalities such as viewing information and uploading images for AI analysis. ***Affiliate user:*** Users who already have an account on our platform. By logging in, they gain access to download datasets and view research papers. ***Researcher:*** Registered users with extended privileges, including uploading datasets, models, and research papers. ***Administrator:*** Users with full control over the platform, responsible for managing users, datasets, and models, and ensuring system security.

The UML Use Case Diagram provides an overview of the interactions between different actors and the system, highlighting their roles and the functionalities they can access. The General Use Case Diagram (see Fig. 1(a)) encompasses interactions for non-administrator roles. Whereas, the Administrator Use Case Diagram (see Fig. 1(b)) focuses specifically on administrative functionalities and system management.

Principle Use Cases' Descriptions: In this section, we delve into detailed descriptions of key use cases that outline essential functionalities of the platform:

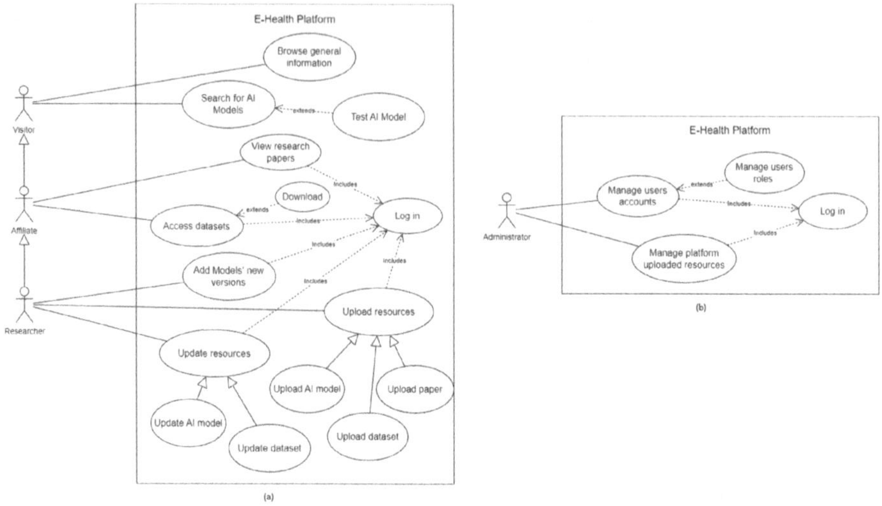

Fig. 1. (a) General Use Case Diagram, (b) Administrator Use Case Diagram

1) Search (and Test) uploaded AI-models: To facilitate the effective utilization of AI models, this use case outlines the process by which users can search for and test the AI models available on the platform.

- ***Functional requirements:*** Any user can search for available AI models on the platform, and their descriptions. He can also upload medical images (or data) to test an AI model and receive its results.
- ***Involved actors:*** Visitor, Affiliate user, Researcher.
- ***Main scenario:*** The following steps outline the process for searching and testing AI models: The user logs into the platform (if not a visitor). The user navigates to the section for browsing available AI models. The user searches for a specific AI model or browses through the list of available models. The user views the description and details of the selected AI model. The user decides to test the model and navigates to the section for uploading medical images. The user uploads a medical image or dataset to be analyzed. The platform processes the uploaded image using the selected AI model. The platform displays the results of the AI model's analysis to the user.

Figure 2 depicts these steps for a visitor.

2) View (and download) resources: This use case describes the steps involved for registered users to browse, view details, and download resources available on the platform, such as datasets and research papers.

- ***Functional requirements:*** Registered users can browse and access available resources such as datasets and research papers. They can download these resources for further analysis and study.
- ***Involved actors:*** Affiliate users and Researchers.

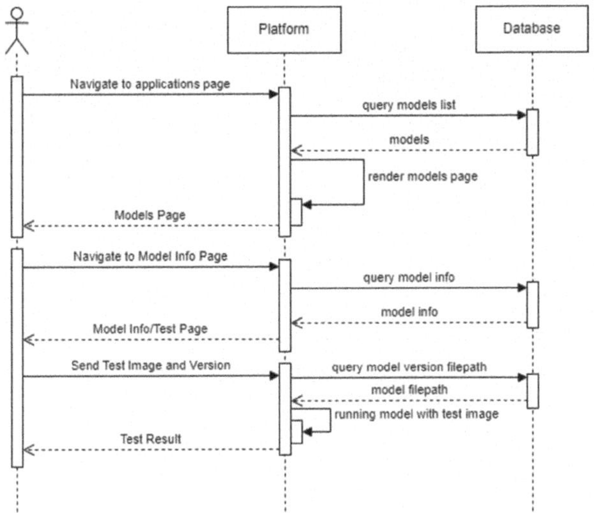

Fig. 2. Searching and testing AI model Scenario by a visitor.

- ***Main scenario:*** The following steps outline the process for viewing and downloading resources from the platform: The user logs into the platform. The user navigates to the section for browsing available resources. The user searches for a specific resource or browses through categories. The user views the details and metadata of the selected resource, including its description, format, and usage terms. The user decides to download the resource. The platform verifies user permissions and initiates the download process. The user receives the downloaded resource for local use.

3) Upload AI models: This use case outlines the process for researchers to upload their AI models to the platform, ensuring that each model is accompanied by the necessary metadata.

- ***Functional requirements:*** Researchers can upload AI models to the platform. They should provide descriptions and relevant metadata for each model.
- ***Involved actors:*** Researcher.
- ***Scenario:*** The following steps outline the process for uploading a new AI model on the platform: The user logs into the platform. The user navigates to the section for uploading AI models. The user fills out the required information for the AI model, including the title and description. The user uploads the AI model file. The platform adds the AI model to its repository.

Figure 3 illustrates these steps by a sequence diagram. While Fig. 4 outlines the process of updating an existing model on the platform.

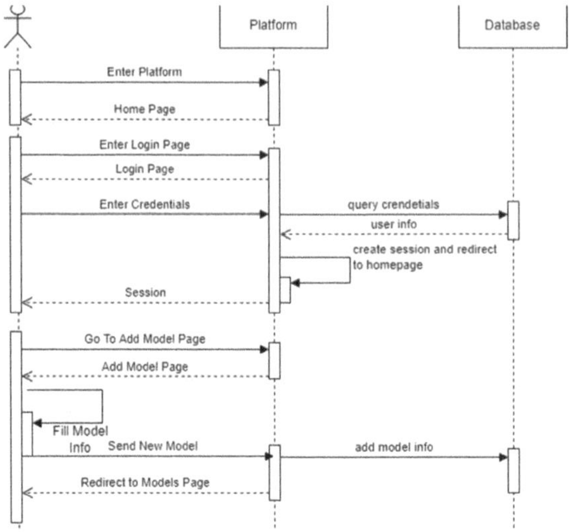

Fig. 3. Model Upload Scenario

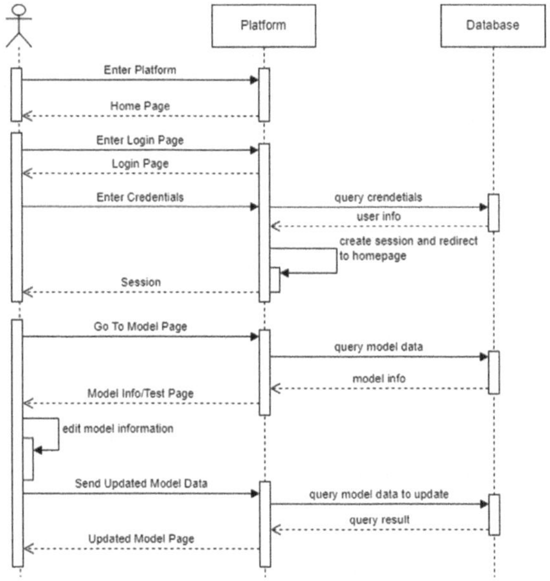

Fig. 4. Model Update Scenario

4.2 Platform Architecture

The architecture of our AI-powered healthcare platform showed, in Fig. 5, consists of three main components: (i) the user browser interface, (ii) the Application Server, and (iii) the storage back-end (Data Server). The user browser interface provides a user-friendly environment for browsing AI models, uploading medical images, and accessing resources. The Application Server, which forms the core of the platform, handles business logic and facilitates communication between the interface and the Data Server. It includes key modules such as Router, Authentication Middleware, Controller, Model, Flask API, and View. The Data Server manages data and file storage, utilizing MySQL Database and File Storage system. This architecture ensures a robust, scalable, and efficient platform that meets functional and non-functional requirements.

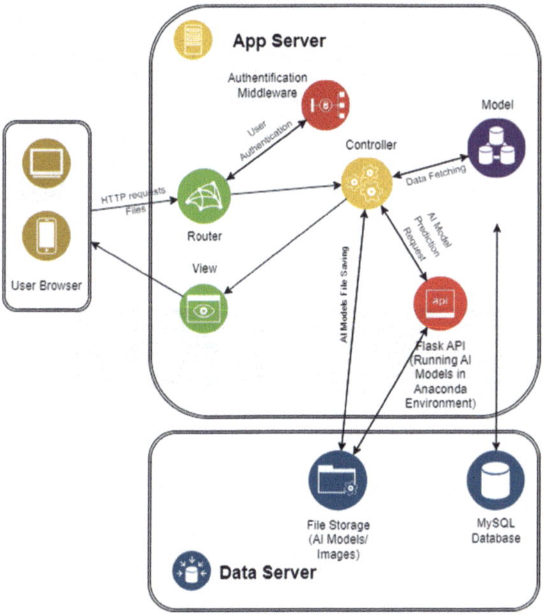

Fig. 5. Platform Architecture

4.3 Platform Realization

Our platform is built using modern technologies and frameworks which are: HTML, CSS, JavaScript, NodeJS, MySQL, Flask, Visual Studio Code. And The following figures represent screenshots from different pages from our platform. Figure 6 illustrates the home page view of the platform.

Fig. 6. Home Page

The “Published Papers” page “AI Applications”, and Add Application, and “Test Application” page are respectively shown in Figs. 7, 8 and 9.

Fig. 7. Publications Page

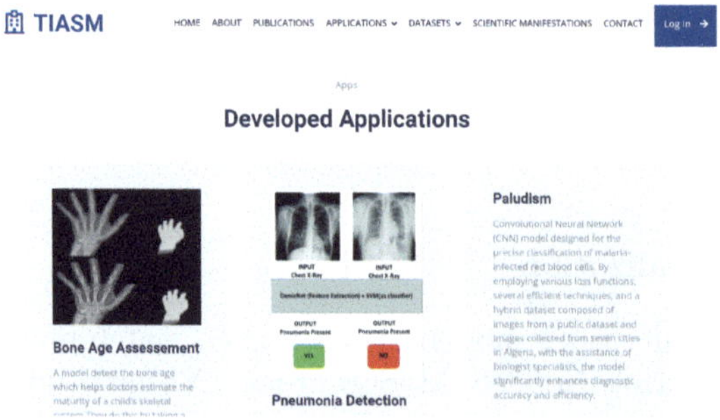

Fig. 8. Apps Pages

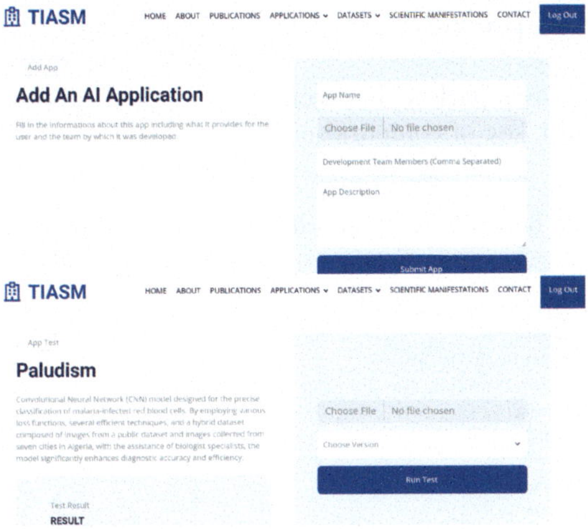

Fig. 9. Add App, and Test App Pages

Figure 10 provides an of the datasets page, uploading a new dataset, and download a specific dataset.

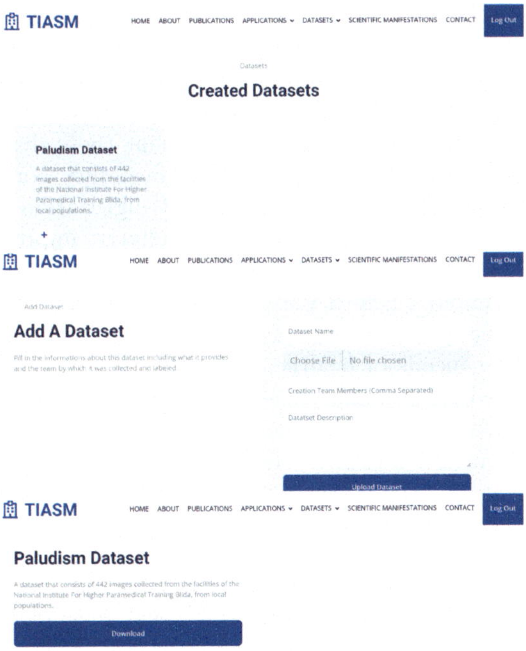

Fig. 10. Datasets, add dataset, and download dataset Pages

5 Results Comparison

Table 1 illustrates a comparison between our proposed system and other platforms using the most key of comparaison.

Table 1. Comparison between our proposed system and best previous related systems.

Platform	Functionalities	Security	Scalability	Performance
[23]	Aid physicians in diagnosing Aid physicians in selecting therapies	No	No	Yes
[15]	Improve clinical processes	No	No	No
[13]	Clinical decision support Healthcare analytics	No	No	No
[9]	Automating clinical tasks Developing augmented reality functionalities	No	No	No
[2]	AutoML Custom training	No	No	No
[11]	AI-assisted virtual care	Yes	Yes	No
[19]	Early cancer detection	No	No	Yes
[10]	Detecting gastrointestinal illness	No	Yes	Yes
[21]	Predictive fall prevention	Yes	No	Yes
Our System	**Store Medical Papers Store and Test Models Store and Download new Datasets**	**Yes**	**Yes**	**Yes**

6 Conclusion

In conclusion, the importance of accessible and effective healthcare as a fundamental human right cannot be overstated. Recent global crises, including the COVID-19 pandemic, have underscored the critical need for robust healthcare systems and international collaboration. However, significant challenges persist, particularly in terms of affordability, accessibility, and the availability of comprehensive health information. It is clear that addressing these challenges requires a multi-faceted approach, involving individual responsibility, community support, and technological innovation. The emergence of digitization and AI presents promising opportunities to enhance healthcare delivery, improve efficiency, and personalize care. Our paper introduces a groundbreaking initiative: the development of an AI-powered e-health Platform specifically tailored to address the unique healthcare challenges prevalent within the Algerian context. By focusing on the specialized needs of healthcare professionals and offering a tailored environment for collaboration and research, our platform aims to revolutionize healthcare practices and improve outcomes. With robust security measures in place to safeguard sensitive healthcare data and a vibrant community of experts facilitating knowledge sharing and innovation, our platform represents a significant step towards advancing healthcare in Algeria, Africa, and beyond.

Acknowledgment. This work is the result of collaboration within the interdisciplinary team known as TIASM (Techniques de l'IA pour le Soutien de la Médecine).

References

1. Baissi, F., Abdelbaki, E.A., Kahloul, L., Mohammedi, A., Ammari, A.: Paludism diagnosis using deep learning. In: Kerrache, C.A., Tahari, A., Kassimi, D., Chakraborty, C. (eds.) Proceedings of the International Conference on Emerging Intelligent Systems for Sustainable Development (ICEIS 2024), pp. 246–260. Atlantis Press International BV, Dordrecht (2024). https://doi.org/10.2991/978-94-6463-496-9_19
2. Berg, G.: Image classification with machine learning as a service:-a comparison between azure, sagemaker, and vertex AI (2022). https://doi.org/10.3390/bioengineering10080891
3. Brennen, S., Kreiss, D.: Digitalization and digitization. Cult. Digit. **8**(2), 10–19 (2014). https://doi.org/10.1002/9781118766804.wbiect111
4. Dakhia, Z., Belouaar, H., Belaiche, L., Kahloul, L.: Precision unleashed: enhancing vital sign predictions using long-short-term memory networks. In: 2023 International Conference on Innovation and Intelligence for Informatics, Computing, and Technologies (3ICT), pp. 397–401. IEEE (2023). https://doi.org/10.1109/3ICT60104.2023.10391537
5. Fellah, K.M., Tigane, S., Kahloul, L.: Diabetic retinopathy detection using deep learning. In: International Symposium on Modelling and Implementation of Complex Systems. pp. 234–246. Springer (2022). https://doi.org/10.1007/978-3-031-18516-8_17
6. Gassmann, O., Frankenberger, K., Csik, M.: The St. Gallen business model navigator. Int. J. Prod. Dev. **18**, 249–273 (2013). https://bmilab.com/s/The_StGallen_Business_Model_Navigator-raee.pdf
7. Henriette, E., Feki, M., Boughzala, I.: The shape of digital transformation: a systematic literature review (2015). https://aisel.aisnet.org/mcis2015/10
8. Karhunen, J., Raiko, T., Cho, K.: Unsupervised deep learning: a short review. In: Advances in Independent Component Analysis and Learning Machines, pp. 125–142 (2015). https://doi.org/10.1016/B978-0-12-802806-3.00007-5
9. Kim, S.H., Jin, J., Sevinchan, M., Davies, A.: How do automated reasoning features impact the usability of a clinical task management system? Development and usability testing of a prototype. Int. J. Med. Inform. **174**, 105067 (2023). https://doi.org/10.1016/j.ijmedinf.2023.105067
10. Kunst, N., Burger, E.A., Coupé, V.M., Kuntz, K.M., Aas, E.: A guide to an iterative approach to model-based decision making in health and medicine: An iterative decision-making framework. PharmacoEconomics, pp. 1–9 (2023). https://doi.org/10.1007/s40273-023-01341-z
11. Lin, S.: A clinician's guide to artificial intelligence (AI): why and how primary care should lead the health care AI revolution. J. Am. Board of Family Med. **35**(1), 175–184 (2022). https://doi.org/10.3122/jabfm.2022.01.210226
12. Płaciszewski, K., Wierzba, W., Ostrowski, J., Pinkas, J., Jankowski, M.: Use of the internet for health purposes-a national web-based cross-sectional survey among adults in Poland. Int. J. Environ. Res. Public Health **19**(23), 16315 (2022). https://doi.org/10.3390/ijerph192316315
13. Radley, D.C., Collins, S.R., Baumgartner, J.C.: 2020 scorecard on state health system performance. The Commonwealth Fund, 2020–09 (2020). https://www.commonwealthfund.org/sites/default/files/2023-06/Radley_2023_State_Scorecard_APPENDICES.pdf

14. Remaigui, I., Kahloul, L., Benharzallah, S.: A new deep learning architecture for pneumonia detection in pediatrics. In: 2022 19th International Multi-Conference on Systems, Signals & Devices (SSD), pp. 1098–1104. IEEE (2022). www.doi.org/10.1109/SSD54932.2022.9955773
15. Retiene, R.: Health-related activities of big tech. Munich Personal RePEc Archive (2022). https://mpra.ub.uni-muenchen.de/115080/
16. Salem, T., Belaiche, L., Kahoul, M., Kahloul, L., Belouaar, H.: One-stage deep learning approach for the detection of fractures in cervical spine. In: 2023 Third International Conference on Theoretical and Applicative Aspects of Computer Science (ICTAACS), pp. 1–5. IEEE (2023). https://doi.org/10.1109/ICTAACS60400.2023.10449644
17. Sarker, I.H.: Deep learning: a comprehensive overview on techniques, taxonomy, applications and research directions. SN Comput. Sci. **2**(6), 420 (2021). https://doi.org/10.1007/s42979-021-00815-1
18. Sebti, R., Zroug, S., Kahloul, L., Benharzallah, S.: A deep learning approach for the diabetic retinopathy detection. In: Ben Ahmed, M., Boudhir, A.A., Karaş, İR., Jain, V., Mellouli, S. (eds.) Innovations in Smart Cities Applications Volume 5: The Proceedings of the 6th International Conference on Smart City Applications, pp. 459–469. Springer International Publishing, Cham (2022). https://doi.org/10.1007/978-3-030-94191-8_37
19. Stephens, K.: Freenome, siemens healthineers collaborate for breast cancer research. AXIS Imaging News (2021). https://axisimagingnews.com/imaging-news/company-news/freenome-siemens-healthineers-collaborate-for-breast-cancer-research
20. Stolterman, E., Fors, A.C.: Information technology and the good life. In: Kaplan, B., Truex, D.P., Wastell, D., Wood-Harper, A.T., DeGross, J.I. (eds.) Information Systems Research. IIFIP, vol. 143, pp. 687–692. Springer, Boston, MA (2004). https://doi.org/10.1007/1-4020-8095-6_45
21. Strubhar, A.J., Tan, P., Storage, L., Peterson, M.: Concurrent validity of the virtusense® gait analysis system for the quantification of spatial and temporal parameters of gait. Int. J. Exercise Sci. **11**(1), 934–940 (2018). https://digitalcommons.wku.edu/ijes/vol11/iss1/12
22. Tihinen, M., Iivari, M., Ailisto, H., Komi, M., Kääriäinen, J., Peltomaa, I.: An exploratory method to clarify business potential in the context of industrial internet – a case study. In: Afsarmanesh, H., Camarinha-Matos, L.M., Lucas Soares, A. (eds.) PRO-VE 2016. IAICT, vol. 480, pp. 469–478. Springer, Cham (2016). https://doi.org/10.1007/978-3-319-45390-3_40
23. Van Melle, W.: MYCIN: a knowledge-based consultation program for infectious disease diagnosis. Int. J. Man-Mach. Stud. **10**(3), 313–322 (1978). https://doi.org/10.1016/S0020-7373(78)80049-2
24. Xin, Y., et al.: Machine learning and deep learning methods for cybersecurity. IEEE Access **6**, 35365–35381 (2018). https://doi.org/10.1109/ACCESS.2018.2836950
25. Zerari, A., Djedidi, O., Kahloul, L., Carlo, R., Remadna, I.: Paediatric bone age assessment from hand X-ray using deep learning approach. In: Senouci, M.R., Boulahia, S.Y., Benatia, M.A. (eds) International Conference on Computing Systems and Applications, pp. 373–383. Springer (2022). https://doi.org/10.1007/978-3-031-12097-8_32

A Hybrid Job Recommendation Approach Based on BERT and WordNet

Assia Brek(✉), Hassiba Laifa, and Yahya Issa Diakhite

Larbi Tebessi University - Tebessa, Tébessa, Algeria
{assia.brek,laifa.hassiba}@univ-tebessa.dz, i17771.etu@iscae.mr

Abstract. Job searching can be complex and time-consuming, requiring candidates to sift through numerous job offers to find those that best match their skills and aspirations. Recommendation systems are crucial in streamlining this process by suggesting relevant job opportunities based on user profiles. This research addresses the challenge of improving job recommendations by leveraging the semantic information contained in job documents, such as descriptions and resumes. We propose a hybrid approach that combines BERT, a language model proficient in understanding text context, with WordNet, a lexical database rich in semantic relationships. This integration aims to enhance the relevance of recommendations by capturing detailed semantic nuances. Evaluation results indicate that incorporating WordNet with BERT generally improves recommendation performance by 10%. However, the effectiveness varies across different profiles, with some cases experiencing reduced relevance due to semantic noise introduced by WordNet. Despite this, the approach demonstrates significant potential for enhancing job recommendations using advanced natural language processing techniques.

Keywords: Job Recommendation · Semantic Matching · BERT · WordNet

1 Introduction

An effective job recommendation system plays a crucial role in matching job seekers with suitable employment opportunities. Traditional methods, while useful, often struggle with understanding the complex relationships between job descriptions and candidate profiles.

In several works, matching job offers and candidates' profiles is described as detecting lexical similarity between document entities, counting common words in both documents or measuring a word's importance to a document within a collection. However, these methods are not adequate since there can be semantic heterogeneity in the texts. An example would be matching a job that requires "object-oriented programming" skills and a resume that offers the skill "java".

BERT (Bidirectional Encoder Representations from Transformers) enhances job recommendation systems by improving the understanding of job descriptions and candidate resumes through contextual word representations [16]. It

T. Marir et al. (Eds.): ICCSA 2024, ISEM 45, pp. 329–340, 2025.
https://doi.org/10.1007/978-3-031-90758-6_24

enables better semantic matching, accurately aligns skills and experience, and personalizes recommendations based on user preferences.

BERT's embeddings are context-dependent but do not explicitly encode lexical relations such as synonyms, hypernyms, or domain-specific similarities [2]. For example, BERT might not recognize that "software engineer" and "programmer" can often be used interchangeably in job contexts. Therefore, Combining BERT with additional tools or knowledge bases, such as WordNet or domain-specific databases, can help mitigate this limitation and improve the effectiveness of job recommendations.

WordNet as large lexical database of English, can enhance recommendation results with BERT by providing lexical relationships and semantic understanding. It helps in disambiguating terms, expanding the context of keywords, and improving semantic matching between user preferences and items. This combination allows BERT to deliver more accurate and relevant recommendations by leveraging both contextual embeddings and semantic knowledge.

In this paper, we propose a hybrid job recommendation approach that leverages the power of integrating WordNet with BERT to recommend job offers. Unlike other existing approaches, we propose first enriching the document content using WordNet before embedding it. We integrated WordNet during the fine-tuning phase, not during the initial BERT training. We enriched the job description with synonyms retrieved from WordNet. This enrichment offers a richer semantic context and reduces word ambiguity.

The rest of this paper is organized as follows: Sect. 2 outlines the related work. Section 3 presents the proposed recommendation approach. Section 4 describes the used dataset. Section 5 provides the obtained results and the evaluation of the approach. Finally, Sect. 6 presents the conclusions and future lines of work.

2 Related Work

Job recommender systems primarily rely on various technologies and algorithms to recommend jobs to job seekers and candidates to recruiters. These systems suggest appropriate job offers to users based on their interests, experience, and skill set. According to several studies in the literature, job recommendation approaches can be categorized into four main types: Collaborative filtering(CF), Content-based recommendation (CBR), Knowledge-based recommendation (KBR) and Hybrid approaches [8].

Content-based recommendation (CBR) methods primarily use and calculate the similarity between candidate information and job offer content. A common source of candidate information is their resume. Therefore, these methods generally rely on the ability to cross-reference information between job offers and resumes in a semantically efficient manner [1].

Collaborative filtering (CF) methods rely on user behaviour data, typically stored in a matrix representing user-item ratings. In online recruitment, this matrix often captures click behaviour, where a rating of 1 indicates that a job seeker clicked on a job offer to view the details, and 0 otherwise. In the absence of such interaction data, the sequence of previous jobs can be used to populate the rating matrix, requiring job categories to represent the discrete items in the matrix [5].

Knowledge-based recommendation (KBR) methods enhance matching capabilities by using domain-specific expertise to identify contextual similarities rather than just matching terms [14]. AnnoJob [4] is a recommendation system that relies on calculating semantic similarity between job offers and resumes exploiting knowledge graphs, such as WikiData, to identify semantic relatedness.

Hybrid approaches essentially combine multiple methods to achieve better results. The combination of CBR and CF approaches is discussed in [10], where the authors propose a two-step process to calculate a matching score that simultaneously considers information extracted from the resume and other applications. In [9], a hybrid information filtering engine integrates several recommendation engines: demographic-based, knowledge-based, content-based, concept-based, and ontology-based, combined with the behavioural and collaborative recommendation, distributed across different integrated modules.

Recently, AI has significantly improved job recommendation systems by using advanced techniques like deep learning and natural language processing (NLP). These innovations enable more personalized and accurate matches between job seekers and employers by better analyzing resumes, job descriptions, and user behaviour. AI-driven systems can quickly adapt to changes, ensuring that recommendations remain relevant and tailored to individual needs.

BERT, introduced in [6], has revolutionised the field of NLP by providing deep contextualised embeddings for words in a sentence. This model captures the context-dependent meaning of words, making it highly effective for tasks like job recommendation. First, it can extract relevant features from job descriptions and user profiles, enabling a deeper understanding of the text content and user preferences. Second, BERT can classify text sequences, allowing for the prediction of the relevance of job postings to users based on contextual and semantic information [11]. Work in [13] demonstrated the effectiveness of BERT in understanding job titles and descriptions, leading to better alignment between job postings and candidate profiles. Moreover, in [12], they applied a refined version of a BERT variant, the Siamese Sentence-BERT (SBERT) model, to recommend vacancies to job seekers. They described their methodology by explaining the use of multilingual SBERT with a bi-encoder and a cosine similarity score as the output layer. The aim of using SBERT is to compute peer-to-peer semantic similarity.

While BERT provides powerful contextual embeddings, it can be further enhanced by integrating external knowledge bases like WordNet.

WordNet, a large lexical database of English, groups words into sets of synonyms (synsets) and provides short definitions and usage examples, as well as the semantic relationships between these sets [15]. Research in [17] has shown that incorporating WordNet into job recommendation systems can help in disambiguating terms and improving the matching process. By aligning BERT's embeddings with WordNet's hierarchical structure, the system can better understand the semantic similarities between different job descriptions and candidate profiles, leading to more accurate recommendations. Likewise, a study in [7] proposed a hybrid approach where BERT embeddings were refined using WordNet to address the challenges of synonymy and polysemy in job descriptions. This approach demonstrated a significant improvement in matching accuracy compared to using BERT alone, particularly in cases where job titles or skills had multiple meanings or closely related.

3 Methodology

We will exploit WordNet to enrich job descriptions and candidate resumes. WordNet will allow us to add semantic synonyms of the terms used in descriptions and resumes. This approach aims to improve the accuracy of recommendations by considering different possible linguistic terms, thereby increasing the chances of matching qualified candidates with appropriate job offers.

There are several approaches to integrate WordNet with BERT for job recommendation, such as the two methods cited in [3]. WordNet synonyms were integrated into a BERT model to enrich text representations using the concatenation approach and the fusion approach.

Concatenation Approach: Combines BERT embeddings with WordNet embeddings by concatenating them into a single vector. This method leverages both BERT's contextual information and WordNet's semantic relationships.

Fusion Approach: Integrates WordNet information directly into BERT's processing, such as augmenting input embeddings or modifying the attention mechanism. This method blends BERT's contextual understanding with WordNet's semantic features for improved text representation.

In our case, we proposed to enrich the text first with wordNet and then embed the enriched text. Enriching the text with wordNet before embedding ultimately improves the quality of vector representation by providing a richer semantic context and reducing word ambiguity.

3.1 The Input: Profile and Job Descriptions

The input data consists of profile descriptions and job descriptions. It is important to note that, for our study, the profiles used in our system are in the form of job search queries, with a future perspective of using traditional resumes. A detailed description of the used dataset as inputs will be delivered in the next section

3.2 Text Enrichment with WordNet

Text descriptions are enriched with synonyms using WordNet to improve context and term diversity.

- **Step 1**: Identification and selection of keywords: first, the text is split into words (tokens), followed by identifying the POS tags for each word to filter out non-content words like determiners (e.g., "the", "a") and prepositions (e.g., "in", "on"). Next, TF-IDF (Term Frequency-Inverse Document Frequency) is applied to identify the most important and unique words in the document. Then we select words with high TF-IDF scores.
- **Step 2**: Use WordNet to search synonyms for each identified word.
- **Step 3**: Selecting relevant synonyms: while WordNet can provide multiple synonyms for a given word, not all of them are relevant in the context of a job description. To avoid introducing too much noise and maintain the clarity of the text, we limit the number of synonyms added. We opted for a maximum of one synonym to control the amount of additional information.
- **Step 4**: Text enrichment: The selected synonyms are then integrated into the original text. Each word is followed by its chosen synonyms, creating an enriched version of the text that retains the original words while adding lexical alternatives.

Example 1: Consider a simple job description: *"The candidate should have experience in programming and software development"*.
The enriched text will be as follows: *"The candidate should have experience in programming* ***coding*** *and software development* ***software engineering***".

3.3 Tokenization (BERT Tokenizer)

Once the text is enriched with synonyms, we use the BERT-specific tokenizer, which includes steps like word segmentation, subword management (WordPiece tokenization), and adding special tokens needed for BERT ([CLS], [SEP]).

The enriched text is converted into a sequence of tokens. For example, a sentence like "Developer with experience in Python and Java" could be tokenized to "[Developer, with, experience, in, Python, and, Java]"

Token sequences are adjusted to match the maximum length accepted by BERT, with padding or truncation if necessary.

3.4 Embedding (BERT Initial Embeddings)

Once the text is tokenized, each token is transformed into an initial embedding vector based on BERT's pre-trained embedding tables, which are vector representations capturing both contextual and syntactic information. These initial embedding include information about the tokens and their positions in the sequence. Position embedding is then added to the token embeddings to incorporate the positional information, which is crucial for capturing the sentence structure.

3.5 Contextual Encoding (BERT Transformer Layers)

Initial embedding pass through multiple layers of transformers. Each layer applies attention mechanisms and nonlinear transformations to refine the vector representations. BERT's attention mechanisms capture long-distance relationships between tokens, producing final contextual embeddings rich in semantic and contextual information.

3.6 Profile and Job Description Embeddings (Contextual Vectors)

Profile and job descriptions are now represented by dense context vectors. Contextual embedding generated by BERT for job search queries and job descriptions serves as the final representation of profiles and jobs. After passing through BERT's transformer layers, the final context vectors for each token are extracted. These can be aggregated, such as by averaging, to create a global representation, or the [CLS] token can be used to represent the entire sequence.

Figure 1, presents the embedding of example 1 mentioned above.

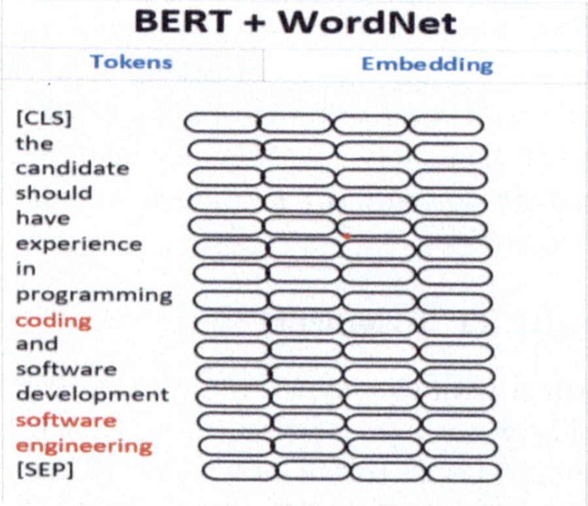

Fig. 1. Example 1 sentence embedding.

3.7 Cosine Similarity Calculation

To measure the similarity between the context vectors of profiles and job descriptions, we use cosine similarity. This measure evaluates the angle between two vectors in the vector space, capturing their semantic similarity. The cosine similarity between two vectors A and B is defined as

$$\text{cosine similarity} = \frac{\mathbf{A} \cdot \mathbf{B}}{\|\mathbf{A}\|\|\mathbf{B}\|} \tag{1}$$

For each pair of vectors (profile and job description), the cosine similarity is calculated by normalizing the vectors and dividing their dot product by the product of their norms.

3.8 Ranking Job Offers

The offers are ranked according to their cosine similarity with the search profile. The most relevant jobs, identified by their high cosine similarity with the search profiles, are then recommended to users.

Job recommendations are presented to users as a list, ranked in descending order of similarity. Each recommendation includes details about the job, such as title, company, location, and a summary of responsibilities and requirements.

4 Dataset

We evaluate our approach on a dataset from Kaggle[1] as CSV file where a total number of inputs are 19077 job offers composed of 9 attributes: Job_id, job_description, Job_title, Skills, Industry, Experience, Job_location, Company.

Before using this dataset, it undergoes several main steps. First, the dataset is cleaned to ensure data quality and consistency. Second, data visualization is utilized to analyze the data through graphs, extracting additional information that the data may contain.

The chart in Fig. 2 shows the 20 most common industries in job postings, ranked from highest to lowest based on their frequency of appearance.

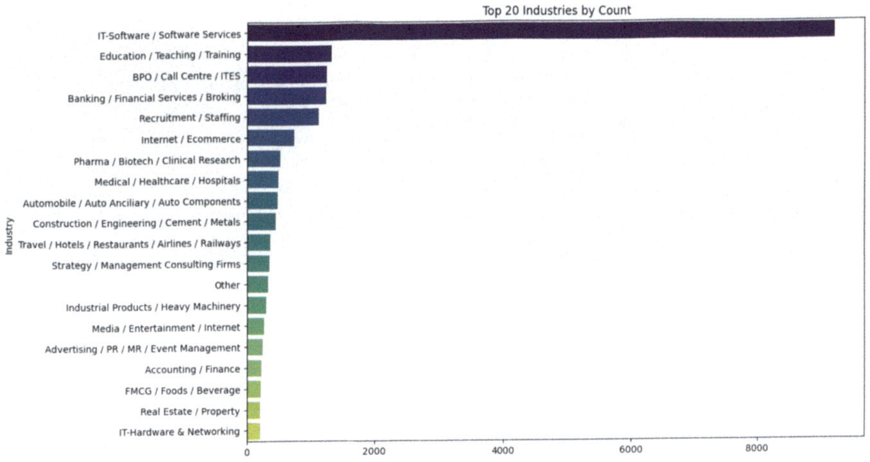

Fig. 2. Top 20 industries.

Figure 3 presents the word cloud of skills. Which highlights the most important and frequently required skills in job offers. The most frequent skills appear larger and more visible.

[1] https://www.kaggle.com/datasets.

Fig. 3. Word cloud of skills.

For the user profile dataset, we couldn't find any available datasets to use. Therefore, we proposed using simple sentences that can serve as queries to search for jobs online. Figure 4 presents the used queries, where each one presents a user profile.

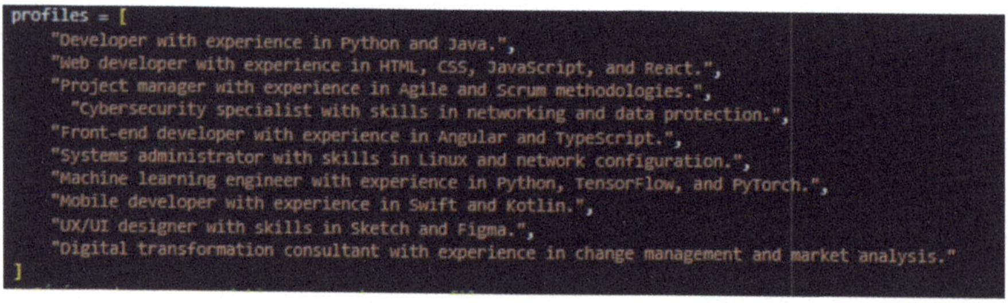

```
profiles = [
    "Developer with experience in Python and Java.",
    "Web developer with experience in HTML, CSS, JavaScript, and React.",
    "Project manager with experience in Agile and Scrum methodologies.",
     "Cybersecurity specialist with skills in networking and data protection.",
    "Front-end developer with experience in Angular and TypeScript.",
    "Systems administrator with skills in Linux and network configuration.",
    "Machine learning engineer with experience in Python, TensorFlow, and PyTorch.",
    "Mobile developer with experience in Swift and Kotlin.",
    "UX/UI designer with skills in Sketch and Figma.",
    "Digital transformation consultant with experience in change management and market analysis."
]
```

Fig. 4. Users profile.

5 Results

To evaluate the quality of the recommendation results, we used the Normalized Discounted Cumulative Gain (NDCG) measures to measure *how relevant the results are* and *how good the ordering is.*

The NDCG as a measure of ranking quality is used to measure the order correctness of the selected ranked job list. The NDCG metric involves finding the normalized discounted cumulative gain. Where the gain is just the relevance score for each item recommended, and cumulative gain at K is the sum of gains of the first K items recommended. Moreover, Discounted cumulative gain weighs each relevance score based on its position. The recommendations at the top get a higher weight, while the relevance of those at the bottom gets a lower weight.

$$DCG@k = \sum_{i=0}^{k} \frac{G_i}{\log_2(i+1)} \tag{2}$$

The normalized discounted cumulative gain is the DCG with a normalization factor in the denominator. The denominator is the ideal DCG score when we recommend the most relevant items first.

$$NDCG@k = \frac{DCG@k}{IDCG@k} \tag{3}$$

The ground truth data is needed to evaluate our approach using NDCG measure. We couldn't find any available ground truth dataset in the job recommendation field. Therefore, we created it manually. First, we manually selected a given profile's top 10 relevant offers. Then, we assign a relevant value for each offer. The numbers 0,1,2,3 are assigned as a designation of the degree of relevance, for example: 3: very relevant, 2: Relevant, 1: slightly relevant, 0: not relevant. The relevance value of each offer is based on the cosine similarity score of recommendation:

- If the similarity is $>= 0.8$ relevance (3)
- $0.6 <$ similarity < 0.8 relevance (2)
- $0.4 <$ similarity < 0.6 relevance (1)
- Similarity < 0.4 relevance (0)

```
# Ground truth data for multiple profiles
ground_truth_profiles = {
    1: [
        (1, 20915501025, 3,'Sr Back end Developer'),
        (1, 190815502644, 3,'Open Source Developer') ,
        (1, 120715500018, 1,'Web Usability Analyst/Information Architect'),
        (1, 30915501203, 1,'Web UI Developer'),
        (1, 170715504541, 0,'PHP Developer') ,
        (1, 81215006897, 1,'Urgent Openings for UI Developer for Product Development Project') ,
        (1, 170316900198, 0,'Technical Architect - Dot Net Developer - OOPS / OOAD') ,
        (1, 241216000556, 0,'.Net Sr. Developer With Plsql') ,
        (1, 211216001564, 1,'Senior Frontend / UI Developer - Javascript/product Development'),
        (1, 71016002174, 1,'Sr Portal Developer')
    ],
    2: [
        (2, 281115900362, 0,'Senior Design Engineer (crash and Safety)') ,
        (2, 130216900602, 3,'Cognitive Computing') ,
        (2, 280316900414, 1,'Multimedia Automation Engineer - Audio / Video') ,
        (2, 260515500625, 0,'PLM/PDM/PBOM professionals'),
        (2, 240316002069, 1,'Software Developer - C++, VC++') ,
        (2, 231215500061, 2,'Data Specialist') ,
        (2, 150416501222, 0,'Auto Cad Designer') ,
        (2, 160915501674, 3,'Team Lead - Machine Learning'),
        (2, 40816002263, 1,'Systems Engineer- Infotainment, C++') ,
        (2, 200116001813, 3,'Excellent Opportunity in Cognitive Computing'),
    ],
    3: [
```

Fig. 5. Ground truth data.

Each recommendation will be annotated with: profile id, job id, relevance value, and job title. Figure 5 illustrates a part of the generated ground truth. Table 1 presents the results of evaluating the proposed recommendation approach using the NDCG measure. We compare the recommendation results based on BERT alone with those obtained by integrating WordNet with BERT.

Table 1. Comparison of NDCG of BERT and BERT + WordNet

Profils	BERT	BERT + WordNet
1	0.7443	**0.9610**
2	0.6617	**0.9259**
3	0.5518	**0.9395**
4	0.7759	**0.7798**
5	0.5476	**0.6550**
6	0.6929	**0.6987**
7	0.7060	**0.8735**
8	**0.7547**	0.6791
9	**0.9299**	0.6876
10	0.6661	**0.8950**
NDCG Mean	0.7031	**0.8095**

The curve in Fig. 6 summarizes the results presented in Table 1.

The NDCG values obtained for different profiles using the BERT+WordNet model demonstrate remarkable performance, with an average of 0.8095.

BERT alone achieves an average NDCG of 0.7031, or 70.31%, indicating that the base model already provides relevant recommendations.

The increase in average NDCG for BERT + Wordnet compared to BERT is about 0.10 points (about 10%) indicating that the addition of WordNet allows to capture of additional semantic relationships which improves overall the relevance of the recommendations.

The most significant improvements are observed for profiles 1, 2, 3, and 10, where the integration of WordNet results in a substantial increase in NDCG. Profiles 4 and 6 show marginal improvements, while profiles 8 and 9, unlike the other profiles, experience a decrease in the score with the addition of WordNet compared to BERT alone. This could indicate that the semantic relationships added by WordNet disrupt the recommendations for these specific profiles.

For profiles 8 and 9, the integration of WordNet did not enhance the relevance of the results, possibly due to the introduction of irrelevant semantic relationships. This suggests that BERT alone captures the concepts of these profiles more effectively, and the semantic "noise" from WordNet may reduce the model's accuracy, particularly if the data involves technical terms or specific jargon.

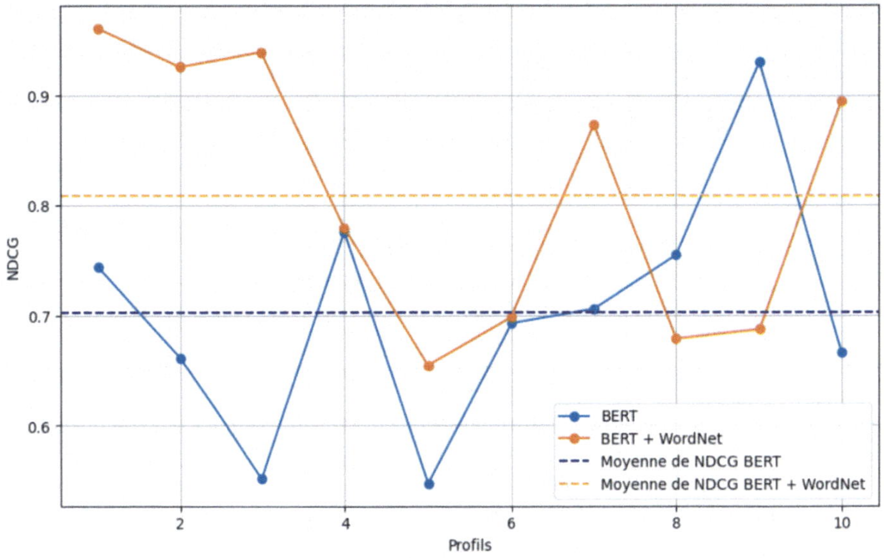

Fig. 6. Comparison of NDCG of BERT and BERT + WordNet.

6 Conclusion

In this study, we proposed a hybrid job recommendation approach that integrates BERT with WordNet to enhance the accuracy and relevance of job recommendations. By enriching job documents with semantic information from WordNet and leveraging BERT's advanced language understanding capabilities, our method aims to provide more precise and contextually relevant job matches. The experimental results indicate that this approach generally improves the performance of job recommendation systems, offering a more nuanced understanding of job descriptions and candidate profiles.

However, the effectiveness of the integration varies among different user profiles. While the addition of WordNet often enhances recommendation relevance, it can sometimes introduce semantic noise, which may negatively impact the accuracy of recommendations for certain profiles. This highlights the need for further refinement of the approach to mitigate potential drawbacks and maximize its benefits across diverse user scenarios.

References

1. Almalis, N.D., Tsihrintzis, G.A., Kyritsis, E.: A constraint-based job recommender system integrating FoDRA. Int. J. Comput. Intell. Stud. **7**(2), 103–123 (2018)
2. Ambalavanan, A.K., Devarakonda, M.V.: Using the contextual language model BERT for multi-criteria classification of scientific articles. J. Biomed. Inform. **112**, 103578 (2020)

3. Barbouch, M., Verberne, S., Verhoef, T.: WN-BERT: integrating wordnet and BERT for lexical semantics in natural language understanding. Comput. Linguist. Netherlands J. **11**, 105–124 (2021)
4. Brek, A., Boufaida, Z., et al.: AnnoJOB: semantic annotation-based system for job recommendation. Acta Informatica Pragensia **12**(2), 200–224 (2023)
5. De Ruijt, C., Bhulai, S.: Job recommender systems: a review. arXiv preprint arXiv:2111.13576 (2021)
6. Devlin, J.: BERT: pre-training of deep bidirectional transformers for language understanding. arXiv preprint arXiv:1810.04805 (2018)
7. Du, Y., et al.: Enhancing job recommendation through LLM-based generative adversarial networks. In: Proceedings of the AAAI Conference on Artificial Intelligence, vol. 38, pp. 8363–8371 (2024)
8. Freire, M.N., de Castro, L.N.: e-recruitment recommender systems: a systematic review. Knowl. Inf. Syst. **63**, 1–20 (2021)
9. Heggo, I.A., Abdelbaki, N.: Hybrid information filtering engine for personalized job recommender system. In: Hassanien, A.E., Tolba, M.F., Elhoseny, M., Mostafa, M. (eds.) The International Conference on Advanced Machine Learning Technologies and Applications (AMLTA2018), pp. 553–563. Springer International Publishing, Cham (2018). https://doi.org/10.1007/978-3-319-74690-6_54
10. Hong, W., Zheng, S., Wang, H.: Dynamic user profile-based job recommender system. In: 2013 8th International Conference on Computer Science & Education, pp. 1499–1503. IEEE (2013)
11. Hossain, M.S., Arefin, M.S.: Development of an intelligent job recommender system for freelancers using client's feedback classification and association rule mining techniques. J. Softw. **14**(7), 312–339 (2019)
12. Lavi, D., Medentsiy, V., Graus, D.: conSultantBERT: fine-tuned Siamese sentence-BERT for matching jobs and job seekers. arXiv preprint arXiv:2109.06501 (2021)
13. Lin, C.Y., Huang, C.K.: Employee turnover intentions and job performance from a planned change: the effects of an organizational learning culture and job satisfaction. Int. J. Manpow. **42**(3), 409–423 (2021)
14. Martinez-Gil, J.: An overview of knowledge management techniques for e-recruitment. J. Inf. Knowl. Manage. **13**(02), 1450014 (2014)
15. Miller, G.A.: WordNet: a lexical database for English. Commun. ACM **38**(11), 39–41 (1995)
16. Panchasara, S., Gupta, R.K., Sharma, A.: AI based job recommedation system using BERT. In: 2023 7th International Conference On Computing, Communication, Control And Automation (ICCUBEA), pp. 1–6. IEEE (2023)
17. Zhou, Y., Bollegala, D.: Learning sense-specific static embeddings using contextualised word embeddings as a proxy. arXiv preprint arXiv:2110.02204 (2021)

Performance Assessment of Tracking Algorithms for Student Monitoring in University Campus: A Comparative Study

Ihsene Zaidi[1](✉), Said Brahimi[2], and Sofiane Boulehouache[1]

[1] Department of Computer Science, University of 20 Août 1955, Skikda, Algeria
i.zaidi@univ-skikda.dz

[2] Department of Computer Science, University of 08 Mai 1945, Guelma, Algeria

Abstract. This paper focuses on exploring the applicability, strengths, and weaknesses of deep learning tracking algorithms to track students' movement and the practical application of it in complex environments such as universities. We focus on three tracking algorithms: BoT-SORT which is recognized for its accuracy and complex matching methods; ByteTrack which is established for accurate and efficient tracking; and DeepSORT which incorporates appearance learning to improve on the traditional tracking algorithm. Then, with the help of real-life data, the effectiveness of each algorithm is discussed at the same time in terms of the given criteria. we used these algorithms to monitor and populate the number of students coming to the classes and exiting the classes. Therefore, the evaluation shows that ByteTrack and BoT-SORT are more accurate in preserving identities and performing better in different tracking scenarios compared to DeepSORT. The findings presented are useful in enhancing the physical security and organizational management of a university by incorporating modern technologies in tracking systems.

Keywords: Multiple Object Tracking · Student Monitoring · ByteTrack · BoT-SORT · DeepSORT · YOLOv8

1 Introduction

Due to advancing technology and high demands towards security and safety, the role of tracking systems is rather important in today's world for different applications. Surveillance technologies are widely used to monitor people in diverse environments, including marketplaces, shopping malls, parks, companies, educational institutions, and many other contexts [1–6]. On a university campus, the ability to accurately track the movement of students is particularly important as it can significantly enhance safety, and improve campus management, and the utilization of resources. Nevertheless, tracking students in a dynamic and busy environment has several difficulties and aspects. Due to the high mobility of students most of the time, individuals are hidden behind other people making it difficult to monitor them all the time. Also, students often wear different clothes and accessories, which change daily, making it very difficult for the tracking

T. Marir et al. (Eds.): ICCSA 2024, ISEM 45, pp. 341–351, 2025.
https://doi.org/10.1007/978-3-031-90758-6_25

system to maintain consistent identities. Besides the variations in lighting, weather conditions, and layout of the university campus, the performance of tracking is also affected. Without forgetting the fact that students often move in groups, and interact with each other adding complexity to accurate tracking.

A general tracking process is initiated by associating a serial number with every object that is discerned in the first frame. When it identifies these objects, the system makes estimations on the future positions of these objects via motion models such as the Kalman Filters [7] and also extracts features to create embeddings that capture object characteristics. Such embeddings act as discriminative features which would help in the discernment of objects. Data association techniques, including the Hungarian Algorithm or deep learning-based methods, are then employed to match detections throughout frames to keep the continuity of object IDs. Finally, the system is in control of the initiation of new tracks that are newly detected, the current update of the tracks with the current detection, and the end or kill-off tracks for objects that are no longer detected. Figure 1 shows the flow chart of tracking system steps.

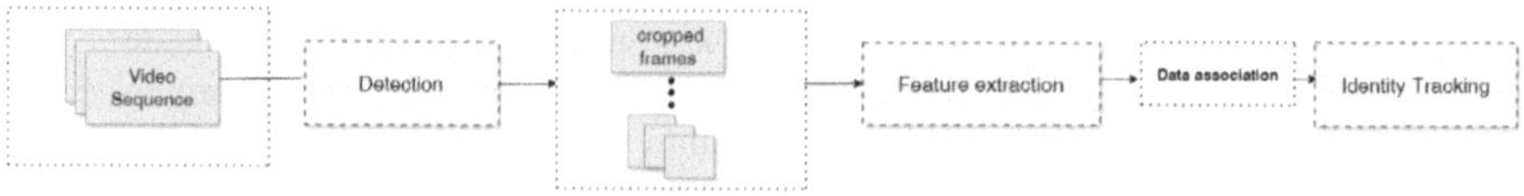

Fig. 1. Tracking system.

This research is intended to provide insights into how various tracking approaches can anticipate student movements and enquire about the performance of such tracking systems in high levels of occlusion and interaction contexts. Furthermore, it is intended to assess the capability and susceptibility of these tracking approaches in such environments. Three tracking algorithms were selected in this research due to their relevance to this use case: Bot-Sort [8] has high tracking accuracy in the object's motion as it applies advanced matching techniques to maintain track continuity even in crowded scenes. ByteTrack [9] attains a solid trade-off of tracking accuracy and computational cost, making it suitable for real-time application in environments like university campuses. Finally, the popular tracking algorithm DeepSort [10] extends traditional Sort [11] with deep learning techniques to improve object identity preservation over time. We will then go with a comprehensive case study, use real-world scene data with the algorithms, and analyze their performance on different metrics. Findings from our research have the potential to significantly impact the development of tracking systems and guide their implementation in educational institutions. Findings could provide recommendations that universities should implement to improve security and operational efficiency on campuses.

This paper is structured as follows: Sect. 2 presents the compared algorithms, and Sect. 3 outlines our experiment and presents the evaluation results. Finally, Sect. 4 concludes with a discussion of future research directions.

2 Presentation of the Compared Algorithms

Object tracking has come a long way over the years. Initially, many researchers had treated tracking as an optimization problem. More recent developments have combined these optimization techniques with deep learning models, particularly convolutional neural networks (CNNs) trained heavily on large datasets to produce resilient visual descriptors. This section delves into both optimization-based and deep learning-based tracking methods.

2.1 Optimization-Based Tracking Algorithms

Optimization-based tracking algorithms try to solve a cost function that indicates what object state or motion is the most likely one. These methods include the Kalman Filter [7], which improves a state transition model for object positioning, along with other algorithms such as Camshift [12], Particle Filter [13], Multiple Hypothesis Tracking (MHT) [14], and the IOU Tracker [15]. These methods are usually lightweight and quick to execute, which also makes them easy to use in the absence of much data. But are likely to have a challenge in high-density scenarios where they will not be able to operate at real-time speed.

2.2 Deep Learning-Based Tracking Algorithms

Deep learning-based tracking algorithms leverage the power of neural networks to enhance tracking accuracy and robustness. Algorithms like ByteTrack [9], FairMot [16], CenterTrack [17], Bot-Sort [8], DeepSort [10] are generally more accurate and resilient to variations in lighting, occlusions, and scale changes. However, these methods often come at the cost of increased computational complexity and a higher demand for extensive training data. In this study, we will focus on ByteTrack, Bot-Sort, and DeepSort to understand their performance and limitations in tracking students entering and exiting a university door.

ByteTrack [9] considers both high-confidence and low-confidence detection boxes to improve data association accuracy and manage occlusion. Initially, it matches high-confidence detection boxes with tracklets if their motion is similar by using a Kalman filter, or it uses appearance similarity (re-id feature distances). Subsequently, ByteTrack attempts to match the remaining tracklets with low-confidence detection boxes. A successful match indicates the detection likely represents an actual object rather than background noise. Conversely, unmatched low-confidence boxes are typically considered background noise or false positives and are removed, thus reducing the number of false positives. ByteTrack relies on high-performance detectors and motion models but is also adaptable to integrate additional features (such as appearance descriptors). While efficient, ByteTrack may encounter challenges in highly dynamic scenes, resulting in more frequent assignment of new IDs to objects.

Bot-Sort [8] enhances DeepSort by improving the state vector within the Kalman filter (KF) to better align the predicted frame with the target's actual motion. Instead of estimating the aspect ratio, which can cause inaccuracies in bounding box width, bot-sort directly estimates the width of the bounding box alongside the height. This technique is efficient in cases where the object's shape changes dynamically and causes Id switches. Another significant improvement is the integration of camera motion compensation (CMC). By estimating the camera's motion, Bot-Sort ensures that the predicted frame of the target closely matches the target's movement, even when it is not moving at a constant speed. Additionally, Bot-Sort can also be extended with a reidentification module (Bot-Sort-Reid). However, in high-density scenes, accurately estimating camera motion becomes challenging, leading to more frequent track fragmentation.

DeepSort [10] enhances the sort algorithm [11] to reduce identity switches by combining motion and learned appearance features. It uses a deep appearance descriptor, which is a convolutional neural network (CNN) pretrained on a person re-identification dataset, to retrieve key appearance features from the detections of each frame. For the motion features, deep sort makes use of the Kalman filter to estimate the tracklet location in the current frame and utilizes the Mahalanobis distance to calculate the tracklet and detections' spacio-temporal dissimilarity. With this distance, we can create a barrier that will only allow associations that are likely to be relevant and thus save time. Thus, this combination preserves deep sort's speed but it barely handles occlusions and identity changes in crowded scenes and low-quality images.

3 Proposed Work

We evaluate ByteTrack [9], DeepSort [10], and Bot-Sort [8] using a video from the Pdestree dataset [18], which contains footage of pedestrians walking outside public spaces. This dataset was obtained using UAVs at two different university campuses in Portugal and India, students consented to be involved in the data collection. It has many tracking cases of people with different occlusion and densities of the crowd, which makes the set highly appropriate for testing tracking algorithms. Figure 2 below is a frame captured from the video used in our study.

In our experiment, we applied these three trackers to monitor students entering and exiting a university door. Our objective is to track the students, count the number of entering and exiting, and assess the performance of each tracker based on the results obtained.

3.1 Experiment

In this experiment, the implementation was carried out on a server featuring an Intel Xeon E5-1620 v2 processor (3.7 GHz, 4 cores, 8 threads) and an NVIDIA Quadro P2200 GPU with 5 GB of VRAM. The GPU, running on NVIDIA driver version 535.183.01 and CUDA 12.2, was used for object detection and tracking tasks, leveraging PyTorch's GPU acceleration for efficient model inference. The system had 15 GB of RAM, with approximately 7.1 GB free during execution, and operated on Ubuntu 22.04.3 LTS with the Linux 6.8.0-40-generic kernel.

We utilize PyTorch framework for algorithm implementation and Ultralytics YOLOv8 for object detection [19]. Yolov8 is an advanced version of the YOLO [20] (you only look once) that predicts bounding boxes and the probability of classes all at

Fig. 2. Frame captured from the Pdestree dataset.

once with the help of cells in a grid. Yolov8 further updated the architecture of its previous versions and other factors related to training planning were improved for enhanced accuracy.

We systematically applied each tracker to the selected video, Fig. 3 shows examples of detected and tracked students.

We evaluated the performance of ByteTrack [9], Deepsort [10] and Bot-sort [8] using Trackeval toolkit [21]. Trackeval is the official evaluation kit used when it comes to multi-object tracking (MOT) that provides support for multiple tracking evaluation metrics, such as the higher-order tracking accuracy (HOTA) [22] and the CLEARMOT metrics [23]. Its flexible architecture enables the evaluation of tracking performance in a unified and standardized manner and provides fair comparability between different tracking algorithms, in addition to more detailed insights into performance across different measures such as accuracy, overlapping, and fragmentations. The key metrics we will focus on include:

IDF1 [24] (Identification F1 Score) calculates the accuracy of the entire predicted trajectory and ground truth entire trajectory, aiming to optimize the correct identification of objects over their whole paths. It is calculated as:

$$IDF1 = \frac{2IDTP}{2IDTP + IDFP + IDFN} \tag{1}$$

Where true positives (IDTP) represent correctly matched object identities, false positives (IDFP) are incorrectly matched identities, and false negatives (IDFN) are missed matches.

MOTA (Multiple Object Tracking Accuracy) calculates the average detection errors of the false negatives (FN) and false positives (FP) along with the identity switches (IDSW),

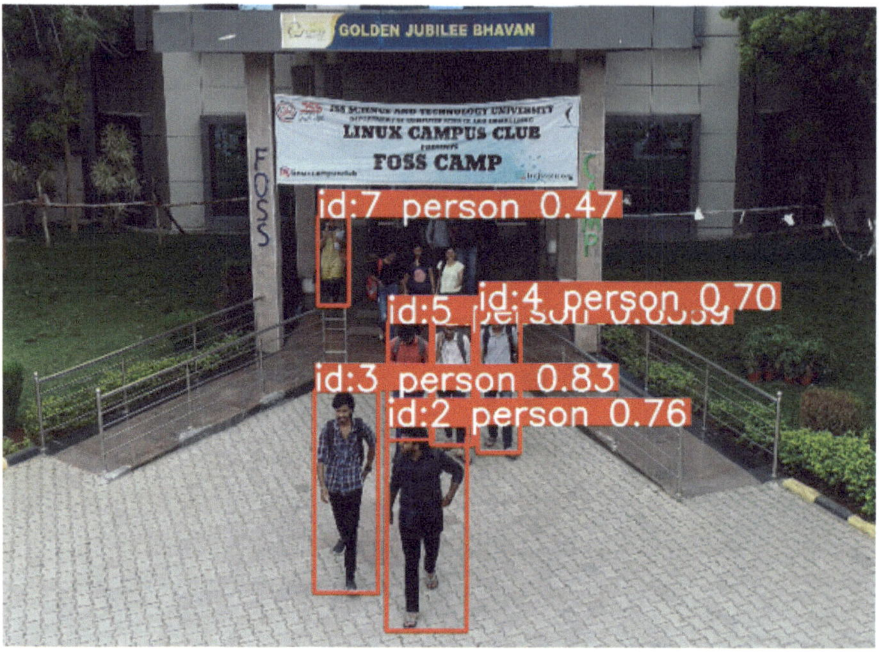

Fig. 3. Frame captured from the video showing tracked students.

adding them to the total and then dividing by the number of ground truth detections (GT). It is calculated as follows:

$$MOTA = 1 - \frac{FN + FP + IDSW}{GT} \quad (2)$$

MODA (Multiple Object Detection Accuracy) measures the accuracy of object detection in videos, similar to the MOTA score but without considering ID switches (IDSWs). It focuses solely on the detection aspect, ignoring the complexities of tracking object identities over time. It is calculated as follows:

$$MODA = \frac{TP - FP}{(TP + FN)} \quad (3)$$

HOTA (Higher Order Tracking Accuracy) evaluates the performance of the tracker for the detected orders (levels) or associations. This means it determines the ability of the tracker to perform in longer sequences and different complex scenarios. It is calculated as follows:

$$HOTA = \sqrt{DetA \times AssA} \quad (4)$$

Detection Accuracy (DetA) reflects the accuracy of object detection focusing on both identifying objects correctly and localizing them accurately within the image. It is calculated as follows:

$$DetA = \frac{TP}{(TP + FP + FN)} \quad (5)$$

Association Accuracy (AssA) relates to the capability of correctly linking the identified objects with their corresponding identities in the subsequent frame. It calculates the accuracy of association for each true positive (x) and then averages it to obtain the overall association accuracy. The formula is as follows:

$$AssA = \frac{1}{TP}\sum\nolimits_{x\in TP}\frac{TPA(x)}{TPA(x)+FNA(x)+FPA(x)} \tag{6}$$

Detection Recall (DetRe) refers to the ability of a tracker to identify all of the ground-truth detections, misses happen when trackers do not predict detections that are in the ground-truth list. It is determined as shown below:

$$\text{DetRe} = \frac{TP}{(TP+FN)} \tag{7}$$

Detection Precision (DetPr) refers to how a tracker can avoid issuing out false alarms which are additional detections that are not real. It is calculated as follows:

$$\text{DetPr} = \frac{TP}{(TP+FP)} \tag{8}$$

Association Recall (AssRe) quantifies the ability of trackers to overlay with the ground-truth trajectories and unify the same target into one consistent track. It is calculated as follows:

$$AssRe = \frac{1}{TP}\sum\nolimits_{x\in TP}\frac{TPA(x)}{TPA(x)+FNA(x)} \tag{9}$$

Association Precision (AssPr) measures the ability of the tracking system to avoid assigning multiple objects (merging) to the same predicted track. A low association precision indicates that the predicted track extends over multiple ground-truth objects. The formula is:

$$AssPr = \frac{1}{TP}\sum\nolimits_{x\in TP}\frac{TPA(x)}{TPA(x)+FPA(x)} \tag{10}$$

ID Switches (IDSW): counts the number of times an object's identity changes between consecutive frames. The formula is:

$$IDSW = \sum\nolimits_{f=1}^{F}\sum\nolimits_{i=1}^{N} 1\left(prID_{i,f-1} \neq prID_{i,f}\right) \wedge \left(gtID_{i,f-1} \neq gtID_{i,f}\right) \tag{11}$$

For each object i at each frame f, the indicator function$1(\cdot)$ returns 1 if the predicted identity of the object changes from the previous frame (indicating an ID switch) and the ground truth identity also changes, otherwise, it returns 0.

MTR (Mostly Tracked Ratio): calculates the ratio of the object's durability of its tracking by recording the percentage of time it was managed accordingly across the whole span of its trajectory. It is calculated as:

$$MTR = \frac{\textit{Number of Mostly Tracked Objects}}{\text{Total Number of Ground Truth Objects}} \times 100 \tag{12}$$

An object is considered "mostly tracked" if it is sustained all along its trajectory (e.g., more than 80% of its lifespan).

MLR (Mostly Lost Rate) displays the percentage of lost steps in a tracked object as a result of poor tracking. It is calculated as:

$$MLR = \frac{Number\ of\ Mostly\ Lost\ Objects}{\text{Total Number of Ground Truth Objects}} \times 100 \tag{13}$$

A mostly lost object is an object not tracked for a significant portion of its lifespan (e.g., more than 80% of its total trajectory duration).

PTR (Partly Tracked Ratio) provides information about the proportion of objects that are tracked for a significant portion of their lifespan (e.g., between 20% and 80% of its total trajectory duration), but not consistently enough to be considered "mostly tracked." indicating moderate tracking performance. It is calculated as:

$$PTR = \frac{Number\ of\ Partly\ Tracked\ Objects}{\text{Total Number of Ground Truth Objects}} \times 100 \tag{14}$$

Frag (Fragmentation) measures how many times the trajectory of a single object is interrupted and then reestablished, indicating periods where the tracker loses and then reacquires the object. We calculate the sum of each continuous segment where the object is correctly tracked without losing its identity (fi) as follows:

$$Frag = \sum\nolimits_{i=1}^{N} (f_i - 1) \tag{15}$$

3.2 Result Analysis

To better understand the performance differences among DeepSORT, ByteTrack, and BoT-SORT, we analyze the provided metrics. Table 1 below presents the findings, while Fig. 4 provides a simplified visualization of these results for easy comparison:

ByteTrack and BoT-SORT are superior in performance because of enhanced data association procedures and specific changes compared to DeepSORT. Thus, by applying two types of detection boxes – the high-confidence ones and the low-confidence ones, ByteTrack can detect and track more objects and their movements, at the same time assessing the real conditions within which it operates. This kind of approach helps in minimizing cases of false negatives and at the same time enhances the quality of tracking.

BoT-SORT improves the state vector in the Kalman Filter (KF); the bounding boxes' width and height are predicted instead of the aspect ratio. Such differentiation results in better estimation of the movement and location of objects, particularly in dynamic conditions that improve the tracking quality and eliminate the switching of identities which explains the highest IDF1.

On the other hand, DeepSORT which employs a deep appearance descriptor for the data association task performs poorly in comparison. As it pertains to the deep appearance descriptor, it is useful when identifying objects by their appearance but is less capable of handling situations where the objects occlude each other or interact with other objects. Additionally, DeepSORT's method involves higher affinity of identity switches and incomplete tracking in contrast to ByteTrack and BoT-SORT.

Table 1. Performance results of DeepSORT, ByteTrack, and BoT-SORT.

	DeepSort	ByteTrack	BoT-SORT
IDF1	55.869	71.661	72.197
HOTA	42.01	56.539	55.286
MOTA	38.2	52.136	52.241
MODA	19.772	52.203	52.316
DetA	41.26	53.191	51.79
AssA	42.856	60.11	59. 024
DetRe	50.42	64.52	63.126
DetPr	64.766	71.64	69.58
AssRe	52.49	65.343	64.54
AssPr	63.784	83.473	80.95
IDSW	34	7	8
MLR	18.18	2.9412	2.9412
PTR	68.182	61.765	58.824
MTR	13.63	35.294	38.235
Frag	38	24	23

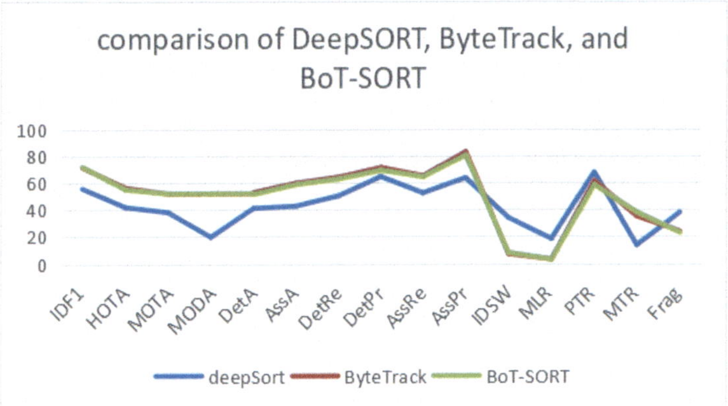

Fig. 4. Comparison of DeepSORT, ByteTrack, and BoT-SORT performances.

Altogether, ByteTrack has stable detections in the different confidence levels and BoT-SORT has better state vector updating in the Kalman Filter. However, DeepSORT falls short of such a system because of difficulties in handling multiple tracks and keeping the identity of objects for a longer time.

4 Conclusion

This work attempted to assess ByteTrack, DeepSORT, and BoT-SORT for tracking students' entry and exit through a university door. Based on the results presented in this paper, ByteTrack and BoT-SORT perform better and are more stable than DeepSORT especially in preserving consistent person IDs. ByteTrack seems to outperform the others regarding the capability to handle different detection confidence levels while BoT-SORT appears to enhance the state-vector estimation process. DeepSORT, while a good model, seems to struggle in dynamic and crowded scenes.

Based on these results, we can define the advantages and limitations of each tracking algorithm, which may help researchers understand better the capabilities of these algorithms in dynamic and populated environments such as university campuses. However, the present study has some limitations like the use of only one dataset for the experiment and the general environmental conditions that are depicted in the videos. Further studies can look into the applicability and efficiency of the algorithms under different working settings and other databases to supplement and confirm our results, moreover, we plan to make some recommendations and adjustments to the tracking algorithms according to this research.

References

1. Hu, H., Hachiuma, R., Saito, H., Takatsume, Y., Kajita, H.: Multi-camera multi-person tracking and re-identification in an operating room. J. Imaging **8**(8) (2022). https://doi.org/10.3390/jimaging8080219
2. Kumar, S., Vishal, Sharma, P., Pal, N.: Object tracking and counting in a zone using YOLOv4, DeepSORT and TensorFlow. In: Proceedings of the International Conference on Artificial Intelligence and Smart Systems ICAIS 2021, pp. 1017–1022 (2021). https://doi.org/10.1109/ICAIS50930.2021.9395971
3. Papaioannou, S., Markham, A., Trigoni, N.: Tracking people in highly dynamic industrial environments. IEEE Trans. Mob. Comput. **16**(8), 2351–2365 (2017). https://doi.org/10.1109/TMC.2016.2613523
4. Alam, T., Hadi, A.A., Najam, R.Q.S.: Designing and implementing the people tracking system in the crowded environment using mobile application for smart cities. Int. J. Syst. Assur. Eng. Manag. **13**(1), 11–33 (2022). https://doi.org/10.1007/s13198-021-01277-7
5. Matos, H., Santos, H.: People tracking in a smart campus context using multiple cameras. In: CEUR Workshop Proceedings, vol. 3601 (2023)
6. Nadeem, A., et al.: Tracking missing person in large crowd gathering using intelligent video surveillance. Sensors **22**(14), 1–25 (2022). https://doi.org/10.3390/s22145270
7. Kalman, R.E.: A new approach to linear filtering and prediction problems. J. Fluids Eng. Trans. ASME **82**(1), 35–45 (1960). https://doi.org/10.1115/1.3662552
8. Aharon, N., Orfaig, R., Bobrovsky, B.-Z.: BoT-SORT: robust associations multi-pedestrian tracking, no. 2 (2022). http://arxiv.org/abs/2206.14651
9. Zhang, Y., et al.: ByteTrack: multi-object tracking by associating every detection box. In: Avidan, S., Brostow, G., Cissé, M., Farinella, G.M., Hassner, T. (eds.) ECCV 2022. LNCS, vol. 13682, pp. 1–21. Springer, Cham (2022). https://doi.org/10.1007/978-3-031-20047-2_1
10. Wojke, N., Bewley, A., Paulus, D.: Simple online and realtime tracking with a deep association metric. In: Proceedings of the International Conference on Image Processing ICIP, vol. 2017, pp. 3645–3649, September 2018. https://doi.org/10.1109/ICIP.2017.8296962

11. Bewley, A., Ge, Z., Ott, L., Ramos, F., Upcroft, B.: Simple online and realtime tracking. In: Proceedings of the International Conference on Image Processing ICIP, vol. 2016, pp. 3464–3468, August 2016. https://doi.org/10.1109/ICIP.2016.7533003
12. Bradski, G.R.: Real time face and object tracking as a component of a perceptual user interface. In: Proceedings of the 4th IEEE Workshop on Applications of Computer Vision, WACV 1998, vol. 1998, pp. 214–219, October 1998. https://doi.org/10.1109/ACV.1998.732882
13. Gordon, N.J., Salmond, D.J., Smith, A.F.M.: Novel approach to nonlinear/non-Gaussian Bayesian state estimation. IEE Proc. Part F Radar Signal Process. **140**(2), 107–113 (1993). https://doi.org/10.1049/IP-F-2.1993.0015
14. Reid, D.B.: An algorithm for tracking multiple targets. In: 1978 IEEE Conference on Decision and Control, which included the 17th Symposium on Adaptive Processes, pp. 1202–1211 (1978). https://doi.org/10.1109/CDC.1978.268125
15. Bochinski, E., Eiselein, V., Sikora, T.: High-speed tracking-by-detection without using image information. In: 2017 14th IEEE International Conference on Advanced Video and Signal Based Surveillance, AVSS 2017, August 2017. https://doi.org/10.1109/AVSS.2017.8078516
16. Zhang, Y., Wang, C., Wang, X., Zeng, W., Liu, W.: FairMOT: on the fairness of detection and re-identification in multiple object tracking. Int. J. Comput. Vis. **129**(11), 3069–3087 (2021). https://doi.org/10.1007/s11263-021-01513-4
17. Zhou, X., Koltun, V., Krähenbühl, P.: Tracking objects as points. In: Vedaldi, A., Bischof, H., Brox, T., Frahm, J.M. (eds.) ECCV 2020. LNCS, vol. 12349, pp. 474–490. Springer, Cham (2020). https://doi.org/10.1007/978-3-030-58548-8_28
18. Kumar, S.V.A., Yaghoubi, E., Das, A., Harish, B.S., Proenc, H.: The P-DESTRE : a fully annotated dataset for pedestrian detection, tracking, re-identification and search from aerial devices, pp. 1–11 (2020)
19. ultralytics/CITATION.cff at main · ultralytics/ultralytics · GitHub. https://github.com/ultralytics/ultralytics/blob/main/CITATION.cff. Accessed 14 June 2024
20. Redmon, J., Divvala, S., Girshick, R., Farhadi, A.: You only look once: unified, real-time object detection. In: Proceedings of the IEEE Conference on Computer Vision and Pattern Recognition, vol. 2016, pp. 779–788, December 2016.https://doi.org/10.1109/CVPR.2016.91
21. TrackEval/docs/MOTChallenge-Official at master · JonathonLuiten/TrackEval · GitHub. https://github.com/JonathonLuiten/TrackEval/tree/master/docs/MOTChallenge-Official. Accessed 04 July 2024
22. Luiten, J., et al.: HOTA: a higher order metric for evaluating multi-object tracking. Int. J. Comput. Vis. **129**(2), 548–578 (2021). https://doi.org/10.1007/s11263-020-01375-2
23. Bernardin, K., Stiefelhagen, R.: Evaluating multiple object tracking performance: the CLEAR MOT metrics. Eurasip J. Image Video Process. **2008** (2008). https://doi.org/10.1155/2008/246309
24. Ristani, E., Solera, F., Zou, R., Cucchiara, R., Tomasi, C.: Performance measures and a data set for multi-target, multi-camera tracking. In: Hua, G., Jégou, H. (eds.) ECCV 2016. LNCS, vol. 9914, pp. 17–35. Springer, Cham (2016). https://doi.org/10.1007/978-3-319-48881-3_2

Author Index

T. Marir et al. (Eds.): ICCSA 2024, ISEM 45, pp. 353–354, 2025.
https://doi.org/10.1007/978-3-031-90758-6

The manufacturer's authorised representative in the EU is Springer Nature Customer Service Centre GmbH, Europaplatz 3, 69115 Heidelberg, Germany. If you have any concerns regarding our products, please contact ProductSafety@springernature.com

Printed and bound by CPI Group (UK) Ltd, Croydon, CR0 4YY
30/09/2026
02225584-0012